THE CHANGING COUNTRYSIDE

Proceedings of the First British–Dutch Symposium on Rural Geography

University of East Anglia, 3–5 September, 1982

Edited by

Gordon Clark, Jan Groenendijk and Frans Thissen

The editors would like to thank Paul Cloke and Joost Hauer for
organising the symposium, and Malcolm Moseley for his hospitality
during the symposium in Norwich and for his assistance in finding
a publisher. Jan ter Haar (re)drew many of the figures and
Jacqueline Sehlmeier, Carolyn Brown, Jean Burford and Christina
Skinner provided valuable assistance.

Lancaster and Amsterdam

September 1983

© Individual Authors, 1984

ISBN 0 86094 155 8

Published by Geo Books
          Regency House
          34 Duke Street
          Norwich NR3 3AP
          England

Printed in Great Britain by Short Run Press Ltd

Contents

                                                                        page

List of Contributors   v

List of Figures   vii

1.   Introduction   1
     Gordon Clark, Jan Groenendijk and Frans Thissen

     SECTION ONE: THE NATURE AND DEFINITION OF RURAL AREAS

2.   Rural geography in Britain: some aspects of rural   11
     problems and policies   *Bruce Proudfoot*

3.   Proposal for a theoretical basis for the human   17
     geography of rural areas   *Joeke Veldman*

4.   A multivariate approach to a rural typology of Dutch   27
     regions at different spatial scales   *Joost Hauer*

5.   The intra-regional differentiation of peripheral   41
     regions in the Netherlands   *Wim Ostendorf*

6.   Urban field developments and the changing position   51
     of rural areas   *Lambert van der Laan*

7.   The concept of resources in rural geography   59
     Chris Park

8.   Towards a policy-oriented analysis in rural areas: 67
     a reflection on living conditions   *Tom van der Meulen*

     SECTION TWO: RURAL STANDARDS OF LIVING

9.   Mobile services and the rural accessibility problem   79
     Malcolm Moseley and John Packman

10. Access in a remote rural area   *Paulus Huigen*   87

11. The downward development of small service centres   99
     in rural areas: a theoretical exploration   *Henk de Haard*

12. Rural education services: the social effects of   111
     reorganisation   *Michael Tricker*

13. Spatial mobility problems of the elderly and disabled   121
     in the Cotswolds   *Robert Gant and José Smith*

14. Incomes of the elderly in rural Norfolk   *Roger Gibbins*   137

15. Public spending in rural areas   *Ad van Bemmel*   145

16. Population trends and community viability on the   153
     Danish small islands   *Tony Martin*

17. A comparison between objective and subjective   163
    measures of quality of life   *Jaap Gall*

SECTION THREE: RURAL PLANNING

18. Planning small villages in the Netherlands: 171
    a comparative approach   *Herman Kiestra*

19. Key-village and related settlement policies in   187
    Scotland   *Douglas Lockhart*

20. Optimising the settlement pattern: review and   199
    research design   *Dinny de Bakker*

21. Planning policy and socio-economic changes in   207
    post-war Montgomeryshire   *David Grafton*

22. Land speculation and the under-use of urban-fringe   221
    farmland in the Metropolitan Green Belt   *Richard Munton*

23. Rural settlement planning: too great an expectation? 233
    Patrick Hanrahan and Paul Cloke

SECTION FOUR: POLITICAL INFLUENCES IN THE COUNTRYSIDE

24. Geography and nationalism: the case of rural Wales   243
    Gareth Edwards

25. Politics and the countryside: the British example   251
    Andrew Gilg

26. The implementation of land-use policy in the urban   261
    fringe: the North Middlesex Green Belt estates
    1920-1950   *Elizabeth Sharp*

27. Small rural communities and the reorganisation of   277
    Dutch local government districts   *Jacob Groot*

28. A key to settlement growth in rural areas: local   283
    administrators, their powers and the size of their
    territories   *Jan Groenendijk*

SECTION FIVE: RURAL HOUSING

29. Housing in small villages: a classification of   297
    contexts   *Frans Thissen*

30. Changing migration to rural areas in the Netherlands 309
    Oedzge Atzema

31. Housing and conservation in the Lake District: 317
    a study in ambiguity   *Gordon Clark*

32. Public-sector housing in rural areas in England   327
    David Phillips and Allan Williams

33. Alternative tenures in rural areas: the role of   339
    housing associations   *Patricia Richmond*

List of Contributors

British Contributors

Gordon Clark, Department of Geography, University of Lancaster, Lancaster, LA1 4YR.

Gareth Edwards, Department of Geography, St. David's College, Lampeter, Dyfed SA48 7ED.

Robert Gant and José Smith, Department of Geography, Kingston Polytechnic, Kingston-upon-Thames KT1 2EE.

Roger Gibbins, postgraduate student at University of East Anglia, now at Northamptonshire County Council, County Hall, Northampton NN1 1DN.

Andrew Gilg, Department of Geography, University of Exeter, Exeter EX4 4RJ.

David Grafton, Department of Geographical Sciences, Plymouth Polytechnic, Plymouth PL1 8AA.

Patrick Hanrahan and Paul Cloke, Department of Geography, St. David's College, Lampeter, SA48 7ED.

Douglas Lockhart, Department of Geography, University of Keele, Keele ST5 5BG.

Tony Martin, Department of Geography, University of Strathclyde, Glasgow G1 1XH.

Malcolm Moseley and John Packman, School of Environmental Sciences, University of East Anglia, Norwich NR4 7TJ.

Richard Munton, Department of Geography, University College, 26 Bedford Way, London WC1H OAP.

Chris Park, Department of Geography, University of Lancaster, Lancaster LA1 4YR.

David Phillips and Allan Williams, Department of Geography, University of Exeter, Exeter EX4 4RJ.

Bruce Proudfoot, Department of Geography, University of St. Andrews, St. Andrews, Fife KY16 9AL.

Patricia Richmond, Department of Geography, University of Exeter, Exeter EX4 4RJ.

Elizabeth Sharp, Department of Geography, University College, 26 Bedford Way, London WC1H OAP.

Michael Tricker, Public Sector Management Group, University of Aston, Birmingham B4 7DU.

Dutch Contributors

Oedzge Atzema, University of Nijmegen, Department of Geography,
    Nijmegen, The Netherlands.

Dinny de Bakker, State University of Utrecht, Department of
    Geography, Utrecht, The Netherlands.

Ad van Bemmel, State University of Utrecht, Department of
    Geography, Utrecht, The Netherlands.

Jaap Gall, State University of Leiden, Department of Social
    Psychology, Leiden, The Netherlands.

Jan Groenendijk, Free University, Institute for Geographical
    Studies and Urban and Regional Planning, Amsterdam, The
    Netherlands.

Jacob Groot, Agricultural University, Department of Sociology and
    Sociography, Wageningen, The Netherlands.

Henk de Haard, Provincial Physical Planning Agency of Overijssel,
    Zwolle, The Netherlands.

Joost Hauer, State University of Utrecht, Department of Geography,
    Utrecht, The Netherlands.

Paulus Huigen, State University of Utrecht, Department of
    Geography, Utrecht, The Netherlands.

Herman Kiestra, Provincial Physical Planning and Housing
    Department Gelderland, Arnhem, The Netherlands.

Lambert van der Laan, Free University, Institute for Geographical
    Studies and Urban and Regional Planning, Amsterdam,
    The Netherlands.

Tom van der Meulen, Provincial Physical Planning Agency of Zeeland,
    Middelburg, The Netherlands.

Wim Ostendorf, University of Amsterdam, Sociaal-Geografisch
    Instituut, Vakgroep Stad en Land, Amsterdam, The Netherlands.

Frans Thissen, Free University, Institute for Geographical Studies
    and Urban and Regional Planning, Amsterdam, The Netherlands.

Joeke Veldman, State University of Utrecht, Department of
    Geography, Utrecht, The Netherlands.

List of Figures

		page
1.1	Map of COROP regions in the Netherlands	3
3.1	A conceptual model for regionalising	20
3.2	The interaction between physical-spatial structure and the socio-spatial system in large cities	21
3.3	The interaction between physical-spatial structure and the socio-spatial system in peripheral rural areas	22
3.4	The interaction between physical-spatial structure and socio-spatial system in peri-urban rural areas	23
4.1	The COROP regions classified in diminishing order of rurality on the basis of the original measurement values of the structure variables	36
4.2	The COROP regions classified in diminishing order of rurality on the basis of the original measurement values of the process variables	36
4.3	The Frisian municipalities classified in diminishing order of rurality on the basis of the original measurement values of the structure variables	38
4.4	The Frisian municipalities classified in diminishing order of rurality on the basis of the original measurement values of the process variables	38
5.1	The differentiation of Dutch regions based on a socio-morphological typology	44
10.1	Locational profile of a settlement	91
10.2	Distance in time from Gaastmeer to Sneek (Zuidwest-Friesland)	92
11.1	Number of service sequences per value of call frequency, given the size of centre	101
11.2	The functional hierarchy of centres in Lüchow-Dannenberg based on service sequences, given a call frequency of one	108
11.3	The functional hierarchy of centres in Lüchow-Dannenberg based on service sequences, given a call frequency of four	108
12.1	Location of case study villages	113
13.1	North Cotswolds: location of health and social service facilities used by the survey population	122
13.2	Availability of services	123
13.3	Availability of mobile services	123
13.4	Availability of public transport	124

16.1 Distribution of a) small islands and b) linked   154
     islands

16.2 Ferry traffic and retail turnover on Anholt, 1970-75   159

18.1 Study areas   173

19.1 Angus District: proposed rural settlement policy   191

19.2 Western Rural Area, Central Area: proposed   194
     rural settlement policy

21.1 Rural Districts with remote rural characteristics   208
     in 1971

21.2 Mid-Wales and the Borders: topography and major   210
     settlements

21.3 Socio-economic scores in Montgomeryshire parishes,   212
     1961

21.4 Socio-economic scores in Montgomeryshire parishes,   214
     1971

21.5 Relationship between 1961 and 1971 index scores   215
     for Montgomeryshire parishes

21.6 Residuals from regression   216

22.1 Area of the approved Metropolitan Green Belt in   224
     1982 and location of the study areas

25.1 Changes in planning fashion   258

26.1
 (a) Number of properties approved for acquisition as   264
     green belt in the Greater London area as a whole,
     in total and with L.C.C. grant aid only

 (b) Area approved for acquisition as green belt in the   264
     Greater London area as a whole, in total and with
     L.C.C. grant aid only

26.2 The North Middlesex study area   266

26.3
 (a) Number of properties approved for acquisition as   268
     green belt in Middlesex, in total and with L.C.C.
     grant aid only

 (b) Area approved for acquisition as green belt in   268
     Middlesex, in total and with L.C.C. grant aid only

28.1 Population living in declining settlements   288

28.2 Areas without growing settlements   292

29.1 Growth/size types of village by COROP region   299

29.2 Types of COROP region by housing stock, 1971   303

29.3 Aspects of housing in COROP regional types, 1971   304

29.4 Small village policies in the provinces   306

30.1 Decrease in regional mobility, 1973-79   310

30.2 Change in proportion of in-migrants in regional   310
     mobility, 1973-79

30.3 Migration surpluses of municipalities in the   313
     Randstad region, 1970-74 and 1975-79

31.1 The English Lake District   318

32.1 Local authority housing in England   329

32.2 Local authority housing in South Hams   331

32.3   Council housing stock and proportion sold in   332
       South Hams

32.4   Areas in which resale conditions apply to   334
       council house sales

33.1   A guide to the voluntary housing movement in   341
       the U.K.

33.2   Location of housing association projects in Devon   345

33.3   Housing association projects in East Devon   346

33.4   Housing association projects and council   347
       housing stock in East Devon

Chapter 1

Introduction

Gordon Clark,

Jan Groenendijk and Frans Thissen

This book derives from a British-Dutch Symposium which was held at
the University of East Anglia in Norwich between 3 and 5 September
1982. The symposium's theme was 'Living conditions in peri-urban
and remoter rural areas in north-west Europe' and the participants
were largely drawn from the membership of the two sponsoring bodies,
the Rural Geography Study Groups of the Institute of British Geographers and the Royal Dutch Geographical Society (KNAG). Some of
the Dutch contributors are from other disciplines and are collabora-
ting with geographers in a working party on socio-spatial research
in rural areas. Most of the symposium papers have been specially
re-written for this book which, in the editors' opinion, gives a
clear view of some of the main avenues of current research in rural
geography in the United Kingdom and the Netherlands.

There are interesting similarities between the work in the two
countries and also some differences. There is a common desire for
geographical research which will be useful for public policy,
though less clarity on how results and, in some cases, political
advocacy should be transformed into political action. In both
countries, the countryside is more frequently analysed from social
and political stances than from a purely economic one. One difference between the two sets of contributions is the greater team-work
evident among Dutch geographers, particularly those working in, or
influenced by, the 'Utrecht' group. The British research tends to
be more individualistic and varied in its themes and settings. All
the geographers are very clearly aware that the countryside is
changing in both countries under the impact of a number of processes; suburbanisation, the changing economic base of the countryside, migration and the role of governments (particularly through
planning systems) in shaping rural areas. Consequently the 32
papers in this book have been collected together into five sections
to reflect these concerns:
- the nature and definition of local areas
- rural standards of living
- rural planning
- political influences in the countryside
- rural housing.

Each section contains both British and Dutch papers, and sometimes papers contribute to the debate on more than one theme.

Proudfoot approaches the definition of rural geography through a review of the work of those who would call themselves rural geographers. He notes the trend from regional to thematic studies, the continuing belief that research should be policy-relevant and the limited theoretical context of rural research. He stresses the need to appreciate the continuing importance of the varying physical endowment of the countryside while still advancing our knowledge of the changing functions and social structure of rural areas.

Veldman proposes a not unrelated theoretical base for rural geography. It starts from the geographical milieu of rural areas, based on a distinctive physical-spatial structure that is dominated by area-bound forms of land use. For peri-urban and peripheral rural areas he proposes conceptual models that are based on the interaction between the socio-spatial system and the physical-spatial structure. The interdependence of the socio-spatial system and the physical-spatial structure is the central hypothesis of the Zuidwest-Friesland project of a research group at the State University of Utrecht whose work is also reflected in the papers by Hauer, van Bemmel, Huigen and De Bakker. Finally, he sees a need for a multidisciplinary approach to socio-spatial research in rural areas and for integrated policies and management.

Hauer approaches the definition of rural areas by constructing a typology of regions at two spatial levels, that of the so-called COROP regions (see Figure 1.1) and that of the municipality. Taking the central hypothesis of the Zuidwest-Friesland research group as his starting point, he looks for structure variables and process variables that indicate rurality. At both levels a rural dimension can be identified which is quite similar in the classifications based on both types of variables.

Ostendorf tests the internal consistency of a typology of municipalities in peripheral areas which is based on their degree of urbanisation. It is clear that intra-regional differentiation in peripheral regions due to suburbanisation is not pronounced, and it seems that other processes of differentiation are at work in these regions. The relationship between socio-spatial and physical-spatial characteristics and the use of different spatial scales are again crucial elements in the research design. The difference is that Ostendorf's starting point is urban and, for him, the daily activity region is important in peripheral rural areas as well as in urban ones.

Recent developments in rural areas are clearly related to changes in urban society. Van der Laan uses the concept of the 'urban field', to study a functional specialisation of rural areas paralleling spatial changes in urban society.

Park examines in more detail the physical distinctiveness of rural areas and how rural resources present by their nature unusual problems of management. He examines the role of conflict over access to resources as a major facet of developments in the countryside and a major theme of rural geography research. A resource management approach to the countryside is seen as fruitful both in practical and conceptual terms.

For Van der Meulen, research should be 'action research', carried out in discussion between actor (policy maker) and researcher. Research provides a potential feedback between actor (organisation) and environment. Geographical knowledge is tested

2

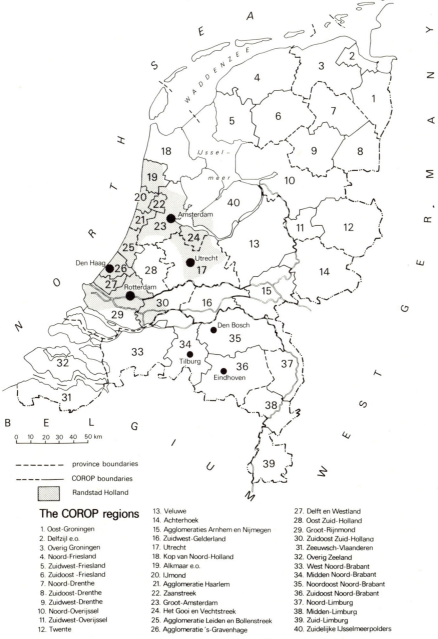

The COROP regions

1. Oost-Groningen
2. Delfzijl e.o.
3. Overig Groningen
4. Noord-Friesland
5. Zuidwest-Friesland
6. Zuidoost -Friesland
7. Noord-Drenthe
8. Zuidoost-Drenthe
9. Zuidwest-Drenthe
10. Noord-Overijssel
11. Zuidwest-Overijssel
12. Twente
13. Veluwe
14. Achterhoek
15. Agglomeraties Arnhem en Nijmegen
16. Zuidwest-Gelderland
17. Utrecht
18. Kop van Noord-Holland
19. Alkmaar e.o.
20. IJmond
21. Agglomeratie Haarlem
22. Zaanstreek
23. Groot-Amsterdam
24. Het Gooi en Vechtstreek
25. Agglomeratie Leiden en Bollenstreek
26. Agglomeratie 's-Gravenhage
27. Delft en Westland
28. Oost Zuid-Holland
29. Groot-Rijnmond
30. Zuidoost Zuid-Holland
31. Zeeuwsch-Vlaanderen
32. Overig Zeeland
33. West Noord-Brabant
34. Midden Noord-Brabant
35. Noordoost Noord-Brabant
36. Zuidoost Noord-Brabant
37. Noord-Limburg
38. Midden-Limburg
39. Zuid-Limburg
40. Zuidelijke IJsselmeerpolders

Figure 1.1 Map of COROP regions in the Netherlands

3

by its policy-oriented application. This approach, he argues, is especially relevant in the countryside.

RURAL STANDARDS OF LIVING

In the study of rural standards of living, the work of Moseley on accessibility as a component of quality of life has been particularly influential for several researchers. The chapter by Moseley and Packman assesses one way of improving accessibility by taking services to the more isolated 'customers' rather than requiring them to travel to the office providing the services.

In his study of Zuidwest-Friesland Huigen extends Moseley's ideas by viewing access in relation to possible behaviour rather than actual behaviour. Research should focus on individuals or small groups and should take the time dimension explicitly into consideration. His research is directed at the development and application of a simulation model of individual access to services and employment.

However, the common experience in Western Europe has been for service provision to be diminished in small settlements as shops, for example, were closed. De Haard's chapter considers exactly what happens when one service facility closes down in villages of various sizes given that most people travel to villages on multipurpose trips. He proves that the loss of a service from a small centre has a proportionately much more damaging effect on that centre's attractiveness than it would have in a larger centre. He also shows how the functional hierarchy of rural settlements is radically altered.

In both the U.K. and the Netherlands, the closure of small village schools is a lively political issue. Tricker studies the effect of such closures on the standard of education of pupils and on the wider social role of schools for parents and villagers. He concludes that, with certain clearly defined exceptions, the harmful effects of school closure have geen overstated.

Gant and Smith study a different aspect of quality of life when they focus on the mobility inside and outside the home of the elderly in part of Gloucestershire. They show how much the family context of the elderly affects their quality of life. They identify public transport, the location of houses and services, and family support as key elements in helping those with restricted mobility to lead full and independent lives.

Personal income also has some part to play in the standard of living of the elderly and Gibbins examines this complex and difficult subject in rural Norfolk. He is able to show that the incomes of the elderly in the countryside differ from those in the towns by virtue of there being both more low-income elderly and more high-income earners.

This polarisation in standards is also found between whole urban and rural areas in the Netherlands. Van Bemmel compares public spending in rural and urban areas and, although precise measurement is tricky, the disadvantages of low spending and poorer services in rural areas are starkly obvious.

Martin's work on the Danish islands tackles an area where accessibility is particularly difficult and where the loss of services is most acutely felt. He charts the ways service provision has changed and assesses the future viability of these vulnerable communities.

Gall argues that, although the 'objective' measures of quality of life which have been used in the other papers are valuable, it is also important to study the country person's subjective impression of his quality of life. Gall attempts to measure subjective quality of life and compare it with objective indices.

RURAL PLANNING

One way in which governments have sought to increase their control over rural areas, partly to help raise living standards, is through the use of planning controls. One of the most intractable problems in both countries has been the future functions of villages, especially small villages. Kiestra examines the future of Dutch villages in the face of depopulation in some areas and intense urban in-migration in others and notes how the use of planning controls is conceived rather differently at the various levels of local government.

Lockhart's study of villages in the Tayside and Central Regions of Scotland again highlights the contrast between the peri-urban village and remoter areas. He also stresses the clear limitations of planning for growth and for the control of the distribution of the population within the current British system of planning.

De Bakker also is concerned with a policy of concentrating services in selected centres. He outlines a research design for evaluating such a policy in Zuidwest-Friesland which explicitly recognises the possible conflict of interests at different levels of local government.

Grafton's study of mid-Wales has a similar theme and his evaluation concludes that the concentration of housing and industry in one town has polarised the region and hindered attempts to provide a more equitable standard of living in the remoter areas.

Much rural planning is occasioned by urban pressures on the countryside, one of the most obvious being the need for building land. Munton reports on a project to assess the Green Belt around London which was designed to restrict the expansion of that city. He examines the powerful effect of land speculation and how green-belt legislation has mostly succeeded in restricting the effects of speculation in land to a narrow band on the immediate fringe of the built-up area of London. In this case planning powers have been able to achieve some of their objectives even within a framework of private landownership.

Hanrahan and Cloke, however, widen the debate by looking at the general effectiveness of British rural planning. They explain its failure to achieve its aims as being partly a result of misunderstanding the countryside's diversity and partly because planning is not structured to meet effectively the needs of the deprived. They regret the lack of co-ordination between the agencies of Government and deprecate the extent to which private capital is outwith planning control. They conclude that rural planning as currently constituted in the U.K. can never be an agent of major social change.

POLITICAL INFLUENCES IN THE COUNTRYSIDE

The view in some of the previous chapters that rural planning has
the wrong objectives and priorities would be explained by Edwards as
being partly a function of where major planning decisions are taken.
He analyses the recent history of rural Wales within a core (London) - periphery (rural Wales) model. He examines the achievements
of the Welsh Nationalist Party (Plaid Cymru) in partially reversing
the exploitative nature of Welsh rural development and puts forward a programme for the simultaneous achievement of nationalist
and socialist aims in rural Wales.

Gilg also adopts a macro-scale framework for his review of
British Government policies affecting the countryside in the post-
war years. He emphasises the primary importance of the political
persuasion of the Westminster government in affecting the balance of
objectives. He also identifies a few policies which have been pursued irrespective of which party was in power, particularly agricultural expansion and curbing the physical growth of cities.

The other papers in this section adopt a narrower canvas.
Sharp's study of the London Green Belt in the inter-war years ex-
plores in considerable detail a theme noted in several other chapters, namely the uneasy nature of the relationship between different
tiers of local government. Groot describes what happens when the
units of local government are amalgamated into larger areas. He
suggests that in the most peripheral and thinly populated areas the
effects of this policy might be markedly negative. Groenendijk, on
the other hand, studies existing local government units of different sizes in his comparison of rural planning in Norfolk and the
northern Netherlands. The small size of Dutch local government
areas and their considerable powers lead to a more dispersed use of
space than in the centralised British system which tends to favour
a greater clustering of development.

RURAL HOUSING

Many of the chapters in this section explore themes from the other
sections in the context of rural housing. Thissen classifies
COROP region and their planning policy for small villages. He
shows how influential are suburbanisation, concentration of development and the policy of the provinces with respect to small villages.
These affect housing provision whose scale, type and location are
identified as major influences on the social development of the
countryside. Atzema pursues the theme of residential suburbanisation in more detail by mapping trends in migration. Recent changes
in migration flows are marked and he attempts to explain these.

The pressure of incomers on rural housing markets is also found
in the English Lake District which is the site of Clark's study of
attempts to favour local people in need of housing over incomers.
He focusses on the lack of clarity in planning objectives and the
centralised nature of the British planning system as key influences
in shaping this attempt by rural planners to engineer social
changes through the private housing market.

The chapters by Phillips and Williams and Richmond focus on the
public sector of housing in Devon, south-west England. Phillips
and Williams explain the traditionally low levels of provision of
council houses in the countryside. The need for subsidised housing
to rent is shown to be rising at the same time as national legislation is making it more difficult for local authorities to meet

6

their population's needs. Recently, housing associations have been
promoted as a suitable local agency to use national funds to provide more, subsidised housing to rent. Richmond measures their
impact and notes how the limited provision of new houses is only
weakly related to measures of local housing need.

THE FUTURE

The symposium was felt by those who attended to be a most successful meeting. The editors hope that this collection of papers will
help to further links between British and Dutch geographers and
that future symposia will be possible.

SECTION ONE:

THE NATURE AND DEFINITION OF RURAL AREAS

Chapter 2

Rural geography in Britain:
some aspects of rural problems and policies

Bruce Proudfoot

There is a long tradition of studying rural areas in British geography and a wide variety of themes has been examined. Until at
least the mid-1950s a considerable part of British geography dealing with contemporary conditions in Britain was concerned essentially with rural areas, as was most historical geography. Much of
the geography of contemporary Britain was set in a regional framework which, with the exception of the Midland industrial areas and
the coalfields, was based either on rural areas, or on regions defined in terms of physical features and the agricultural responses
to these, or on the characteristics of rural life in the broadest
sense. The first Land Utilisation Survey, both in its classification of land uses and perhaps even more clearly in its county reports, showed the strong rural and regional bias in British geography at that time. Fieldwork for the Survey was carried out in
the early 1930s and maps and reports were published continually
until the final analytical volume on *The land of Britain: its use
and misuse* was published by the Director of the Survey in 1948
(Stamp, 1948). Many of those associated with the Survey were also
involved in government planning during and immediately after World
War II and the involvement of geographers concerned with rural
areas in matters of public policy has continued since that time.

With the virtual demise of studies of regions in British geography during the 1960s there have been few analyses of rural areas
during the last two decades which have considered in detail more
than one or two aspects of rural activities, whatever the scale of
enquiry. Put another way, studies have become much more themati-
cally oriented. Considerable attention in both pure and applied
research has been paid to land use, land-use conflicts, land budgets, conservation, recreation in rural areas, rural population
change, transportation, accessibility, and, more recently, rural
deprivation. Apart from the intrinsic interest of such topics, they
have also been regarded as 'relevant', and perhaps politically significant, not only by the researchers themselves, but also by funding agencies and by many rural and urban people. Although there
has been some discussion of broad spatial patterns of agricultural
activities, particularly of land use, types of farms, and to a
lesser extent of economic inputs and outputs, there have been few
examinations of the extent to which appropriate economic or social
theory exists which might help in understanding functional activities in rural areas. This is surprising because in urban geography on the one hand and in development or Third World studies on
the other, there has been extensive consideration of theory

and theoretical explanations of observed conditions. Those
theoretically-based quantitative studies which have been carried out,
mainly if not entirely by agricultural economists, have dealt with
particular aspects of individual sectors within rural economies.
However, some insights into the whole rural economy of particular
areas have been gained by studies of, for example, the income-
employment multiplier effects of establishing a new pulp mill in
Lochaber (Scotland) and of investment by the Highlands and Islands
Development Board in the fishing industry (Greig, 1971; 1972) and
the impact, within an input-output framework, of tourism in Anglesey,
Wales (Archer, 1973; 1977).

Rather than describe in detail individual contributions to
those topics already noted as being of interest to rural geographers
in Britain, the present paper addresses some broader issues, largely by raising questions rather than providing answers. This is not
to suggest that such topics are unimportant: rather, geographers
tackling such topics are increasingly involved in discussions of
rural public policy and management of human and physical resources
in the countryside. Some indication of recent work can be obtained
from reviews in *Progress in Human Geography* (e.g. Cloke, 1980) and
from reviews and lists of publications in the *Countryside planning
yearbook* edited since 1980 by Gilg. It is regrettably true that in
spite of the large volume of research published on rural areas over
the last few years, there has been little discussion of the kinds
of theoretical arguments which might 'explain' individual findings
or might be used as a framework for the study of policy formation
or implementation.

Such discussion might well start with the fundamental premise
that the physical endowment of rural areas is unequal. Frequently,
all-embracing arguments and policies assume that they are equally
endowed. The Southern Uplands of Scotland, mid-Wales and Dartmoor
are not equally suitable for human use whether it be for grass or
grain production, forestry, game and wildlife management, or recreation. This is clearly shown by maps of such climatic factors as
length of growing season, total rainfall and its annual variability
(Gregory, 1964). These upland areas are in very general terms more
similar than they are to, for example, the Midland Plain, East
Anglia or the chalk and clay scarplands of southern England. All
such long-recognised regional sub-divisions of Britain differ in
terms of their physical endowments, that is, their *natural advantages* for specific purposes. Surprisingly little is known about
these differences, and even less about the differences in terms of
the *comparative advantages* of the individual regions as they are
presently organised for various types of production. Yet much
agricultural and land-use policy decision-making assumes just such
knowledge. The effects of, for example, a hill-sheep subsidy may
be to minimise the physical differences between individual upland
areas just as a guaranteed price for barley may minimise the physical differences between lowland areas. In neither case have the
costs of production necessarily been equalised, although the risk
of economic loss or failure may have been reduced. The effects may
have been sufficient to create a virtually isotropic economic plain
from a varied physical environment. In terms of the national,
rather than the local, economy and in terms of individual producers,
are these the most efficient or most equitable procedures? The
same questions should be raised in the context of the Common Agricultural Policy and the overall economy of the European Community.
To distinguish between the spatial variations in the input of a
particular subsidy and other, often essentially independent, developments in the agricultural geography of areas is crucial, as
Bowler (1979) has indicated in his analysis of spatial responses to
production grants.

A wider range of questions must also be raised concerning such instruments of public policy as agricultural subsidies. What was the original purpose of particular subsidies? To what extent were they to benefit farm incomes or living conditions in rural areas? Whom were they designed to benefit? Whom have they benefited? Historically, the goals of agricultural policy in Britain, as elsewhere, have been related to farm interests and have been focussed on increasing or stabilising farm production and income. A broader goal has sometimes been to guarantee for the national population particular quantities of farm produce at a particular price. To varying extents, these policies will have income re-distributive effects both within and outwith agriculture. One might question how relevant it is to discuss the impact of policies concerned with individual producers in spatial or regional terms, since any definition of spatial units, except at a micro-level, is going to include producers and production units with their varying responses to, and benefits from, subsidies. Here is surely a fruitful area for further enquiry, not least because there are clear spatial variations in the input of grant aid in England and Wales which were scarcely envisaged by the policy makers (Bowler, 1979).

Some agricultural policies have been geared to the equalising of farm incomes and the desire to assist farmers in the more difficult physical environments where, for a variety of physical, socio-economic, and historical reasons, small farmers in particular have suffered the lowest incomes. Yet in spite of governmental assistance over many years, such problem areas persist. Perhaps, instead of arguing that the economic system is to blame for the persistence of rural problems in such remote, peripheral and physically difficult areas as the west of Ireland or the Highlands and Islands of Scotland, it should be admitted that it is neither possible to produce agricultural goods there at prices comparable with production from more favoured areas, nor to support populations under any kind of farming system at standards of living equal to those found elsewhere without massive subsidies. The solution is to subsidise people to live there - if that is acceptable to society at large - rather than to subsidise agriculture. Problems of alternative or conflicting land uses might also be more critically examined in the context of limited environmental opportunities and broader societal requirements.

The more remote and difficult rural areas encapsulate in exaggerated form many of the features of all rural areas - static or declining populations; out-migration; scattered and ageing populations which lead to difficult service provision with high unit costs; difficulties of accessibility; low incomes; seasonal variation in employment and income; chronic under- and un-employment. However, it should be emphasised that at present there is considerable spatial variation in such features and that historical trends within and between individual areas are not necessarily consistent. During the last two decades out-migration in some areas has been balanced to some extent by in-migration so that population totals are no longer falling, but detailed evidence on a national scale is lacking except for the period 1966 to 1971 (Grafton, 1982). During this period net migration losses in remote rural areas of England and Wales were caused by relatively low levels of immigration rather than high levels of outmigration. The proportions of outmigrants aged 15-44 did not differ significantly between the 'least remote' and 'most remote' Rural Districts although there was a high proportion of immigrants of this age-group into the 'least remote' Districts. Advance analysis of the 1981 census returns shows a reversal of depopulation in many remoter areas but the explanation of this change is not yet clear and is likely to vary between different areas (OPCS, 1981). In some areas there is a high

proportion of elderly, retired people among the in-migrants and
this emphasises the overall ageing of the population and the attendant social problems of, for example, accessibility and health care
provision. In the remotest areas, especially of the northern and
western uplands, populations are still too small to be viable in
terms of social provision even with the recent increases, and the
increasing number of older people emphasises the prevalence of low
incomes.

Low-income people in rural areas are usually able to purchase
locally an adequate, if not abundant, supply of lowest-order goods.
They can often obtain public services at subsidised rates, particularly transport and public utilities. An increase in income is
likely to lead to a greater demand for higher-order goods and for
better quality services, which will not necessarily be available
locally. If general observations derived from urban consumer patterns are any guide, then increases in income will lead to proportionately greater increases in spending on clothing, transport and
recreation. These observations, together with general notions of
the threshold and range of goods and services and of the hierarchical arrangement of the provision of goods and services in central
place theory, would suggest that increasing incomes among the rural
poor would not necessarily increase expenditures locally. They
might increase them at higher points in the hierarchy; the village
shop will benefit less than the regional supermarket, and public
transport will attract even fewer passengers as private transport
becomes more widely used. In this context, services might be maintained on a wider basis, and perhaps more cheaply, by providing
state aid to shops rather than to shoppers, as in the Norwegian Aid
Programme in Sparsely Populated Areas (Kirby *et al.*, 1981). Similarly, Moseley's (1977) suggestions for a low-cost package of peripatetic public services might be more effective in providing
equality of access to a widely scattered population.

However, there are so few studies of expenditure patterns in
rural areas (as distinct from studies of shopping behaviour) that
it is difficult to be certain of the likely outcome of any public
policy aimed at assisting rural areas. The likely effects of new
industries introduced into rural areas, including the growth of
tourism and recreation and the spread effects of growth centres
upon their rural hinterlands, have still to be demonstrated in
detail (compare Bohlin and Ironside, 1977; Ironside and Williams,
1980; and Coppock and Duffield, 1975). Data are most abundant at
regional level, but sparse at individual level and scanty on cash
flows within local or regional economies. Crude employment figures,
however, clearly show that employment in most rural areas is now
predominantly in manufacturing and service industries, and not in
primary production. Agricultural subsidies, which are the most
frequently used instrument of public policy in rural areas, are
likely to help directly only a small proportion of the rural population, and their local multiplier effect is unknown. Employment
opportunities in non-farm activities are spatially sporadic. There
has been substantial growth of non-farm employment in some areas
such as East Anglia which has been particularly well-documented by
geographers (Sant and Moseley, 1977). In other rural areas such as
some in Northern Ireland and northern England, there has almost
certainly been a decline in non-farm employment following a decline
in a dispersed textile industry. Almost everywhere there has been
the demise of small firms processing agricultural produce, for
example, small maltings, breweries and distillers. These changes
are not associated so much with rural factors as with widespread
industrial change. The geography of rural areas should, perhaps,
be studied less in isolation and more in its wider context.

14

Discussion so far has been concerned with essentially economic problems, with only passing mention of the social characteristics of rural areas. Just as there is considerable physical and economic variety within British rural areas, so there is social variety, some of it based on long-standing interrelations between traditional societies and economies, such as fishing or crofting where each term is regularly used to embrace both society and economy. Increased intercourse between urban and rural societies has resulted from economic expansion, increased mobility and an explosive development in communication such that isolation from the media and unawareness of conditions beyond the local area have virtually disappeared. Nevertheless a rural geographer should not be surprised that a recently published series of essays by social anthropologists on identity and social organisation in British rural cultures should be entitled *Belonging* (Cohen, 1982). People in specific locations perceive differences between themselves and those in other locations, and their behaviour reflects this sense of difference. The location means something to them which it might not mean to others. People identify with locations and their capacity to do so explains to them why they behave as they do, that is, why they behave differently from others, and it may also consciously incline them to behave as they do. This does not mean that locations are isolated, nor that the sense of belonging to a particular location is the only sense of belonging which individuals possess. Nevertheless, the association of people with place and the sense of belonging to that place are powerful forces which may develop rapidly as witnessed by the association of migrants from urban areas with their new peri-rural or rural surroundings. Indigenous inhabitants do not regard the newcomers as 'belonging', but treat them as outsiders and not surprisingly tensions arise. Often these are most publicly exposed in the conflicts over development and the creation of new employment opportunities, and in the conflicts between agricultural intensification and rural conservation. Village preservation societies are often dominated by newly immigrant villagers.

More generally, many rural communities feel threatened not only by commuters or second-home dwellers, but also by unemployment; by changes in land use often resulting from changes of landownership or public policy; by the creation of recreation facilities; and by out-migration and the consequent changes in the community and its facilities, such as the closing or the village school or shop. Often communities respond, in part, by emphasising the tightly structured intricacy of local social life. Sometimes there is dissent and opposition of a broadly political kind which is most often directed against central government which is, rightly or wrongly, held to be the cause of whatever problems are currently recognised.

What then are the tasks of the rural geographer? At least one task must be to examine more fully the actual functioning of rural areas - the ways in which individual farms and other firms operate, individual responses to the physical environment and to more rapidly changing political, social and economic forces. He should examine actual cash and income flows within rural economies and the broader economies of which they are a part as well as the relations between behaviour and aspirations and the association of people with place. These last are all legitimate aspects of a genuinely *social* rural geography.

The results of such inquiries are likely to be diverse and often contradictory. In as much as they offer greater understanding, they may provide a better base than exists at present for the formulation and implementation of public policy.

REFERENCES

Archer, B.H. 1973. *The impact of domestic tourism*. Bangor Occasional Papers in Economics No.2, (University of Wales Press, Cardiff).

Archer, B.H. 1977. *Input-output analysis: its strengths, limitations and weaknesses*. (Travel Research Association, 8th Annual Conference, Arizona).

Bohlin, K.M. and Ironside, R.G. 1976. Recreation expenditure and sales in the Pigeon Lake area of Alberta: a case of 'trickle-up'? *Journal of Leisure Research,* 8, 275-88.

Bowler, I.R. 1979. *Government and agriculture: a spatial perspective*. (Longman, London).

Cloke, P. 1980. New emphases for applied rural geography. *Progress in Human Geography,* 4, 182-217.

Cohen, A.P. (ed) 1982. *Belonging: identity and social organisation in British rural cultures*. (Manchester University Press, Manchester).

Coppock, J.T. and Duffield, B.S. 1975. The economic impact of tourism: a case study in Greater Tayside. in *Tourism as a Factor in National and Regional Development,* Helleiner, F.M. (ed), Occasional Paper No.4 (Department of Geography, Trent Univ., Peterborough, Ontario).

Gilg, A.W. (ed) 1980-3. *Countryside planning yearbook,* 1-4, (Geo Books, Norwich).

Grafton, D.J. 1982. Net migration, outmigration and remote rural areas: a cautionary note. *Area,* 14, 313-8.

Gregory, S. 1964. Climate. in *The British Isles: a systematic geography,* Watson, J.W. with Sissons, J.B. (eds), (Nelson, Edinburgh), 53-73.

Greig, M.A. 1971. The regional income and employment multiplier effects of a pulp mill and paper mill. *Scottish Journal of Political Economy,* 18, 31-48.

Greig, M.A. 1972. *A study of the economic impact of the Highlands and Islands Development Board's investment in fisheries*. (HIDB, Inverness).

Ironside, R.G. and Williams, A.G. 1980. The spread effect of a spontaneous growth centre: commuter expenditure patterns in the Edmonton Metropolitan Region, Canada. *Regional Studies,* 14, 313-32.

Kirby, D., Olsen, J.A., Sjøholt, P., and Stølen, J. 1981. *The Norwegian aid programme to shops in sparsely populated areas: an assessment of the first four years*. (Norwegian Fund for Market and Distribution Research, Oslo).

Moseley, M.J. 1977. A look at rural transport and accessibility. *The Village,* 32-35.

O P C S (Office of Population Censuses and Surveys) 1981. First results from the 1981 Census. *OPCS Monitor,* July, 1981.

Sant, M.E.C. and Moseley, M.J. 1977. *Industrial development in East Anglia*. (Geo Books, Norwich).

Stamp, L.D. 1948. *The land of Britain: its use and misuse*. (Longmans, London).

Chapter 3

Proposal for a theoretical basis
for the human geography of rural areas

Joeke Veldman

INTRODUCTION

To include rural geography legitimately as a scientific part of
human geography, it has to be proven that rural regions have so
many mutual similarities that they are a regional type to which one
can apply nomothetic statements. At the same time the differences
from non-rural, regional types have to be sufficiently great to be
able to distinguish rural from urban. Seen from an external viewpoint, a regionalised unit can be compared with a 'formal region'.
But seen from an internal viewpoint every regionalised unit has a
functional character. Associations and interactions cause complex
and developing differences within the unit, which form the basis
for internally induced changes (Jones, 1980). Following Hoekveld
et al. (1978), Ottens (1976), and Driessen (1978), the internal
similarities and the external differences can be found, first, in
the socio-spatial system, the users of space being the elements;
second, in the physical-spatial structure or the space used; and,
third, in the character of the interaction between those two – the
use of space.

Rural areas are a regional type by the nature of the industries
and the use of space resulting from them. As opposed to urban
areas, the countryside is dominated by area-bound forms of land use
(Veldman, 1979, 178; Van Bemmel *et al.*, 1979, 6), as exemplified
by those activities where the extent and/or the quality of the
results are directly influenced by the land used. These activities
include agriculture, cattle rearing, market gardening and forestry
as well as reservoirs, recreation in the open air, military training areas and rubbish dumps (De Vries Reilingh, 1967, 146). Clout
(1972, 3) would add nature conservation and mineral extraction.
The entire physical-spatial structure includes also land left to
nature and water surfaces. Hence inland fishing is also an area-
bound activity. So the physical-spatial structure is not exclusively the passive result of the inhabitants' activities analysed in
the framework of space users, space use and used space. It may be
an independent factor in the interaction between space and space
users. This is the basic difference between rural areas and urban
areas. For, the dominance of area-bound activities and/or surfaces
left to nature directly influences the functioning of the socio-
spatial systems in small villages, hamlets or isolated dwellings
which are surrounded by relatively large areas of uninhabited
space and consequently a relatively low density of settlement. That
is why Clout (1972, 1) could describe rural areas as "less-densely
populated areas which are commonly recognized by virtue of their
visual components as 'countryside'"!

The difference between urban and rural areas has been immensely strengthened by such phenomena as the Industrial Revolution at the national level and by the 'undevelopability' of some peripheral rural areas. At the present time there is great interest in the qualities of the rural physical-spatial structure, especially for nature, ecology, quiet and scenic beauty. That is why society attaches greater significance to rural areas than in former days.

SOME EFFECTS OF THE RURAL PHYSICAL-SPATIAL STRUCTURE ON THE SOCIO-SPATIAL SYSTEM

Thus, rural areas have their own geographical milieu based on the physical-spatial structure, which results from area-bound activities and land left to nature and water. The effects on the socio-spatial system and on the life and employment of the rural inhabitants have social relevance. In this context the direct connection between the rural physical-spatial structure and the relatively low density of population has already been mentioned. It is meaningful to arrange the problems of the socio-spatial systems which are connected with the physical-spatial structure on the basis of the scale at which one approaches the rural geographical milieu. One can discern three levels of study:
- the macro-level of (inter-)national scale. The elements of research are the regions showing growth, stagnation or decline with regard to each other in population, prosperity and development. This is the problem of imbalance between urban concentrations and lagging rural areas as it has been formulated in the regional economic policies of most Western states. The key problems at this level are the quantity and quality of non-agricultural employment, the decision making on rural planning goals, the effect of policies and the management of rural areas.
- the meso-level or regional scale. At this scale the elements of research are the villages and the village areas. The research is directed at the functioning of the settlement pattern. Key problems on this level include regional organisation of the supply of central goods and services, open-air recreation, the realisation of rural planning goals and landscape management.
- the micro-level or local scale. The elements of research are the households, firms and other institutions and individuals. The relations between these are expressed in the welfare of village communities and the small-village problem. An aspect of that is the organisation into large administrative units of small numbers of citizens (see Groot's paper in this book).

In reality the division between the scales is not sharp. This is expressed clearly in the regional effects of some national regulations. Uniformity of postal rates is a concealed subsidy to the rural areas since the real costs of postal delivery are much higher there than elsewhere. On the other hand, a single national minimum number of pupils for keeping open kindergarten and elementary schools is a disadvantage for rural areas because of their low population densities.

The interrelationships between the different scales also apply to rural research and rural policy. At the macro-level individual actions are just a statistical mass. That is why attention is devoted to spatial-economic problems. At the local level households and individuals are the elements of research. That is why social problems get most attention at that level. At the regional level both approaches have to be brought into balance.

18

TRANSFORMATIONS OF RURAL AREAS

Referring to the transformations of rural areas, Mauret (1974) makes a distinction between pre-industrial and industrial societies. Before 1800, pre-industrial village communities were diverse. In addition to farmers, craftsmen of all kinds lived and worked there. Social relations were predominantly territorially bound.

During the Industrial Revolution in the nineteenth century, craft industries failed to survive in the countryside. Out-migration resulted so that just agricultural employment remained (Franklin's (1969) agrarianisation of the countryside). The rural-urban dichotomy was complete; the rural areas were occupied by isolated village communities where all members showed a territorially bound spatial behaviour.

After the Second World War, a structural crisis in agriculture developed. Some farmers had to change their profession, although they might not be compelled to leave their houses. Besides, many city-dwellers were looking for a residence in the rural environment - residential suburbanisation. After 1970 this evolution was strengthened by industrial suburbanisation; a lot of companies sought a rural location for their expansion.

All languages of Western Europe have a term for the concept of 'countryside'. It refers to the rural-urban dichotomy which finds its expression in the physical-spatial structure of area-bound land-use forms, and in the socio-spatial system with a predominantly agricultural composition and territorially bound spatial behaviour. Countryside is a term for this twofold unity of a community form and a land-use form, the former influenced by the latter. That is why village communities are of great intensity too. There is a basic difference from other forms of society. Considered spatially a settlement pattern arises with villages, towns and cities in a characteristic hierarchy of numbers, sizes, functions and market areas.

After the Second World War the characteristic twofold unity of community form and land-use form was severed completely. The village's communal relations had already been weakened. Compulsory education, the organisation of farmers into larger groups, increasing use of non-factor inputs in agriculture, the increased mobility from private and public transport and the mass media had all contributed to a greater individualisation of behaviour and attitudes. The basic differences between the communities of the countryside and the wider society disappeared completely as a consequence of the individual mobility caused by the car. That is why non-agricultural active households are able to come and live in villages. Socio-spatial relations are spread over a much wider area with many other settlements within reach. Fundamental differences in rural and urban mentalities no longer exist. In the Netherlands 'countryside' has come to be used as a historical concept. It has been replaced by the term 'rural areas' (for example, in the annual report of the Dutch State Planning Service). The term rural areas refers only to a rural physical-spatial structure, that is, the coincidence of area-bound forms of land use and the interaction between physical-spatial structure and socio-spatial systems.

These processes in rural areas are often described as urbanisation (Prillevitz and De Groot, 1968; Burie, 1966; CBS, 1964; Van Engelsdorp Gastelaars et al., 1980). A distinction is made between physical, mental and socio-economic urbanisation for successive changes in the physical environment, culture and social structure. One may object to the way in which the term urbanisation is used.

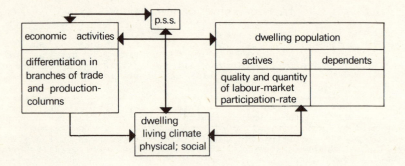

Figure 3.1   A conceptual model for regionalising

The term does not just have one meaning as it may refer to a situation or to a process, and because it may indicate socio-spatial,
socio-economic and socio-cultural qualities. Another impediment is
that urbanisation suggests a one-sided approach to changes in rural
areas. Endogenous developments are hidden by the use of the term
(Mathieu and Bontron, 1973). Moreover the three forms of urbanisation are not always and everywhere co-extensive.

TYPES OF RURAL AREAS

For regionalisation of rural areas, one could choose an approach
through any.of these three elements; space users, space use and
used space. This study considers the interaction between space
users and used space or, in other words, between the socio-spatial
system and the physical-spatial structure. This encompasses the
entire spatial dimension. The conceptual model for a regionalising
human geography is one which aims to recognise the character of
these interactions (Figure 3.1).

    Considered regionally, one may take the view that people live
where they work; one lives where one can get an income. This
statement is not applicable for incomes not bound to a place, such
as interest and rent, pensions and redistributive incomes. But to
region-bound households the dwelling place is derivative from the
working place. In a broader sense the population of a region is a
reflection of the nature and size of the regional activities. It
is not so strange, therefore, that the regional population has
characteristics related to the regional economic structure; for
example, its size and density, average income, average educational
level and age and sex structures.

    The character of the interaction may be urban, peripheral rural,
semi-peripheral rural or peri-urban rural. In general, activities
in urban areas (Figure 3.2) are characterised by space being a
catalyst without being directly productive. The activities are
mainly secondary, tertiary and quaternary in character including
housing. This has led to a considerable concentration of people,
objects and buildings. Another characteristic of urban activities
is heterogeneity. Most kinds of trade and industry are represented.
That is why urban labour markets are differentiated both in terms
of training and employment. Accordingly the active population has
a broad choice of career and education. As a rule, average income
is higher than the national average.

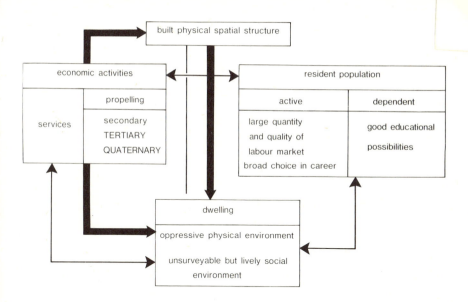

Figure 3.2   The interaction between physical-spatial structure and
the socio-spatial system in large cities

The physical-spatial structure is the detailed concrete
expression of this concentration of industries and housing. As a
consequence of an imperfect control of the built environment, the
physical environment is burdened by over-concentration, noise and
polluted air, notwithstanding the great differentiation of milieux
for living. The social milieu is characterised by its 'unsurvey-
ability' (i.e. lack of clear structure). Nevertheless, the big
cities are centres of arts, culture and spiritual life.

In urban areas the spatial result of residential and economic
activities is a dominance of the socio-spatial system in the interdependence with the physical-spatial structure. This presents a
contrast with rural areas.

SEMI-PERIPHERAL AND PERIPHERAL RURAL AREAS

In peripheral rural areas the physical-spatial structure consists of
some agricultural land and much land left to nature. The population
is small and lives at low density. The physical-spatial structure
is dominant in the interaction with the socio-spatial system. In
semi-peripheral rural areas much less land is left to nature. The
most important (propelling) activities are agriculture and similar
area-bound occupations. The physical-spatial structure consists of
large, almost uninterrupted areas that are directly productive.
Therefore other forms of land use are almost excluded. The agricultural changes lead to problems which are insoluble because of the
characteristics of, and the appreciation for, the physical-spatial
structure. The last is dominant in the inter-dependence (Figure 3.3).
The origin of the landscape and settlement pattern are in the distant past. They were achieved as a function of an agricultural
land use in a period of a now outdated technological-organisational

21

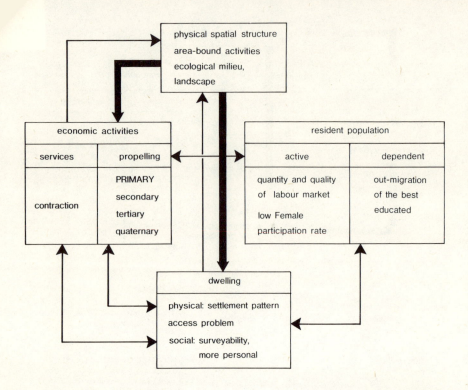

Figure 3.3 The interaction between physical-spatial structure and
the socio-spatial system in peripheral rural areas

system. These kinds of rural area are peripheral being out of the
daily reach of the labour and housing markets of the big cities and
city regions. That is why there are few urban commuters with a
rural background. Economic development is strongly coupled to
agricultural changes. Buttel (1980) draws attention to four connected structural changes in farming:
  i) a tendency for big, specialised production-units and farm
     enlargement;
 ii) a far-reaching and continuing mechanisation aimed at increas-
     ing labour productivity;
iii) an increasing use of bio-chemical, non-factor inputs aimed at
     raising production per animal and per hectare;
 iv) a strengthening of the agribusiness complex and interregional
     commerce aimed at expansion.
The consequences of these developments for the peripheral rural
areas are:
- fewer workers and many independent farmers going out of business;
- a shrinking labour market lagging both in quality and quantity
  which makes it unattractive for non-agricultural firms;
- agribusiness (the supplying, producing and processing industries)
  spans a relatively large part of the production process. The
  differentiation of professions and so the choice of a career is
  broadened. However, the tendency for fewer and bigger plants
  leads to less employment and longer journeys to work for the
  commuting workers;

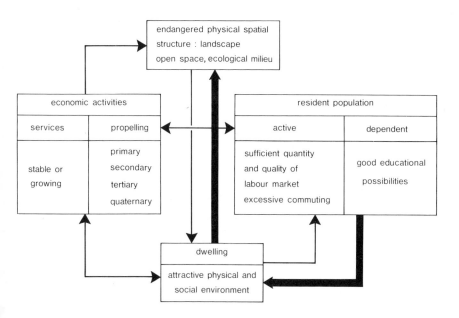

Figure 3.4   The interaction between physical-spatial structure and
the socio-spatial system in peri-urban rural areas

- an outdated settlement pattern in the sense of too many small
villages to keep up a sufficient carrying capacity for public and
private services. The tertiary activities show a tendency to
concentrate in larger villages and shops;
- pressure from the government and public opinion on farmers to
protect scenic beauty and the ecology. Residing in peripheral
rural areas is judged positively by many people. It is not
exceptional for households with a free choice of housing to
settle down in these areas. The originally rural inhabitants
accept the greater strains in getting central services and the
lag in supplying activities which are beyond their access. They
feel compensated for this by both the qualities of the physical
environment (quiet, pure air, nearer to nature) and by those of
the social environment (clarity of structure (i.e. 'survey-
ability') more personal, less stressful). All these trends are
greater in the peripheral areas.

PERI-URBAN RURAL AREAS

These rural areas are situated within the daily reach of the housing and labour markets of the big cities. They offer space for
residential and industrial suburbanisation. As in the cities, the
socio-spatial system is dominant in the interdependence with the
physical-spatial structure (Figure 3.4).

In a morphological sense, the peri-urban rural areas are rural
since the landscape is characterised by area-bound activities.
However, an important part of the peri-urban areas belongs to the
rural parts of city regions. The physical urbanisation of city
expansion transfers rural land to urban uses. Residential

suburbanisation, by which new residential quarters are built near villages, does not do that. Industrial suburbanisation has led to the construction of industrial parks near small towns and villages, particularly near traffic nodes. These developments have been very influential in moulding the character of peri-urban rural areas. If socio-economic forces have free play, the suburbanisation processes cannot be stopped. The area-bound activities produce less value per hectare than the non-area-bound activities, including housing. This is best seen in the price of land. Urban land users bid far higher prices, so that the area-bound activities are driven out. Though the present rural areas are morphologically rural, the interdependence between physical-spatial structure and socio-spatial system is characterised by the dominance of the socio-spatial system. The peri-urban rural areas have a rural character visually, but inwardly they are urban.

As a rule intensive agricultural methods are found closer to cities but that does not mean that they can resist suburbanisation pressing forward. Government policy has as one of its aims the regulation of the processes of suburbanisation. For that purpose the physical-spatial structure is protected by regulations especially on behalf of the ecological milieu, the scenic beauty of land-scapes and open spaces. The conservation of the physical-spatial structure is pursued through urban-fringe zones and green belts formulated in regional and structure plans. The labour market shows all the characteristics of a metropolitan labour market. The industrial economy is broadly based. Quantity and quality of employment are relatively favourable and a wide choice of career is possible. The educational possibilities are also diverse. The growing number of inhabitants offers a viable market for the main-tenance of public and private services. The cost of all this is excessive commuting.

It is clear that the population's wishes in relation to the physical and social milieu are critical in these developments. A favourable physical and social climate is ascribed to rural areas; peace and purity of living in the countryside in a clearly struc-tured community.

SECTORS AND FACETS OF RURAL AREAS

Integrated research in rural areas has special significance for building a theoretical framework. Specialised research is important for applied research and policy. Socio-spatial research in rural areas has to be engaged in a choice of phenomena and factors which partly explain the geographical milieu in question. Together these contribute towards research of *sectoral problems* in rural areas. The sectoral approaches may proceed from:
- the used space, expressing itself in the physical-spatial struc-ture. This research is engaged in problems of land structure, the scenic beauty of the landscape and the ecological milieu, the settlement pattern and the major works of nature and man;
- space use - agriculture and forestry, open-air recreation, non-agricultural activities, public and private services, access and housing problems;
- the space users - demographic changes, shifts in authority and the location power, costs and benefits of living in rural areas, communities and rural deprivation.

Besides this sectoral research directed at problems inherent in rural areas, 'facet research' is characterised by the position from which rural or sectoral problems are observed and interpreted. The facets of the rural question find expression in the diversification

of rural sciences. Rural human geography takes space as an
approach to these problems; rural economics and rural sociology
take space use and space users as their respective starting-points
for research. The epithet 'rural' to these scientific specialisa-
tions makes it clear that they are all handling a regionalised concept. Some of the central questions which may be chosen in 'facet
research' are briefly mentioned below:
- rural geography: the specified field of research is the interaction between the physical-material structure and the socio-
  spatial system in rural areas. One of the consequences of this
  structure for society is the problem of access and hence passive
  and active deprivation (see Huigen's chapter in this book).
  Another is the origins of, and changes in, the scenic beauty of
  landscapes. In this respect, the 'facet research' of historical,
  recreational and agricultural geography may all contribute in an
  important way. A third topic is the qualitative and quantitative
  nature of rural communities as this relates to the physical-
  spatial structure since this affects the costs and benefits of
  living in rural areas. Also important is the optimising of the
  settlement pattern in accordance with a modern technical-
  organisational system (see de Bakker's paper in this book).
  Finally, we may study the problems of spatial planning in rural
  areas as this relates to all these facets.
- the social problems (especially of agriculturally defined areas)
  come within the scope of rural sociology. The viability of
  small villages, the nature of communities and the welfare of the
  inhabitants are basic research goals. An important aspect is
  passive and active rural deprivation. The same goes for the
  shifts in power away from villages.
- agricultural economics studies the prosperity of agriculture:
  the returns to agricultural production systems in relation to
  farm size, the degree of mechanisation and rationalisation, and
  land management.
- ecology deals with the problems of the natural milieu and the
  scientific value of landscape. This science aims at conservation and management of nature and so physical geography is
  important.
- the administrative sciences direct their research especially to
  the division of communes and provinces. The ruling force of
  small communities is considered to be too weak to fulfil properly
  many community tasks.

   In rural areas, both sectoral and facet research and also integrated research are confronted with the methodological problems of
the small numbers involved. That goes especially for peripheral
rural areas. By definition, the research population exists in a
small number of households and/or firms, which in general have
direct, personal and strong mutual relationships. That is why all
elements of the geographical milieu are intensively related.
Changes in one element have direct repercussions for the other
elements. Two important consequences for socio-spatial research in
rural areas result from this, namely:
- a methodological question: what scientifically justified method
  may be used to lay down how far the results of small-scale research may be generalised?
- a pragmatic consequence: socio-spatial research in rural areas
  has to accept a multidisciplinary approach to prevent the conclusions from being disadvantageous for the rural physical-
  spatial structure and for the rural communities by accentuating
  its own facets too strongly.

   Government policy in rural areas also has to be aware of the
unusual sensitivity of the relations between physical-spatial
structure and the socio-spatial system. Therefore, it is

desirable to work towards integrated policies and management in these areas.

REFERENCES

Bemmel, A. van, Huigen, P. and Veldman, J. 1979. *Projektkader, De stand van zaken per 1-9-1979.*(GIRUU, Vakgroep SGS/Toegepaste Geografie, Utrecht).

Burie, J.B. 1966. *Stedelijkheid als driedimensioneel fenomeen. Een poging tot verheldering.* (Sociologisch Instituut, R.U.U., Mededeling no. 35, Utrecht).

Buttel, F.H. 1980. Agricultural structure and rural ecology: towards a political economy of rural development. *Sociologia Ruralis*, XX, 1/2, 44-62.

CBS 1964. *Typologie van de nederlandse gemeenten naar urbanisatiegraad, 31 mei 1960.* (De Haan, Zeist).

Clout, H.D. 1972. *Rural geography: an introductory survey.* (Pergamon Press, Oxford).

Driessen, B.G. 1978. *Grondslagen en programma Stepro. Nota no. 2.* (GIRUU Vakgroep SGS/Toegepaste Geografie, Stepro-Onderzoekenreeks Stedelijke problematiek West Nederland).

Engelsdorp Gastelaars, R.E. van, Ostendorf, W.J.M. and Vos, S. de 1980. *Typologie van nederlandse gemeenten naar stedelijkheidsgraad. Milieu-differentiatie en maatschappelijke ongelijkheid binnen Nederland. Monografie Volkstelling 1971. Deel 15B.* (Staatsuitgeverij, 's-Gravenhage).

Franklin, S.H. 1969. *The European peasantry: the final phase.* (Methuen, London).

Hoekveld, G.A., Jobse, R.B., Weesep, J. van and Dieleman, F.M. 1978. *Geografie van stad en platteland in de Westerse landen.* (Romen, Bussum).

Jones, G.E. 1980. Economic growth and regional development - harmony and disharmony in a differentiation process. *Sociologia Ruralis*, XX, 1/2.

Mathieu, N. and Bontron, J-C. 1973. Les tranformations de l'espace rural: problèmes et méthode. *Etudes Rurales*, 49/50 (janvier/juin), 137-159.

Mauret, E. 1974. *Pour une équilibre des villes et des campagnes: aménagement, urbanisme, paysage.* (Dunod, Paris).

Ottens, H.F.L. 1976. *Het groene Hart binnen de Randstad. Een beeld van de suburbanisatie in West-Nederland.* (Van Gorcum, Assen).

Prillevitz, F.C. and Groot, K. de 1968. Landbouw in een verstedelijkend Nederland, *Landbouwkundig Tijdschrift,* 12-23.

Veldman, J. 1979. Enige gedachten over toekomstige ontwikkelingen in landelijke gebieden. in *Nederland op weg naar een post-industriële samenleving?* Kwee, S.L. et al. (eds), (Van Gorcum, Assen), 177-189.

Vries Reilingh, H.D. de 1967. Verstedelijking en landbouw: de agrarische ruimte in de toekomstige planologische ontwikkeling van Europa. *Landbouwkundig Tijdschrift,* 146-153.

Chapter 4

A multivariate approach to a rural typology
of Dutch regions at different spatial scales

Joost Hauer

INTRODUCTION

The research reported in this paper has been carried out within the context of the Zuidwest-Friesland Project of the Department of Geography of the State University of Utrecht. The construction of a rural typology was one of the research items in the main project. The aim of the typology project was to give a more specific indication of the position of the research area with respect to other Dutch regions at the same spatial scale using a range of rural indicators. In a wider context however, this taxonomic exercise serves a deeper research function. Especially since Cloke (1980) made it clear that present-day rural geography is still very much searching for concepts and methods suitable for rural problems, efforts to reach these goals have been intensified. One of the problems met is that in order to be able to present a more sound theoretical framework, much more insight has to be gained into the behaviour of relevant variables through empirical research designed with this explicit aim (see also Cullingford and Openshaw, 1982). In the Zuidwest-Friesland Project a preliminary conceptual framework was designed to guide the selection and study of supposedly significant variables (see Veldman's chapter in this book). The lines of research needed are mentioned elsewhere (Hauer and Veldman, 1983) and only two small aspects are investigated in this paper. These concern the choice of indicators from a rural point of view and the behaviour of indicators at different spatial levels.

The first analysis deals with regions at the COROP level. The decisive reason why this level of analysis is chosen stems from the fact that COROP region no.5, Zuidwest-Friesland, is the research area of the overall project. Therefore the starting point for this analysis is the division of the Netherlands into 40 regions known as COROP regions (Figure 1.1). This division was introduced about a decade ago and serves as a framework for integrating basic statistics. It is important to note here that this division is based on a nodal regionalisation principle using material from the last census (1971), while administrative boundaries of provinces and municipalities were not to be violated by the COROP boundaries. These points have to be remembered, of course, when interpreting results.

The second analysis deals with the 44 municipalities of Friesland. The choice of this province is evident from the research

27

area of the overall project. Fortunately, the municipality boundaries in Friesland have not yet been changed as has happened in
some other parts of the country where a new administrative system
has been introduced recently. Apart from the general disadvantage
that municipalities are more or less historic entities and not geographic ones, the administrative structure in Friesland differentiates some of the bigger settlements quite neatly.

It must be stressed here that these analyses are of an exploratory character, especially because the choice of indicators rests
on the central hypotheses of the overall research project, as will
be seen in the next paragraph. Besides this, results are only
given for a constant set of indicators at the two spatial levels;
we have not tackled here the problem of looking at different sets
of indicators.

THE CHOICE OF INDICATORS

The aim of the classification project is to develop a typology of
regions according to the extent to which these regions can be
described as rural. As a starting point for the process of selecting indicators, we used the central theme of rural geography as it
is formulated in our research group; the interdependence of the
material-spatial structure of rural areas and the socio-spatial processes taking place in these areas. In relation with this interdependence, it is hypothesised that the more rural regions are, the
more dominant is the position of structure variables, process variables being more dependent.

Looking at rural areas we note that these regions have the
following characteristics:
a) a dominance of activities that need vast amounts of space
   (agriculture, recreation);
b) a settlement pattern with relatively small nuclei and dispersed
   settlement;
c) a low density of population;
d) a regional organisation of service provision.

Especially regarding the more peripherally situated areas, one
could add:
e) non-participation in urban-centred, regional housing and
   labour markets;
f) the low quality of non-agricultural employment.

These characteristics were formulated in the early phase of the
overall research project and will be investigated more fully elsewhere. We use them as theoretical statements from which to derive
the operational indicators. They lead us to two sets of variables;
a set of 9 structure variables and another set of 13 process variables, these reflecting the distinction between the material-spatial
structure and the socio-spatial processes.

The choice and operationalisation of the *structure variables* is
almost self-evident. The following 9 indicators are used in the
analysis.
1. UNB The area not built-up as a percentage of the total area on
       1.1.1976;
2. CUL The cultivated area as a percentage of the total on 1.1.1976;
3. GLS The area of gross living space (i.e. inclusive parks etc.)
       per 1000 inhabitants on 1.1.1976;
4. PRO The number of kilometres of paved roads outside built-up
       areas per square kilometre of area not built up on 1.1.1975;

5. P50 The population living in municipalities with less than
       50 000 inhabitants as a percentage of the total population
       on 1.1.1972;
6. DIS The average distance from the biggest place to the three
       nearest towns with more than 100 000 inhabitants on 1.1.1976;
7. POP The population of the COROP region taken together with the
       population of the neighbouring COROP regions as a percentage
       of the total population of all COROP regions on 1.1.1976;
8. FAM The number of one-family houses as a percentage of all
       dwelling units in 1971;
9. OLD The number of houses dating from before 1945 as a percentage
       of all dwelling units in 1971.

The operational definitions are given here as they were used in
the COROP-level analysis. In the municipality-level analysis some
definitions had to be adapted as follows:
5. P40 The population living in nuclei with less than 4000 inhabi-
       tants as a percentage of the total population on 1.1.1972;
7. POP The population of the municipality on 1.1.1976 taken
       together with the populations of neighbouring municipalities
       as a percentage of the total population of the COROP region
       in which the municipality lies.

These nine variables are closely connected with the characteristics of rural areas as given above, especially the first four.
Aspects of the housing market are measured through variables 8 and 9.
Population density is measured indirectly by variables 5, 6 and 7;
direct measurement was rejected on statistical grounds since population density correlates very strongly with some other indicators,
in particular with variable 1. All nine variables, except variable
7 (POP), are believed to covary positively with the extent to
which regions are rural.

The selection of *process variables* led to a choice of variables
which are indicators of processes more strongly associated with
rural areas. The final choice reflected a desire to use a set of
indicators which is representative of those processes that are
selective towards rural regions. The following thirteen indicators
were used in the analysis:
10. INC Average income per inhabitant in 1974;
11. WOR The number of lower-grade employees and workers as a
        percentage of all heads of household in 1971;
12. EDU The working population with primary education as a percent-
        age of the total working population in 1971;
13. GRO The growth of population 1970-1976 as a percentage of the
        population in 1970;
14. MIG The net migration 1973-1976 as a percentage of the
        population in 1973;
15. 019 The number of 0-19 year olds per 1000 inhabitants in 1976;
16. 64+ The number of people 64 years old and over per 1000
        inhabitants in 1976;
17. OUT The number of out-migrants 20-29 years old as a percentage
        of total emigration in 1974;
18. IMM The number of in-migrants of 60 years old or over as a
        percentage of total immigration in 1974;
19. EMP The number of unoccupied houses per 1000 dwelling units on
        1.10.1976;
20. PRI The number of privately owned houses as a percentage of all
        dwelling units in 1971;
21. CAR The number of cars per 1000 inhabitants in 1975;
22. MEM The number of 20-64 year old, male unemployed as a
        percentage of all men in this age group in 1975.

29

Mainly because of data availability some slight adaptations
had to be made for the municipality-level analysis:
13. GRO The growth of population 1972-1976 as a percentage of the
    population in 1971;
14. MIG Net migration 1974-1978 as a percentage of the population
    in 1973.
For variables 15, 16 and 19, the year of measurement is 1978,
while for variable 21 the year of measurement is 1976.

    The selection of these indicators can be justified by the
following reasoning which also gives their hypothesised direction
of covariation with rurality. In rural areas generally we find a
relatively small growth of population (GRO), mainly caused by a
negative net migration (MIG). We see an over-representation of two
groups among the migrants; emigration of the younger age group is
over-represented (OUT) and immigration predominantly comprises
elderly people (IMM). As a consequence, we expected relatively
high proportions of the young (0-19) and the elderly (+64). The
more peripheral rural areas especially are not attractive for non-
agrarian economic activity. The negative selection in the quality
of non-agrarian employment becomes evident when one sees that highly
skilled or trained labour, head offices, laboratories etc. are
almost absent. This line of thought is indicated with variables 11
and 12 (WOR and EDU). It is likely that the more highly trained
will not find suitable jobs on the regional labour market. Many
feel themselves forced to leave the region. Variables 10 and 22
measure the consequences for the region itself; low average income
(INC) and unemployment (NEM). The last variable however is also
incorporated because cyclic and structural developments in the
economy have a stronger and more direct impact on employment in the
predominantly productive industry in rural areas. A high percentage of privately owned houses (PRI) and also a relatively large
number of unoccupied houses (EMP) are considered as characteristics
of rural areas. The same goes for the number of cars (CAR), since
mobility is an absolute necessity in regions where distance plays
such an important role in the organisation of most activities.

    In summary we see that all indicators except 10, 13 and 14
covary positively with rurality; INC covaries negatively, for GRO
we expect lower than average values in peripheral regions and MIG
will mostly be negative. Because of the extreme value of most
indicators in COROP region no.40, the IJsselmeerpolders, this
region has been excluded from all the following calculations at the
COROP level.

                THE STRUCTURE OF THE INDICATOR SETS

Before we enter into the classification of regions on the basis of
the chosen indicators, we would like to investigate the internal
structure of the two groups of variables through correlation and
principal components analysis. The results of these analyses are
interesting in themselves because they give a check on the choice
and hypothesised indicative value of the variables. Before any
comments are given we present the results in tabular form
(Tables 4.1, 4.2, 4.3 and 4.4).

Table 4.1 The Pearson correlation coefficients between the structure
          variables at two spatial levels (C = COROP level,
          M = municipality level)

Variables		1	2	3	4	5	6	7	8	9
1. UNB	C	x	.71	.77	.27	.69	.62	-.35	.74	-.41
	M	x	.06	.35	-.68	.71	.18	-.02	.83	.50
2. CUL	C		x	.56	.19	.50	.47	-.23	.42	-.29
	M		x	-.66	.18	-.12	-.27	.49	.08	.22
3. GLS	C			x	.22	.66	.73	-.54	.73	-.35
	M			x	-.44	.39	.44	-.27	.32	-.06
4. PRO	C				x	.17	-.13	-.01	.34	-.66
	M				x	-.55	-.19	.07	-.49	-.19
5. P50	C					x	.63	-.34	.69	-.32
	M					x	.32	-.23	.66	.59
6. DIS	C						x	-.72	.44	.05
	M						x	-.54	.27	.24
7. POP	C							x	-.43	.04
	M							x	-.21	-.24
8. FAM	C								x	-.55
	M								x	.43
9. OLD	C									x
	M									x

Table 4.2 Principal Components Analysis, the loadings on the first
          three unrotated components of the structure variables at
          two spatial levels

Variables	COROP - level components			Municipality-level components		
	I	II	III	I	II	III
1. UNB	.90	-.04	.26	.84	.40	-.21
2. CUL	.70	-.01	.54	-.27	.85	.20
3. GLS	.90	.11	-.06	.61	-.57	-.38
4. PRO	.33	-.78	-.24	-.71	-.07	.46
5. P50	.82	.06	.14	.85	.18	.04
6. DIS	.75	.58	-.08	.53	-.44	.43
7. POP	-.58	-.45	.59	-.41	.56	-.56
8. FAM	.84	-.19	-.15	.81	.31	-.01
9. OLD	-.49	.76	.15	.55	.43	.55
Eigenvalue	4.72	1.79	.83	3.79	2.03	1.23
Cumulative % of total variance	52.4	72.3	81.5	42.1	64.6	78.3

Table 4.3 The Pearson correlation coefficients between the process
variables at two spatial levels (C = COROP level,
M = municipality level)

Variables	10	11	12	13	14	15	16	17	18	19	20	21	22
10. INC C	x	-.39	-.90	-.24	-.51	-.75	.43	-.01	-.24	.02	-.14	.36	-.49
M	x	-.15	-.45	.24	.36	-.55	.44	.21	-.16	-.29	.01	-.32	-.40
11. WOR C		x	.39	.13	.20	.30	-.48	-.17	-.16	-.10	-.10	-.18	.37
M		x	-.11	.29	.02	.19	-.54	-.23	-.18	-.22	-.58	.12	-.02
12. EDU C			x	.14	.53	.64	-.26	-.26	.30	.14	.20	-.30	.43
M			x	-.22	-.33	.29	-.21	-.15	.01	.26	.21	.22	.21
13. GRO C				x	.47	.36	-.33	.06	.01	-.15	.03	-.34	.08
M				x	.03	.02	-.17	-.06	-.00	-.36	-.14	-.09	-.08
14. MIG C					x	.66	-.34	-.22	.19	-.09	.49	-.37	.04
M					x	-.24	.05	-.11	-.37	-.24	-.24	-.23	-.56
15. 019 C						x	-.72	.16	.01	-.29	.18	-.43	.09
M						x	-.50	-.08	-.04	.10	-.01	.06	.12
16. 64+ C							x	-.37	.51	.45	.11	.31	-.06
M							x	.17	.11	-.09	.33	-.22	.10
17. OUT C								x	-.36	-.44	.01	-.02	-.16
M								x	.06	-.18	.37	-.34	.23
18. IMM C									x	.16	.16	.07	.33
M									x	-.04	-.12	.35	.29
19. EMP C										x	-.01	.04	.17
M										x	.10	.21	.12
20. PRI C											x	-.04	.26
M											x	.04	.42
21. CAR C												x	-.03
M												x	-.04
22. NEM C													x
M													x

Table 4.4 Principal Components Analysis: the loadings on the first
four unrotated components of the process variables at two
spatial levels

Variables	COROP-level components				Municipality-level components			
	I	II	III	IV	I	II	III	IV
10. INC	-.88	-.22	.12	-.27	-.85	.09	.07	.12
11. WOR	.52	-.03	-.52	-.18	.17	-.76	.38	.00
12. EDU	.80	.43	-.07	.16	.63	.16	-.27	-.10
13. GRO	.48	-.17	.18	-.42	-.23	-.39	.48	-.20
14. MIG	.74	.12	.48	-.17	-.64	-.34	-.39	-.06
15. 019	.89	-.20	.12	.14	.60	-.24	-.03	-.52
16. 64+	-.66	.64	.23	.02	-.47	.65	.05	.34
17. OUT	-.01	-.71	.03	.46	-.22	.50	.40	-.37
18. IMM	.08	.74	.20	.31	.30	.04	.59	.59
19. EMP	-.19	.64	-.09	-.36	.45	.22	-.51	.19
20. PRI	.20	.14	.77	.18	.10	.80	-.01	-.21
21. CAR	-.53	.12	-.10	.45	.49	-.15	-.07	.60
22. NEM	.35	.45	-.61	.18	.48	.50	.48	-.13
Eigenvalue	4.17	2.44	1.65	1.06	3.02	2.56	1.62	1.39
Cumulative % of total variance	32.1	50.9	63.6	71.7	23.2	42.9	55.3	66.0

The results of the correlation and principal components analysis are presented here in such a way that comparison between the two spatial levels is immediately possible. A warning is apposite, however. A direct comparison of the coefficients would have been more interesting and, from the point of view of a multi-level design, methodologically more sound if we had analysed all Dutch municipalities and not the Frisian ones only; on the other hand it may be said that our aim, at least at this stage of the project, was not so ambitious as this. Attention should be directed towards the question of whether the indicators give a realistic picture of the COROP regions and, if so, whether these indicators discriminate in a satisfactory way between the smaller Frisian regions with respect to their rurality.

Consider first the internal structure of the sets of variables. In all cases the correlation coefficients are rather low especially in Table 4.3 where we find only three coefficients higher than .70, all three being at the COROP level. In the case of the structure variables (Table 4.1) we find six coefficients higher than .70 at the COROP level and only two at the municipality level. This leads to the conclusion that the amount of redundant information contained in the indicator sets is relatively small, at least when we look at it from the statistical point of view. Therefore the possibility of finding dimensions which could be an artefact of the choice of strongly overlapping variables is almost excluded.

A second remark to be made after inspection of Tables 4.1 and 4.3 is that there is no consistent direction of change of coefficients connected with a shift in the spatial scale of analysis. In both cases the absolute value of the majority of correlation coefficients becomes smaller, as is to be expected in the general multi-level case, but here this majority is less than 60 per cent.

An elementary linkage analysis of the correlation matrices could be presented here, but because of the rather low absolute values of the coefficients and the strong overlap with the principal components analysis, it is better to turn to Tables 4.2 and 4.4. Principal components analysis has been used here because of the exploratory character of the investigation, methodologically this would also imply that rotation is out of the question. Nevertheless varimax rotations were calculated to see if any improvement in the loading-structure could be reached. This did not occur. Taking the structure variables at the COROP level first, we note that with the slight exception of variable 7 (POP) two clear groups of indicators may be distinguished in connection with the first and second components. The loadings of the third component are not interesting. The first component represents more than half of the total variance and combines variables 1, 2, 3, 5, 6 and 8. We may quite safely designate this component as a 'rurality' dimension. This interpretation becomes even clearer when we consider the scores of the COROP regions on the first component. All the regions having a positive score on the rurality dimension, in the north, east, south and south-west of the Netherlands could have been designated as more or less rural using prior knowledge only. Looking at the position of the regions with respect to the second component, we see the more peripheral rural areas as well as the (highly) urbanised areas scoring positively on this dimension. From the literature it is of course a well known fact that extremely rural areas and urban regions show common characteristics, although the theoretical explanation is completely different. In our set of indicators the variables 4 and 9 make the highest contribution to the second dimensions in the way hypothesised earlier. On the basis of these observations, we may expect the first component and both components together to discriminate between COROP regions.

Looking at the structure variables at the municipality level, it is important to stress that the Frisian municipalities are all situated in an area that is typified as specifically rural in the COROP-region analysis. In both cases we want to discriminate according to the extent to which regions are rural; in the COROP analysis we find non-rural regions mainly where we also find larger urban agglomerations. In Friesland urban agglomerations are absent, but because we shift to the lower spatial level of analysis we can still distinguish between rural and non-rural since we now find a series of municipalities whose built-up areas follow quite closely in several cases the administrative divisions. Besides, it may be noted that in all analyses of this kind the measure used is a relative one. In the light of these variables, we can only conclude that region A is more rural than region B, not that region A is rural and region B is non-rural. Rurality must be interpreted in a relative sense.

The components analysis of the structure variables at the municipality level yields results that resemble loosely those at the COROP level. It is also clear that the first component represents a high proportion of the total variance (though not as high as at the COROP level because of the lower correlation coefficients). Again, this may be interpreted as a rurality dimension. Municipalities having a positive score on this dimension are found in the north, north-east and south-west of the province with the exception of most of the towns. The most remarkable point is the changing effect of variables 2 and 4. Variable 4, kilometres of paved roads outside built-up areas, behaves more loosely than expected and seems quite sensitive to the scale of measurement. Variable 2, cultivated area, makes the highest contribution to the second dimension in this case, being largely responsible for the discrimination between municipalities with and without cultivated area, notably between the Wadden Islands and the towns.

The picture resulting from the components analysis of the process variables is less clear. The coherence between variables is less, an elementary linkage analysis would show a considerable measure of chaining and the shift in position of indicators in the loading structures shows a more chaotic picture than with the structure variables. We shall not try to interpret the dimensions, in all cases we may conclude that elements of the hypothesised social structure become apparent in the loadings. We may note the strong position of some variables, notably income and to a lesser extent education and 0-19 year olds, and the interesting behaviour of the two migration variables (17 and 18) when we compare the two spatial levels. For the purpose of this paper, however, it is more interesting to see if one of the main ideas behind this exercise (the expected similarity between the regional typologies based on the structure and process variables respectively) shows up in the results. Therefore, we direct our attention now to the classification of the regions at the two spatial levels.

A TENTATIVE CLASSIFICATION OF REGIONS WITH RESPECT TO RURALITY

In constructing a typology of regions on the basis of a multiple set of indicators we have to choose between several possible procedures. In order to avoid the danger of what has been called the subjectivity of objective measures (Johnston, 1968) several options in the taxonomic procedures have been worked out, although only a few of them are presented here. We discuss the results from a grouping of the regional units on the basis of the component scores and from a cluster analysis performed on the original values of the variables, handling separately the structure and process variables and the two spatial levels.

In all cases, grouping is executed with a hierarchical clustering procedure using Ward's criterion. A hierarchical procedure
implies that at every step in the grouping process one or more
groups are formed or extended after which the (dis)similarity
matrix is adapted before the next step follows. In Ward's method
the (dis)similarity is measured as the squared Euclidian distance
between groups in the variable space. The aim, of course, is to
keep the groups as homogeneous as possible. During the grouping
one could define several stop criteria, the F-ratio is one of the
most widely used. In our analyses we found between 10 and 15
groups in almost all cases when using an F-ratio of 1.00, i.e.
accepting only a relatively small loss of information. However, we
used a much more pragmatic criterion in the end. Because of the
visual presentation of the results in maps and the necessity to be
able to compare the results of a range of different groupings we
chose seven groups in all cases.

The COROP level

A classification into seven groups of COROP regions using the
structure variables was made in two ways. The first uses the
scores on the first two components presented in Table 4.2 and the
second uses the original values. The first two components discriminate according to two distinct groups of variables and represent
a fairly high percentage of total variance. This grouping of
COROP regions and the second classification using the original variables show a striking resemblance, therefore we only show the map
of the second grouping (Figure 4.1). The northern and south-western
parts of the country show up as the most rural, while rural characteristics are also seen in the east, the south and the river area
in the centre. At the other end of the scale, the non-rural areas
in the west come out clearly.

One of the disadvantages of the grouping method is that since
differences at the least rural end of the scale are bigger, the
groups formed in the hierarchical procedure become smaller. If we
formed groups each with the same number of regions, Arnhem-Nijmegen
(COROP region 15) and some regions in Brabant would shift to the
less rural groups. Nevertheless the overall picture is quite clear.

The procedures used on the *process variables* are exactly the
same. The only difference is that in the case of the process
variables we used the scores on the first three components (cf.
Table 4.4). In the results the general pattern is repeated,
although as in the previous case there are local differences. The
results of the classification performed on the original values is
shown in Figure 4.2.

The results of the classification of COROP regions are combined
in Table 4.5 in which the seven groups by each classification are
brought together using the numbers of the COROP regions (for these
numbers see Figure 1.1).

Remembering that our measurement of rurality results in a
relative ordering of regions according to this dimension and concentrating on the question of whether the two sets of indicators
produce similar results, it seems logical to compare the groupings
of the regions by the two classifications.

Out of 39 COROP regions 27 are classified in the same or a
neighbouring group in the four classifications. Of the remaining
12 regions, the most notable exception is region 3, Overig
Groningen. This region is predominantly rural but it also incorporates the biggest town of the northern part of the country. This

Figure 4.1   The COROP regions classified in diminishing order of
             rurality on the basis of the original measurement values
             of the structure variables

Figure 4.2   The COROP regions classified in diminishing order of
             rurality on the basis of the original measurement
             values of the process variables

36

Table 4.5   The four classifications at the COROP level

groups	structure variables		process variables	
	component scores	original values	component scores	original values
1	1, 2, 3, 4, 5,	1, 2, 3, 4, 5, 6, 7, 8, 9, 31 32	1, 2, 8	1, 2, 8, 31
2	6, 7, 8, 9, 18 31, 32	10, 14, 16, 18	4, 5, 6, 7, 9, 14, 16, 31, 32	4, 5, 6, 7, 9, 13, 14, 16, 32
3	10, 11, 12, 14 16, 19, 39	11, 12, 13, 15, 19, 27, 33, 34 39	13, 18, 19, 28	18, 19, 28
4	35, 37, 38	35, 36, 37, 38	10, 15, 34, 35 37, 38	10, 11, 12, 33, 34, 35, 36, 37, 38, 39
5	13, 15, 28, 30, 33, 34, 36	17, 28, 30	3, 11, 12, 22, 29, 30, 33, 36, 39	3, 17, 20, 22, 25, 27, 29, 30
6	17, 20, 22, 24 25, 27, 29	20, 21, 22, 24 25	17, 20, 25, 27	15
7	21, 23, 26	23, 26, 29	21, 23, 24, 26	21, 23, 24, 26

is reflected in its shift of position from group 1, when using
structure variables, to group 5 when using process variables. The
same phenomenon is seen with several other regions although less
markedly: Noord-Overijssel (10) with Zwolle, Zuidwest-Overijssel
(11) with Zutphen, Twente (12) with Hengelo and Enschede, Arnhem
and Nijmegen (15), Delft and Westland (27), West Noord-Brabant (33)
with Breda and Bergen op Zoom, Midden Noord-Brabant (34) with
Tilburg and Zuid Limburg (39) with Heerlen and Maastricht. The
three remaining regions behave in a rather unexpected way: Veluwe
(13), Oost Zuid-Holland (28) and Groot-Rijnmond (29). The main
characteristics of the Veluwe are its natural beauty (it has a
national park) and concentrated urban development (Apeldoorn).
Oost Zuid-Holland lies in the middle of the Randstad and constitutes
the centre of the so-called Green Heart which implies that the
region is much influenced by the surrounding regions. Groot-Rijnmond includes the industrial and harbour area of Rotterdam but also
parts of the Green Heart and the southern isles.

    We may conclude that the similarity between the classifica-
tions based on the structure and process variables respectively is
quite high, especially for the groups at both ends of the rurality
dimension. The disadvantages of the COROP regions however are also
shown clearly. The regions in which rural and non-rural elements
are combined are ordered less consistently using the different sets
of variables.

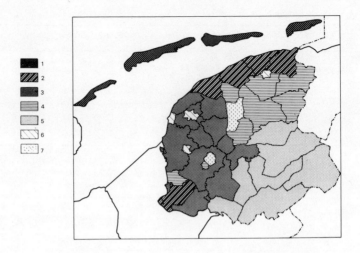

Figure 4.3   The Frisian municipalities classified in diminishing
order of rurality on the basis of the original measurement values of the structure variables

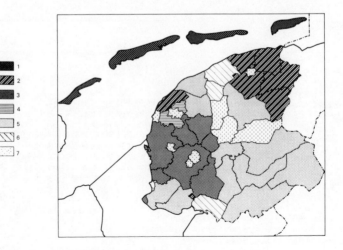

Figure 4.4   The Frisian municipalities classified in diminishing
order of rurality on the basis of the original measurement values of the process variables

The municipality level

The sevenfold grouping of the municipalities using the *structure variables* on the basis of the component scores and the second classification using the original values again produce very similar results. The second classification is shown in Figure 4.3. The most rural parts of the province are found in the north-east, north and south-west, with the exception of some towns. In these parts of the province most towns are independent municipalities - Dokkum, Harlingen, Franeker, Bolsward, Sneek, IJlst, Hindelopen, Stavoren and Sloten. In the east and south-west the municipalities are bigger and the towns are incorporated into them. Here we find Heerenveen, Drachten, Wolvega, Lemmer and Joure.

In Figure 4.4 the classification of the municipalities on the basis of the original values of the *process variables* is shown. In this case the grouping using the scores on the first two components showed enough general resemblance to justify the presentation of only one of the maps.

The differences and similarities of the groupings on the basis of both variables will be described in more general terms here. A tabular ordering, as was made in the case of the COROP regions, is methodologically interesting but interpretation in detail supposes a rather detailed knowledge of the Frisian scene. Therefore in this case we shall restrict ourselves to some general conclusions.

The number of municipalities classified in the same or a neighbouring group (out of seven) is less in this case, 24 out of a total of 44. The remaining municipalities may be divided into three groups:
a) Small towns where the structure indicators point to a higher position on the rurality dimension than the process indicators, notably three small towns in the southwest - Hindelopen, Stavoren and IJlst. We could include in this group the larger municipalities where a town might be responsible for the lower score on the rurality dimension using process indicators. Here we find one municipality in the southwest, Gaasterland, a region also characterised by less rural activities, some agro-industry and tourism, and the municipalities that contain the towns of Joure, Lemmer and Drachten.
b) The second group consists of regions situated around Leeuwarden where the lower score on the rurality dimension using process variables may be an effect of the urban influence of Leeuwarden. This group of municipalities forms a contiguous band in Figure 4.4 between the more rural north-east and west.
c) A miscellaneous group mostly consists of municipalities where we find a striking difference between the two classifications on the process indicators. More detailed investigations have to be made about this group.

SOME CONCLUDING REMARKS

This taxonomic exercise served two purposes. First, we wanted to show a possible way of proceeding from theory towards the selection of indicators and the construction of a rurality dimension on which regions are classified. We found that the maps at both spatial levels gave a realistic general picture given our prior knowledge of the area. Where regions deviated from this expectation or were classified inconsistently this could be explained to some extent by the characteristics of the regional division chosen. A further investigation, however, is necessary especially through looking into the behaviour of individual indicators. The second aim was to find out if the classifications based on the structure variables and those based on the process variables showed a satisfactory similarity; this question is based on the general theory developed by our research group. Although this paper may hardly be seen as a test of this hypothesis, the exploratory investigation shows enough similarity to make a deeper investigation of the matter worthwhile.

When we see the results in a broader context, we may conclude that the measurement of rurality was promising, while the chosen indicators discriminated in a satisfactory way at both the spatial levels. To refine the results, a more detailed study of the behaviour of indicators at other spatial levels is needed. At the same time comparative research might be rewarding in the sense that the behaviour of indicators in urban and rural regions should be compared in order to rethink the choice of variables and to gain more insight into the differences in both contexts.

REFERENCES

Cloke, P.J. 1980. New emphasis for applied rural geography. *Progress in human geography*, 2, 181-218.

Cullingford, D. and Openshaw, S. 1982. Identifying areas of rural deprivation using social area analysis. *Regional studies*, 16, 409-418.

Hauer, J. and Veldman, J. 1983. Rural geography at Utrecht. *Tijdschrift voor economische en sociale geografie*, 74, 397-406.

Johnston, R.J. 1968. Choice in classification: the subjectivity of objective methods. *Annals of the Association of American Geographers*, 58, 575-589.

Chapter 5

The intra-regional differentiation of peripheral regions in the Netherlands

Wim Ostendorf

INTRODUCTION

In this contribution the starting point is the 'resident' or the 'household'. The resident uses his house as a daily starting point to undertake activities and returns to it again. Consequently, the resident is subject to limitations of time and distance and therefore is dependent on a limited spatial area. This area can be called his residential environment. In this residential environment a number of spatial levels can be distinguished based on the nature of daily interaction with its corresponding range of action. The paper will consider two spatial levels, the 'neighbourhood' and the 'daily activity region'.

The neighbourhood is the area, within which people do their daily activities on foot and is limited to one or two kilometres. Within this area we find the daily shopping and visits to church, pub and library; within this area the children go to school and most social life takes place (visits to neighbours, friends, corporate life). Dependent on their degree of being tied to the house, many residents hardly leave their neighbourhood during their daily routine.

The daily activity region occupies a larger area and includes paid employment. Working is the daily activity with the highest spatial threshold. In principle, daily life takes place within the area of the daily activity region (hereafter, the region). Fundamentally, every resident has his or her own neighbourhood and his or her own region. Because of the concentration of employment, public goods and services, many residents' areas coincide. That is why one can speak of regional labour markets or daily systems characterised by a wide network of relations between the land-using units (households, firms, institutions) within this complex and manifest in daily recurrent movements between these units. One can think of a central city surrounded by a collection of connected towns, suburbs and villages.

This approach differs from that of Veldman in this book, who classifies regions as either urban, peri-urban rural or peripheral rural. Peri-urban rural areas are situated within the sphere of urban influence and together with urban areas form a housing and labour market. Peripheral rural areas are situated outside the sphere of urban influence. Apart from the problem that an urban area has to be distinguished from a rural one, a sort of threshold has to be used on the basis of which one can decide whether an area belongs to a sphere of urban influence. In my opinion there will always be service areas. So I prefer to distinguish daily activity regions, even in peripheral rural areas; within these regions further differentiation can be studied.

This is not to say that all regions are more or less equal. On the contrary, some regions are well equipped with a wide range of job opportunities and are supplied with good educational establishments; others are short of employment and have mainly low-quality jobs. This difference results from developments in the economic base and the politico-administrative system. These developments have found an ecological expression in the existence of a polarity between the western parts of the Netherlands (the 'Randstad') as the centre and the remaining parts which are intermediate and peripheral regions.

Central and peripheral regions form different productive environments. It is true that these productive environments are related to the regions through the labour market, but they have a scope and structure of their own which presents itself to the regions' residents as a fact and as such offers certain opportunities. Stagnation in employment is a feature of peripheral rural areas. Apart from the fact that stagnation seems to be a feature of this decade, one can expect similar phenomena in regions which have undifferentiated and stagnating employment; for instance industrial regions like Twente and the Brabantse stedenrij in the Netherlands and the regions of Liverpool and Glasgow in Britain.

This 'centre-periphery constellation' can be seen as differences in degree of urbanisation. It can also be described in terms of a) degree of agglomeration (to be measured by the concentration of jobs, services and residents in each region, that is, in terms of those features which are frequently cited as qualifications for the degree of urbanisation) as well as in terms of b) differences in the characteristics of the population (e.g. differences in welfare, education, life-cycle, etc.).

Within the regions there exists a differentiation between the city which functions as a regional centre and the towns, suburbs and villages that surround it. This differentiation forms the ecological expression of the increasing need for space by societal units (households, firms, institutions) and of the differentiation in the locational requirements of increasingly more specific functions. Of course, these two processes took place and still take place throughout the country, but they had their most marked ecological effect within the built-up area of the older cities. Here the lack of space for extension was most sharply felt. The consequence of this congestion is that households, firms and institutions needing daily contact with one another no longer settle within the city but in its environs. Due to this movement areas near the cities become more and more suburban. The cities and their surrounding (sub)urban zones grow into one (urban) region, with one spatially indivisible labour and housing market, and they are held together through an intensive network of reciprocal interactions.

This suburban migration was and is very selective. In general, one can say that households with an extensive need for space have suburbanised more than households with an intensive spatial occupation, and that rich households have suburbanised more than poor households. In short, the growth of an (urban) region is paralleled by a differentiation between the central city and the more or less suburbanised zone that surrounds it.

42

This 'central city-suburb' pattern presents above all a differentiation of opportunities regarding the way in which the inhabitants organise their everyday lives. Because of the ecological components it comprises, however, it can also be characterised as differences in degree of urbanisation. After all, it can be described not only in terms of a) the degree of congestion (to be measured by the concentration of artefacts in each municipality), and b) in terms of the nature of the housing stock, but also c) according to differences in the characteristics of the population (e.g. welfare, life-cycle, nationality, geographic mobility, etc.).

Also, interregional and intra-regional differentiation are connected. Congestion (e.g. shortage of space for expansion by existing land uses, traffic congestion and lack of parking facilities) is most pronounced in the central regions. Here the lack of space is most urgent, especially in the central cities, because of the high density of land uses and the small-scale infrastructure. The demand for space is large because of the locational benefits. Furthermore, an increasing use of material attributes - made possible by higher incomes - requires increasing use of space. In other words, intra-regional differentiation in terms of the degree of urbanisation is more pronounced in precisely these central, most highly urbanised regions.

DIFFERENCES IN DEGREE OF URBANISATION IN THE NETHERLANDS

The above mentioned ideas formed the starting point for a study of the differences in degree of urbanisation in the Netherlands based on the results of the 1971 Population and Housing Census and are empirically supported by the results of this study (van Engelsdorp Gastelaars et al., 1980). It has been demonstrated that the degree of agglomeration, the degree of congestion and the nature of the housing stock are related to characteristics of the population, and the more central the region is, the stronger this relationship appears to be. Apparently there is an ecological pattern of social inequality at two levels in the Netherlands. At the regional level this inequality is associated with living conditions, that is the residents' position in the job market and 'service market'. At the municipal level (for statistical reasons the municipality had to be used as the unit of analysis) inequality is associated with the way one is housed, that is, the residents' position in the housing market.

In view of the plausibility of the ideas, three typologies based on the degree of urbanisation have been devised. These typologies comprise - completely in accordance with the analysis - indications of a) the degree of urbanisation of the region in which a given municipality is located, and b) the degree of urbanisation of the municipality itself. This is to be expected because an intra-regional typification is significant only in the context of the region in which the municipality is located. The intra-regional analysis was done after the municipal scores had been standardised with respect to the region.

THE SOCIO-MORPHOLOGICAL TYPOLOGY

One of the typologies of areas is primarily social in nature. It is based on socio-morphological criteria, namely, the characteristics of the populations in both the regions and the municipalities. Regions have been clssified by living standards into four groups: central, intermediate, semi-peripheral and peripheral (Figure 5.1). This pattern resembles the ranking by degree of rurality that Hauer

43

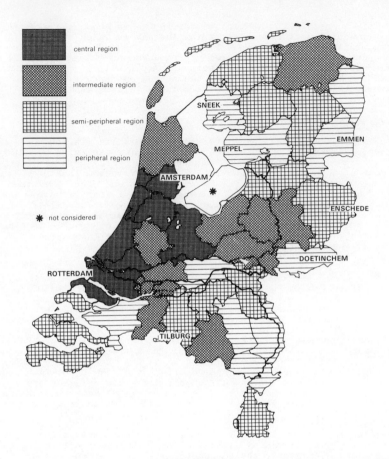

Figure 5.1   The differentiation of Dutch regions based on a socio-
morphological typology.   The selected regions (Table 5.1)
are indicated by the name of the most important centre
in the region (computer map, program KAART, A Mulder
State University, Utrecht)

presents in his study (in this book) of the rurality of regions.
Thus, differentiation by degree of urbanisation merges with a differentiation in terms of rurality.

The municipalities in each of these four classes of region have
been subdivided into urban, suburban, young growing, and rural.
This intra-regional differentiation describes the Netherlands in
terms of those well known ecological components - a) socio-economic
status, b) life-cycle, and c) ethnic status (Shevky and Bell, 1955).
The last component includes not only the proportion of foreigners,
but also the proportions of unemployed, divorced women, non-family
households, and one-person households; this component is labelled
'minorities'. Municipalities with a high score on this minorities
component are urban; municipalities with a high score on socio-

44

economic status are suburban; municipalities with a high score on
the life-cycle component (which points to a high proportion of
young family households) are called 'young growing' areas and the
remainder are rural municipalities. In his contribution to this
book Hauer also finds for the Frisian municipalities a socio-
morphological differentiation based on socio-economic status (Hauer,
table 4.2, his first component). His second component can
be called life-style, and is adapted to the circumstances of a
peripheral region, namely 'greying'; further, the set of variables
seems important for the existing differences. This holds even more
for the absence of the third component, 'minorities'.

The typology, which basically covers aspects of welfare and
well-being, derives its justification from the close correlation
which its component parts possess with such physical-morphological
criteria as the degree of agglomeration, the degree of congestion,
and the nature of the housing stock. The typology has been criticised by Deurloo, Jobse and Thissen (1981) because 'the typification and labelling of the municipalities is not in accordance with
the intra-regional differentiation, especially in peripheral
regions'. A simple rejoinder might be that it is not clear at
which sort of intra-regional differentiation they aim. Furthermore
our study naturally pays attention only to that intra-regional
differentiation which fits into a study of the differences in
degree of urbanisation. Nevertheless, it is interesting to subject
the typology to a test allowing for a judgment of its value. However, not having available ideas and manageable criteria, I am
forced to use the ideas and the data of the study on which the
typology is based. Its validity can be judged on the basis of the
internal consistency of the typology. Maybe this analysis gives
some indications about the poorly understood intra-regional differentiation of peripheral regions.

THE INTERNAL CONSISTENCY OF THE TYPOLOGY

Variables and regions

Does the above-mentioned subdivision into urban, suburban, young
growing and rural municipalities have any validity in peripheral regions? Table 5.1 may help to answer this question.

Variables 1-4 are central to the component minorities; variables 5-7 to socio-economic status; variables 8-10 to the life-cycle
component; variables 11 and 12 point to recent (suburban) immigration; and variable 13, indicating farming population, may be expected
to point to rural municipalities.

The municipalities are, according to the typology, subdivided
into urban, suburban, young growing and rural and belong to different types of regions (see Figure 5.1). First, there are four peripheral regions (Sneek, Meppel, Emmen and Doetinchem) of a rural
nature, with stagnation in employment facilities and often partly
involved in the governmental policy of regional industrialisation.
Two further semi-peripheral regions have been selected (the regions
of Enschede and Tilburg). Here the labour-intensive industrial base
is crumbling away; so, as has been indicated above, employment is
stagnating here too. In contrast, two central regions (the regions
of Amsterdam and Rotterdam) have been chosen. Here one can expect
congestion and as a consequence suburbanisation which functions as
a driving force for intra-regional differentiation.

Table 5.1 shows whether each group of municipalities scores

Table 5.1 Population characteristics of the types of
municipalities in the selected regions

Regions	minorities				socio-economic status			life-cycle			(suburban) immigration		farming population
	1. % unemployed	2. % divorced women	3. % foreigners	4. % non-family and one-person households	5. % high-status employees	6. % low-paid workers	7. % blue collar	8. % families with only young children	9. % families in pre-child stage	10. % settled since 1965	11. % non-native commuters	12. % born in municipality	13. % working at home address
Central													
urban	x	x	x	x	x	x	x	x	x	x	x	x	-.7
suburban	-.4	-.4	-.5	-.5	+1.8	-.7	-.8	+1.4	x	+2.2	+2.5	-.7	x
young growing	-.3	-.3	-.5	-.4	x	x	x	+1.8	+1.4	+2.3	+2.6	-.7	x
rural	-.5	-.4	-.3	-.6	x	x	+1.2	+1.3	x	x	x	x	+3.2
Semi-peripheral													
urban	x	+1.4	+1.3	+1.2	+1.1	x	-.9	-.9	x	x	-.7	-.9	-.5
suburban	-.3	-.5	-.5	-.6	x	-.8	x	+1.2	x	x	+2.1	-.9	+1.8
young growing	x	-.4	x	-.6	-.8	x	+1.1	+1.2	x	x	+1.3	x	x
rural	-.7	-.5	-.7	-.6	-.9	x	+1.1	x	x	x	x	x	+1.9
Peripheral													
urban	+1.2	+1.2	x	+1.2	x	x	-.9	x	+1.1	+1.3	x	-.9	-.5
suburban	x	x	x	x	x	-.8	x	x	x	x	x	-.9	+1.6
young growing	x	-.9	x	x	-.9	x	+1.1	+1.1	x	x	+1.5	x	x
rural	x	-.9	-.8	x	-.9	x	+1.1	x	-.9	x	x	x	x

Source: Population and housing census, the Netherlands, 1971.

- = more than 80% of the municipalities of the type score lower on the variable
  than the region in which the municipalities are located, in which case the
  mean location quotient is given.
+ = more than 80% of the municipalities of the type score higher on the variable
  than the region in which the municipalities are located, in which case the mean
  location quotient is given.
x = the municipalities of the type possess heterogeneous scores (some are higher
  and some lower than the region as a whole).

systematically higher (+) or lower (-) on the given variable than
the region as a whole; the operational threshold is 80 per cent or
more. The remaining cells contain an x, meaning that these municipalities possess heterogeneous scores; some are higher and some
lower than the region as a whole. If the scores are systematically
higher or lower, the mean location quotient on the given variable
(the municipality percentage divided by the regional percentage)
for the group of municipalities has been mentioned.

Expectations

What may be expected from the scores in Table 5.1?
1. In all regions the division into urban, suburban, young
growing and rural municipalities should be clear (that is to say
by filling the boxes in Table 5.1), but this differentiation will
be much less pronounced in semi-peripheral and peripheral regions.
Congestion is much less important in these regions because there
is much more space available, while the demand for space by new and
expanding land-using units is much less. Consequently, there is
less inducement for suburbanisation. The differentiation that becomes evident inperipheral regions will be more dependent on
random and local circumstances.
2. Because of the limited importance of suburbanisation, the
cities in peripheral regions have lost fewer well-to-do residents
and young family households than the cities in central regions.
Consequently, suburban and young growing municipalities will show
a less pronounced profile, while the cities in peripheral regions
will contain a relatively wealthier population and more young-.
family households (Abu-Lughod, 1969).

Results

Urban municipalities in central regions become distinct only indirectly, that is to say, by way of the systematically low scores
of the other types of municipality on variables 1-4. This we may
explain by the fact that Amsterdam and Rotterdam possess such a high
concentration of the population categories involved and influence
the regional score so strongly because of their size, that even the
other cities are below the regional score. Table 5.2, in which
Amsterdam and Rotterdam are split off from the remaining urban municipalities illustrates this point.

Table 5.2 Some population characteristics of the urban
           municipalities in the selected central regions

	1. % unemployed	2. % divorced women	3. % foreigners	4. % non-family and one person households
Amsterdam and Rotterdam	+1.4	+1.3	+1.5	+1.2
Remaining urban municipalities	-.7	-.7	x	-.6

Source: Population and housing census, the Netherlands, 1971
Explanation: see Table 5.1.

The urban municipalities in semi-peripheral and peripheral regions behave as expected; variables 5 and 7 point to the presence of wealthy people (the absence of a poor population); variables 9, 10 and 12 point to a certain growth or population influx in cities in peripheral regions, possibly in connection with their role in the governmental industrialisation policy (Saal, 1972). The limited importance of congestion and the abundant availability of space makes suburbanisation in peripheral regions less urgent and thus contributes to urban growth. Moreover, the concentration of ethnic minorities (variable 3) is virtually absent. Apart from the facts that ethnic minorities naturally occupy a part of the housing market and that their presence and settlement adds to congestion, these categories seem to contribute to the suburbanisation of Dutch family households in central regions. It is said they prevent the desired social homogeneity of neighbourhoods. So, this stimulus to suburbanisation in peripheral regions is virtually absent too.

The suburbs in the central regions behave as expected. In the other regions the high socio-economic status of the residents is much less clear; high-status employees are not systematically present and blue-collar workers are not systematically absent. The cities seem to score higher in this respect. The relative absence of low paid workers (variable 6) is the only indication of the high socio-economic status of suburbs in semi-peripheral and peripheral regions. In peripheral regions even the presence of non-native commuters is not found. Obviously, the well-to-do nature of these municipalities is not to be found exclusively in the influx of wealthy non-native commuters. However, variable 12 points to the influx of non-native population, maybe of farmers (variable 13). Perhaps some of these municipalities serve as the residence of retired farmers and middle-class people of independent means. The existence of such villages, where the local and regional retiring élite settles, has been described by Bergsma (1963) in his study on the Dokkumer Wouden in Friesland, where Murmerwoude possesses this status. In Noord-Holland-Noord, Bergen performs a similar role. If this assumption is right, this type of municipality in peripheral regions consists not only of suburbs inhabited by wealthy non-native commuters, but also of villages at the top of the regional economic hierarchy which are attractive to retiring migrants. Besides, especially in districts with scenic landscapes, the influx of retiring migrants may come from outside the region as well.

The young growing municipalities in central regions need no comment. Neither does the presence of non-native commuters and young-family households in semi-peripheral and peripheral regions. Variable 10 suggests that in peripheral regions the growth of cities might be more important than that of young-growing municipalities. Finally, the blue-collar nature of this type of municipality is striking (variables 5 and 7).

It remains doubtful if we can speak of suburbanisation in relation to young-growing municipalities in peripheral regions. Probably, a key-settlement policy, that aims at housing the lower socio-economic strata, offers a better explanation (Thissen, 1982). After all, it is very difficult in practice to distinguish between suburbanisation and concentration or key-settlement policy; much depends on the definition of the concepts.

Finally, the rural municipalities are characterised by their contrast with the cities, and by their blue-collar (variables 5 and 7) and agricultural nature (variable 13). The last point does not hold true within peripheral regions, where rural municipalities do not form the centres of farming activity.

CONCLUSIONS

1. The above analysis is only to be considered as a test of the internal consistency of the typology. It does not prove anything about other intra-regional differentiating processes in peripheral regions except for processes related to suburbanisation. For that purpose one needs hypotheses about other forms of intra-regional differentiation than those already included in the typology, and also data that fit these hypotheses.

2. The hypothesis that the intra-regional differentiation in peripheral regions is much less pronounced than in central regions has been supported. After all, this does not apply to rural regions only. The same can be said of industrial regions, whose labour-intensive employment can no longer be combined with the high wage level of Western European countries.

3. The hypothesis has been supported that cities in peripheral regions have lost less population due to suburbanisation and, consequently, possess a population with a relatively higher socio-economic status and more family households than cities in central regions.

4. In peripheral regions the use of the label 'suburban' seems less apt. It has been hypothesised that in peripheral regions this type contains municipalities which are at the top of the regional economic hierarchy of villages due to the settlement of retiring migrants.

5. Young-growing municipalities in peripheral regions are inhabited by young-family households and commuters who are not well-to-do. In this connection it is more plausible that this situation is created by a key-settlement policy, which limits new housing to certain villages, than by suburbanisation.

6. The typology fits better central regions, and one has to be careful with the intra-regional distinctions in peripheral regions, because intra-regional differentiation there is less pronounced, and local circumstances might play an important role. So, apart from the distinction between cities and the remaining municipalities, one could leave out of consideration the intra-regional distinctions in degree of urbanisation within peripheral regions. This conclusion is attractive, because in many people's judgement it makes no difference whatsoever where one is in rural areas. It has to be proved how far this judgment is wrong; and this has to be done by those people who presuppose that there really is an important intra-regional differentiation in peripheral regions.

REFERENCES

Abu-Lughod, J. 1969. Testing the theory of social area analysis: the ecology of Cairo, Egypt. *American Sociological Review,* 34, 198-212.

Bergsma, R. 1963. *Op weg naar een nieuw cultuurpatroon.* Studie van de reactie op het moderne cultuurpatroon in de Dokkumer Wouden, (Assen).

Deurloo, M.C., Jobse, R.B. and Thissen, F. 1981. Review of: *Stedebouw en Volkshuisvesting*, Gastelaars *et al.* (eds), 10, 492-493.

Engelsdorp Gastelaars, R. van, Ostendorf, W.J.M. and Vos, S. de 1980. *Typlogieën van Nederlandse gemeenten naar stedelijkheidsgraad. Milieu-differentiatie en maatschappijke ongelijkheid binnen Nederland*. Monografieën Volkstelling 1971, 15 b., ('s Gravenhage).

Saal, C.D. 1972. Dorp en route, waartoe en waarheen? *De Gids*, 4, 270-286.

Shevky, E. and Bell, W. 1955. *Social area analysis; theory, illustrative application and computational procedures*. (Stanford).

Thissen, F. 1982. *Platteland bij de stad, Het geografisch onderzoekmet betrekking tot sociaal-ruimtelijke veranderingen in peri-urbane landelijke gebieden in Nederland*. Paper Nederlandse Geografendagen, (Nijmegen).

Chapter 6

Urban field developments and the changing position of rural areas

Lambert van der Laan

The concept of 'living conditions' can be approached from several angles. The stress placed on different aspects of the concept determines the style of research and policies. First, attention will be paid to the nature of the concept, particularly as it concerns the position of rural areas in relation to urban complexes. The conviction that the study of rural areas has not yet shown an appropriate awareness of a broader change in western urban society forms the basis to this approach. The function and pattern of urban complexes is continually changing and this is affecting rural areas in many ways. Leven summarised these changes in both urban and rural areas in the following way. 'Increasingly, individuals and families can be participant in economic, informational, and even cultural and social aspects of metropolitan life without actually having to live in the metropolis in the sense we know it' (Leven, 1978, 104).

LIVING CONDITIONS AND DEVELOPMENT

Two starting points can be discerned in studying the living conditions of an area. First, living conditions can be seen as the totality of provisions present in a region (e.g. transport facilities, employment opportunities, housing possibilities). Second, the concept can be related to the subjective valuation of a spatial situation. This latter aspect emphasises the alliance between man and environment in particular and is often indicated by the concept of 'liveability'.

The distinction between the two approaches to 'living conditions' is important because it offers different opportunities for research and policies. In the first and more 'objective' approach, changes in the ways of meeting the multidimensional needs of the inhabitants of an area are discussed. From the standpoint of a desire for equity, for example, policies can try to equalise the chance of finding a job in different parts of a country. The more 'subjective' approach to the problem is much less affected by direct policies. The extent to which people really take advantage of chances offered to them is clearly not influenced by policy in the short term. This paper concentrates on the objective part of the concept of living conditions. It must be clear, however, that this should be complemented by a subjective approach in order to embrace the full meaning of the living conditions of a region (see also Van der Laan and Piersma, 1982; and Gall's paper in this book).

Living conditions resulting from social activities have evolved over time and this will affect the future position of rural areas. To understand the complex causes and effects of this changing position, the concept of development has to be defined. In this context some universal ideas about development in both the social and physical sciences need to be linked (Weisz, 1971).

Development refers to the causes and characteristics of the differential increment of the component units of a system over time. This can be looked upon as a morphogenetic process and also as a process of functional differentiation. The former relates to a change in the form of the process in an absolute sense, while the latter deals with relational aspects. For example, consider changes in the sectoral composition of industry. An increasing emphasis on recreation activities in a rural region means both a shifting employment structure in these areas and a functional change of their relations with other regions. A third aspect of development is spatial; namely the (differential) change in the magnitude of variables over space. Morphogenetic and differentiation processes both have a spatial component by which areas attain a specific form and function.

A fourth aspect of development concerns control and the causality of developments. This aspect deals with the ability to steer and dominate the differential growth of the elements of a spatial system such as a rural region. The emphasis will be on the external social dependency of rural areas in the first instance, because this gives an opportunity to place these areas within the sphere of influence of cities. This does not mean that internal control mechanisms are not important. On the contrary, internal factors may ensure that the consequences of continuing urban pressure will be locally distinctive. Besides this, the distinction between internal and external causality is rather diffuse and scale-dependent.

Developments in rural areas reflect changing urban influences and the integration of rural areas into large-scale urban systems. This makes the dispute about changes in non-metropolitan areas highly relevant. The different aspects of development form a basis for subsequent discussion of these changes.

SUBURBANISATION, THE URBAN FIELD AND RURAL AREAS

Much research has pointed to important structural and functional changes in the rural areas. Sternlieb and Hughes (1978) recognised the emergence of rural regions as major growth loci in the United States after 1970. There are also similar observations by Beale (1976), Berry and Gillard (1977) and Lamb (1975) to mention just a few. Some European countries show parallel phenomena (Hall and Hay, 1980; Hoekveld, 1980; Megalopolis, 1980; Vining and Kontuly, 1981).

Traditionally the growth of rural areas has been modelled as a suburbanisation process. Extended commuting in relation to intra-regional migration has played an important role in this. Most of the suburbanisation studies place the changes in rural areas within the framework of processes at the scale of the city-region. The urban fringe and peri-urban areas are the regions at which research has been directed.

Another body of research relates to a higher spatial scale of analysis. The starting point for these studies is the development of rural areas within an urban field. Developments in rural areas are connected to an 'enlargement of the space for urban living that extends far beyond the boundaries of existing metropolitan areas (...) into the open landscape of the periphery' (Friedmann and Miller, 1965, 312-3). Joseph and Smit consider exurban development 'in the countryside which is urban-initiated but not contiguous with any urban area' (1983, 41). However, it is not just a difference of scale which differentiates urban-field studies from suburbanisation studies. More important is the emphasis in the urban-field concept on the emergence of a multi-nodal urban system, the rise of a complex many-sided linkage network within this system and the structuring capacity of the system. These aspects will be discussed briefly later. It must be clear beforehand, however, that shifting attention to the multi-nodal, multi-connected urban field as a system necessitates a parallel shifting of 'our attention from rural/urban differences to the structure of interaction among urban nodes, and from growth at the edge of a city to the allocation of growth among those nodes.' (Simmons and Bourne, 1981, 421). It is within this perspective that the rural situation has to be evaluated. Bourne (1980) has discussed the causes of the emergence of an urban field.

Although various authors sometimes use slightly different terms, the characteristics of the development of an urban field can be traced. This can be related to the various aspects of development as described before.
1. The diminishing role of the traditional metropolis and the emergence of a multi-centred urban region (the spatial-morphogenetic aspect)
2. The rise of a network of social and economic linkages in relation to functional differentiation (the spatial-functional aspect)
3. The changing role of rural areas (a specialisation of the spatial-functional aspect)
4. The structuring capacity of the system (the control aspect).

THE MULTI-CENTRED REGION

As the traditional role of the metropolis becomes increasingly questionable, Sternlieb and Hughes (1978) see a need to alter the basic reference framework. The traditional dominant metropolis gives way to a multi-centred system of cities and becomes merely one of many specialised centres (Friedmann, 1978). This process can be seen within the framework of the urban field. Hayes (1976) uses the term 'dispersed city' for this polycentric phenomenon in relation to the North Carolina urban complex. Also Berry and Gillard (1977) point to the multi-nodal character of today's urban systems and Leven predicts in this context the transition of the expanding post-1945 'car city' to a discontinuously settled urban area, the metropolitan-regional system (Leven 1979, 39-40).

The spatial pattern of the rise of this multi-nodal regional system can be seen as a result of 'cyclic processes' (Cloke, 1978). Cloke coined this concept in describing the pattern of rural change under the influence of urban pressure. Rural areas are affected by this cyclic process in relation to their position in waves of urban growth and 'degrowth'. In due time rural areas increasingly become non-rural. However, particularly due to protective planning, extreme non-rural areas at the edge of urban nodes may become more rural again, thus giving the process its cyclic character. Consequently the spatial pattern of this process is a discontinuously settled urban area. In addition to Cloke's concentric, zonal one-city model, a more complex representation can be suggested. The area outside an 'original urban node' is not an isotropic rural plain. The starting point should be the original population nodes within the rural region. These play an important part in the development of the area because these medium-sized nodes, at some distance from an original large urban centre, often show high rates of growth for employment and population (Lichtenberger, 1976; Fisher and Mitchelson, 1981; Moewes, 1982; Hoekveld, 1980). The resulting spatial pattern has often been called 'dispersed concentration' (Vining and Kontuly, 1981). At first, the rural areas lying between the cities do not take part in the development. In the course of time, continuous urban growth from both an original large urban node and the developing medium-sized nodes causes these rural areas to acquire a specific function in relation to these nodes (see the next section). Quite often spatial policies play an important part in this, as is the case in the Netherlands.

Beale (1976) and Fisher and Mitchelson (1981) describe a more decentralised growth pattern in the rural areas of the United States. In Beale's description 'the non-metropolitan urban towns are experiencing in a micro way the same trends as metropolitan central cities' (1976, 956). This might suggest the different effect of policies in the United States and Europe on the rise of a regional urban system. Alternatively or additionally it might mean the U.S. is a stage ahead of the European situation. The decentralised pattern might then be seen as subsequent to the initial growth.

The shift towards urban nodes within rural regions and a further expansion of these nodes interfers with the dominance of the traditional urban centre and leads to a multi-nodal urban pattern. However, structural changes in the characteristics of areas under the influence of urban complexes do not take place uniformly for all activities. The changes emerge in specific activity-systems and at particular scales. Activities in housing, production and recreation do not have the same scale of spatial effects. Differentiation in activity-spaces can be noticed in this respect. Therefore the cyclic process should also be seen as a sectoral process.

THE FUNCTIONAL NETWORK

There are various functional consequences to the emergence of the multi-nodal urban field. There is a rise of a complex cross-commuting system (Hayes, 1976; Hoekveld, 1982; Fisher and Mitchelson, 1981) and the overlapping of urban spheres of influence which Cloke (1978) pointed to in an analysis of urbanisation patterns in 'intermediate rural districts'. Parallel with multi-nodality there develops a 'finely articulated network of social and economic linkages' (Friedmann, 1978, 42). More and more individual urban nodes become crossroads in a complex network. The centrality of a city is based less on its surrounding region and more on its status in a transactional network.

Because of the functional interdependencies of urban nodes, an urban field functions as an urban entity on a large scale. In this context the distinction often made between rural growth as a result of metropolitan spread and growth caused by internal rural processes is a false one (Taaffe, Gauthier and Maraffa, 1980; Fisher and Mitchelson, 1981). They are the two sides of the same process of urban-field development. Fisher and Mitchelson show an awareness of this and even call it 'a new epoch in the North American urban experience' (1981, 206). They lean on the ideas of Hage (1979), who foresees a situation where 'a number of small towns and cities in the same geo-political area will grow spatially, with specialisation by each

54

city or town, and each town will be connected by a good system of transportation to the other' (Hage, 1979, 99). Functional inter-dependency encourages specialisation. Within the framework of the urban field, subregions accentuate those functions for which they are best adapted. Hence functional selectivity accompanies the emergence of urban fields. This brings us to the changing position of rural areas.

SPECIALISED RURAL AREAS

Many studies have rightly emphasised the problems of suburbanisation in rural areas. However, studies concerning the influence of an evolving urban system on the rural area are rather scarce. Excep-tions to this are Lamb's study of metropolitan impacts on rural America (Lamb, 1975) and the ideas of Hage (1979). Lamb makes clear that because of the growing influence of processes within the urban field, a restructuring of rural areas can be noticed. Rural areas no longer have just an agricultural, mineral or forestry function but also increasingly have a role as a place for housing, recreation and industrial growth (Lamb, 1975). According to Lamb, the principal factors in this process are those relating to accessi-bility, hierarchical position and the presence of amenity resources. In this sense rural areas are increasingly serving specialised functions in specific activity-systems.

Hage emphasises the future role of rural areas in post-industrial society. He sees rural areas developing towards greater specialisation and interdependency. The main factors causing this are government decisions to establish public service organisations, the migration of industry, the development of leisure-time services and educational improvements which influence life style (Hage, 1979). This emerging functional specialisation of rural areas can be seen as a result of integration within the urban field. The latter increasingly steers and controls this specialisation process.

THE CONTROL ASPECT

Control is part of the development process. With reference to Domanski (1978), the concept of 'locational value' can be attached to control. The locational value refers to properties of regions which influence locational decisions. Some locations are preferred for certain activities where value is maximised, e.g. the high quality of the natural environment or infrastructural provisions. Studies of suburbanisation place development processes within the framework of the city-region. The urban-field concept, however, stresses the selection of locational value with a wider framework.

The locational value and therefore the functioning of rural areas are not just related to an urban centre, but increasingly to the position these areas have within the socio-economic network of an urban region at a wider scale. Lamb supports this view. 'In any case, it has become increasingly apparent that the national system of metropolitan centers represents the spatial matrix from which future growth impulses in non-metropolitan America derive' (Lamb, 1975, 2). The outlook for rural areas increasingly depends on the functions these areas can fulfil within an urban field.

At first the development of the urban field widens the range of locational-value choices because more regions are incorporated in the urban field. However, specialisation narrows this choice later.

It is in this way that developmental processes and particularly the functioning of elements of the urban field are steered. Dutch migration illustrates this. Although long-distance migration is not to be expected in so small a country as the Netherlands, the migration flows towards rural regions 'may be akin to long-distance, urban-rural migration in larger countries with respect to the factors which bring them about' (Ter Heide, 1979, 13). The increasing dependency of rural areas on an evolving dispersed urban system is mirrored in new policies for the Dutch rural areas. The implications are such that population re-distribution is no longer directed just at certain rural areas within city-regions, but also has to take into account the dispersed pattern of the urban field (Ter Heide and Eichperger, 1978). This provides a rationale for the emphasis that was given in the Report on Rural Areas (1977) to the nationwide system of areas with rural functions. By this scheme, only those functions are allocated to rural areas which will secure their future functioning in relation to both a multi-nodal urban system and to each other. Hence the locational value of these rural areas is reduced in relation to 'non-rural' functions.

CONCLUSION

Morphogenetic changes in rural areas are increasingly structured by processes within an urban field. Rural areas acquire specific functions in relation to an urban system at a broad spatial scale. This development may be described as dispersed concentrations of specialised growth caused by sectoral and cyclic processes. Of course, the specific historical and spatial conditions in particular areas will create variations on the pattern and processes described.

REFERENCES

Beale, C.L. 1976. A further look at nonmetropolitan population growth since 1970. *American Journal of Agricultural Economics,* 6, 953-962.

Berry, B.J.L. and Gillard, Q. 1977. *The changing shape of metropolitan America.* (Ballinger, Cambridge, Mass.).

Bourne, L.S. 1980. Alternative perspectives on urban decline and population deconcentration. *Urban Geography,* 1, 39-52.

Cloke, P. 1978. Changing patterns of urbanization in the rural areas of England and Wales 1961-71. *Regional Studies,* 12, 603-617.

Domanski, R. 1978. The problems of controllability of spatial systems. *Papers Regional Science Association,* 40, 17-28.

Fisher, J.S. and Mitchelson, R.S. 1981. Extended and internal commuting in the transformation of the intermetropolitan periphery. *Economic Geography,* 57, 189-207.

Friedmann, J. and Miller, J. 1965. The urban field. *Journal of the American Institute of Planners,* 4, 312-320.

Friedmann, J. 1978. The urban field as human habitat. in *Systems of cities,* Bourne, L.S. and Simmons, J.W. (eds), (O.U.P., London), 42-52.

Hage, J. 1979. A theory of non-metropolitan growth. in *Non-metropolitan Industrial Growth and Community Change,* Summers, G.F. and Selvick, A. (eds), (Lexington Books, Lexington), 93-104.

Hall, P. and Hay, D. 1980. *Growth centers in the European system.* (Heinemann, London).

Hayes, C.R. 1976. *The dispersed city: the case of Piedmont, North Carolina.* (University of Chicago, Department of Geography, Research Paper 173, Chicago).

Heide, H. ter 1979. *Implications of current demographic trends for population redistribution policies.* (National Physical Planning Agency, publ. 79-2, Staatsuitgeverij, Den Haag).

Heide, H. ter and Eichperger, C.L. 1978. *Dynamic interrelations between population redistribution policies and demographic developments.* (National Physical Planning Agency, publ. 78-3, Staatsuitgeverij, Den Haag).

Hoekveld, G.A. 1982. Het structuurschema volkshuisvesting, de regionale schaal. in *Bundeling inleidingen Symposium Structuurschema Volkshuisvesting,* (K.N.A.G., Amsterdam), 4-58.

Hoekveld, G.A. 1980. *On the development of the Dutch urbanisation pattern between 1930 and 1971.* Spa, (mimeograph).

Joseph, A.E. and Smit, B. 1983. Preferences for public service provisions in rural areas undergoing exurban residential development - a Canadian view. *Tijdschrift voor Econ. en Soc. Geografie,* 74, 41-52.

Laan, L. van der and Piersma, A. 1982. The image of man: paradigmatic cornerstone in human geography. *Annals of the Association of American Geographers,* 72, 411-426.

Lamb, R. 1975. *Metropolitan impacts on rural America.* (University of Chicago Department of Geography, Research Paper 162, Chicago).

Leven, C.L. 1978. Growth and non-growth in metropolitan areas and the emergence of polycentric metropolitan form. *Papers Regional Science Association,* 41, 101-112.

Leven, C.L. 1979. Economic maturity and the metropolis' evolving physical form. in *The changing structure of the city,* 16, 21-22.

Lichtenberger, E. 1976. The changing nature of European urbanisation. in *Urbanization and counter-urbanization,* Urban Affairs Annual Review, 11, 81-107.

Megalopolis Nordwest Europa, strukturelle Entwicklungen und Tendenzen, Dortmund, ILS, 1980.

Moewes, W. 1982. Was soll aus unseren groszen Städten werden? *Geographische Rundschau,* 11, 502-518.

Simmons, W.J. and Bourne, L.S. 1981. Urban and regional systems/ qua systems. *Progress in Human Geography,* 5, 420-431.

Sternlieb, G. and Hughes, J.W. 1978. *Current population trends in the United States.* (Center for Urban Policy Research, New Brunswick, N.J.).

Taaffe, E.J., Gauthier, H. and Maraffa, T.A. 1980. Extended commuting and the intermetropolitan periphery. *Annals of the Association of American Geographers,* 70, 313-324.

Vining, D.R. and Kontuly, T. 1981. Population dispersal from major metropolitan regions: an international comparison. *International Regional Science Review,* 3, 49-73.

Weisz, P.B. 1971. *The science of Biology.* (McGraw-Hill, New York).

Chapter 7

The concept of resources in rural geography

Chris Park

THE COUNTRYSIDE - CHARACTER AND CONFLICT

The countryside is often seen as 'all things to all men' because it offers a broad spectrum of opportunities to a wide range of users. There is a feeling that 'the countryside stands for all that is important in Britain; it is the expression of the good life away from the stresses and strains of the city, and the symbol of everything which is considered truly British' (Best and Rogers, 1973). Yet is is difficult to completely separate the countryside from urban influences, despite Lefaver's (1978) plea that 'rural issues need to be defined in their own context, and that the policy tools used to solve these issues *must* come from the rural perspective'. The links between town and country are deep-rooted, long-lived and all embracing. However, the countryside offers three important non-urban qualities (Cloke and Park, 1984).
a) it is dominated by extensive land uses such as agriculture and forestry, or large open spaces of undeveloped land;
b) it contains small, low-order settlements which demonstrate a strong relationship between buildings and surrounding extensive landscapes and which are thought of as 'rural' by their residents;
c) it engenders a way of life characterised by a cohesive identity based on respect for the environmental and behavioural qualities of living as part of an extensive landscape.

Moseley (1980) stresses that the distinctive character of the countryside is a function of such rural properties as a pleasant environment, a 'spaced-out' geographical structure (leading, in turn, to problems of accessibility and costly public services) and a distinctive local political ideology which favours the market, the volunteer and the self-helper rather than public provision.

The traditional view of the countryside as a haven of peace and tranquillity, and the 'timeless image of a serene and unchanging countryside' (Selman, 1976) are perpetuated through literature, the media and urban perceptions of the rural 'good life'. However, the reality of the countryside is somewhat different. The countryside is a battleground in the allocation of valuable resources between competing demands (Blacksell and Gilg, 1981; Coppock, 1977), many of which are incompatible, some threatening to impose irreversible changes on the fabric and life of the countryside and others seeking to fossilise rural areas into living open-air museums and sanctuaries of by-gone rustic days. As a result, conflicts are everyday occurrences, and management and planning are essential ingredients in the rural scene.

Rural planning is 'a question of recognising emergent areas of conflict in values, and of taking action to reconcile or otherwise meet that conflict' (Cherry, 1976), and it centres on balancing the needs and aims of various groups of people who seek to use the countryside in the pursuit of different goals. The need for enlightened planning and management in the countryside is clear, and the form these take is critical. Planning can mean many things as Allison (1975) points out. Planning might simply encourage co-ordination of otherwise separate actions; it might offer publicly chosen alternatives or the imposition of policy controls on private and organisational actions; or it might be based on forward projections of recent trends.

The Concise Oxford Dictionary defines a plan as 'a formulated or organised method by which a thing is to be done; a way of proceeding'. In the context of rural planning this means reconciling conflicting values and demands for the common good. How this 'common good' is defined, and who derives most benefit from it, are of course central issues; agreement on such matters seems extremely elusive because of the many vested interests involved. The 'common good' needs also to be considered along several dimensions of both time (the short term or the long term) and space (local, regional or national issues), and in terms which are wider than social and economic gain since they include environmental quality. Much rural planning adopts relatively short time horizons and focusses on local rather than national issues. Moreover, social and economic elements are often viewed as more pressing or more relevant than wider environmental issues.

An ideal planning system for the countryside would seek comprehensive and integrated use of rural resources, it would be forward-looking and cater for the needs and aspirations of as many groups of users as possible, and it would adopt a suitably rural philosophy and perspective. There must increasingly be multi-functional use of limited countryside resources because, as Boote (1976) stresses, 'one cannot have agriculture, forestry, mineral exploitation or urban development taking place in the countryside as single-purpose goals. Often, of course, they are the *dominant* goals, but there are always a number of compatible goals...'.

THE CONCEPT OF A RESOURCE

A resource is a culturally defined notion, rather than a tangible object; anything may be a resource if it offers a means of attaining certain socially-defined goals. Zimmermann (1951) proposed that the environment at large offers a reservoir of 'neutral stuff' which we evaluate on the basis of our biological needs (such as food) and our social requirements (such as fulfilling aspirations). We might, in turn, decide to transform some of that 'neutral stuff' into a usable resource; the ease, cost and speed of doing so being dependent on our knowledge and prevailing technology. Zimmermann (1951) thus argues that 'availability for human use, not mere physical presence, is the chief criterion of resources. Availability in turn depends on human wants and abilities. Resources are not, they become; they are not static, but expand and contract in response to human wants and human actions'.

This inherent flexibility in deciding what is currently a re-source underlies economic principles of resource use, allocation and management (Cottrell, 1978). In general terms a range of fac-tors serve to promote or constrain the viable use of a given re-source. Such factors include environmental issues; economic factors (such as existing market conditions, forecasts of probable demand, prospective development of supplies, trends in market prices and precision of forecasts); political factors and techno-logical factors such as the introduction of new production pro-cesses and the discovery of new reserves (Manners, 1969).

Resources can be classified either by utility or by type. A utility classification distinguishes between *natural resources,* which have some practical value, and *non-utilitarian resources* whose worth is social rather than explicitly practical. Natural resources can be further subdivided on the basis of the extent to which they can be replaced by natural or man-made processes (Dasmann, 1976). Thus inexhaustible resources such as sunlight contrast with non-renewable ones such as coal and oil, whereas other resources are renewable (such as vegetation) or re-cyclable (such as many metals).

Non-utilitarian resources are more important in relation to the quality of life and the environment. In western Europe and the United States in particular, populations with relatively high material standards of living place considerable value on the need for such things as access to unspoiled countryside, unpolluted supplies of air and water, and a high quality environment. Popula-tions in developing countries place lower values on these because they are often more concerned with issues such as food shortages, widespread unemployment and the need to exploit natural resources to achieve a higher material standard of living (Chapman, 1969).

Classification by type would distinguish human and social resources from physical and environmental ones. The latter include mineral resources, water, landscape, wilderness, ecological resources and habitats. These are the resources which figure prominently in many countryside planning debates. Human resources cover rural populations, their life-styles and quality of life, and the complex web of links between people, and between people and environment. The countryside comprises much more than the physical fabric of the landscape. The welfare of rural dwellers, stability of rural communities, viability of the rural economy, and balance between self-sufficiency in economic terms and reliance on income from urban areas are all key elements in countryside planning. Their importance lies in both the ethical need to protect the interests of country folk *per se* and the expedience of maintaining a viable rural economy for the legacy of future generations (Cloke and Park, 1984).

RESOURCE ALLOCATION AND MANAGEMENT

A central task of rural planning is the fair allocation of resources between competing claimants through management policies. Resource management can be defined as 'a process of decision making whereby resources are allocated over space and time according to the needs, aspirations and desires of man within the framework of his techno-logical inventiveness, his political and social institutions, and his legal and administrative arrangements' (O'Riordan, 1971).

The allocation of resources in the countryside is particularly complex because of the need to consider development proposals in

61

the light of their broad social, environmental and institutional
consequences as well as their economic and financial effects
(Sewell, 1973). In many types of resource allocation, market
forces alone are used as a basis for decision making. In an ideal
situation simple economic theory allows market forces to dictate
resource use by competition between supply and demand (Harrison,
1977). This competition is often judged by economic cost-benefit
analyses carried out by each of the major decision-makers involved
in the allocation procedure. The demand for rural resources is
determined by factors such as location (particularly in relation to
urban areas), development potential, land quality, strategic or
political significance, accessibility and economic viability.
Supply will be conditioned by factors such as existing land uses,
landownership patterns, compatibility with other uses, land and
landscape quality and the historical legacy of legislative and
institutional constraints and personality preferences.

The market mechanism is arguably not a suitable basis as the
sole agency for allocating countryside resources. Hines (1973) has
stressed that market forces cannot at the same time encourage both
economic growth and the conservation of natural resources, stimulate all kinds of private economic activity and maximise social
welfare, promote unrestrained economic growth and protect the
environment. Planning intervention is thus required because of the
imperfections of market forces but also because of the limited
availability and irreproducibility of many countryside resources.
As Moore (1978) points out, planning activities are linked by a
concern for providing and allocating public goods for which the
market mechanism alone does not suffice. Intervention demands
planning constraints and rational judgements to take into account
factors such as social, cultural, strategic and environmental impacts of proposed changes in deciding on the allocation and use of
rural resources.

Increasing attention is being paid to the application of
principles of resource management in rural planning (e.g. Statham,
1972; Cloke and Park, 1984), particularly in a climate of monetary
restraint and cautious investment and subsidy. Whitby and Willis
(1978) emphasise the need to identify the most efficient public
choices in rural resource use. This choice must be made in the
face of the continued growth of the public sector, fluctuating
commodity prices, dislocation of the land market, pollution and
declining environmental quality, and the modernisation of derelict
settlement structures.

Planners must start to accept a series of responsibilities in
adopting the principles of resource management (Ashworth, 1974).
First, planners will need to examine the basic premises on which
they base decisions, and to identify which assumptions are based on
the unlimited availability of resources such as energy resources
and land for development. Reappraisal of these assumptions in the
light of scarcity might encourage fundamental shifts in policies,
such as those for transport and settlements which will be based in
the future on more enlightened resource use and conservation.
Second, they will need to recast plans with the aim of maximising
the use of renewable resources including the human resource. Such
reformulation might cast a new light on the significance of air and
water pollution, engender a broader view of agriculture and forestry
enterprises, and favour redevelopment of existing land rather than
sterilisation of productive farmland. A third responsibility is
for planners to take account of the social and physical consequences
of plans with constant feedback between policies, objectives and
outcomes.

IMPLICATIONS AND PROBLEMS

Adoption of a resource management perspective in the countryside might thus encourage co-ordinated and rational rural planning in preference to the fragmented approaches evident in many areas today. A resource management approach would help rural planning in both practical and conceptual terms. It will not be the panacea of rural planning because, although it can offer a useful framework in which to evaluate and organise conflicts and priorities, resource management in the countryside is not free from problems. As Holling and Chambers (1973) stress, 'we need to find directions towards solutions, and not the Utopian solutions themselves'.

One group of problems centres around the type of resource use characteristic of rural areas. Many of the resources currently being used on a large scale are non-renewable ones such as minerals. Krutilla (1967) argues that many of the problems of resource use stem from the irreproducibility of unique phenomena such as natural landscapes and habitats. Although the supply of many rural resources cannot readily be enlarged, their utility to man (and hence the economic value placed on them and the pressure to develop them) is doubtless increasing.

A second suite of problems derives from differing compatibilities of rural resource uses. The countryside is a mosaic of interacting resource-using activities, and inevitably the use of some resources automatically devalues or destroys others. Extraction of mineral resources generally reduces scenic quality, for example. Hall (1980) comments that 'planners have tried in many ways to combine concern for visual aesthetics, economic efficiency and social equity in their plans; planners and planned now have a greater understanding of the limited capabilities of the planning machine to foster all three in abundance'.

The third problem concerns evaluation, both in theory and practice (Lichfield, Kettle and Whitbread, 1975). Natural resources can be evaluated in monetary terms because market forces are generally of central importance in determining resource use and allocation. Non-utilitarian resources, however, cannot be evaluated unambiguously in monetary terms despite attempts to include intangibles such as environmental quality, loss of habitat and extinction of species into conventional cost-benefit analyses (e.g. Helliwell, 1973). Therefore conflicts involving both natural and non-utilitarian resources (such as the conflict between opencast mining and landscape protection) involve many value judgements and, ultimately, ethical decisions.

A fourth set of problems relates to the balance of factors adopted in decision making. This balance is important in terms of scale. Local plans might need to accommodate national or even international objectives and strategies such as the European Community's Common Agricultural Policy or the forthcoming Community Directive on the Environmental Assessment of Projects. Balance also arises through the need to include factors other than economic criteria in resource allocation. Social impacts, environmental repercussions, strategic implications and political acceptability are amongst the more important factors which must be considered.

The final problem area combines ideology and expediency and it concerns allocation procedures and intervention. For most rural resources, allocation is based partly on market forces of supply and demand and also partly on intervention by a planning system and the public responsibility and accountability which this implies. Intervention is an inevitable outcome of the need to balance forces such as national policies and strategic issues, regional priorities and objectives, local development-control interests, political opinion at various levels, social and environmental impacts and economic incentives and constraints. Intervention is also required to introduce welfare and equity considerations into rural planning. It has been argued that 'the aims and objectives of rural planning and management could be redirected towards redressing the inequalities causing (rural) deprivation... Ultimately the problem becomes one of resolving conflicts within the rural resource system, particularly those arising from basic deprivation of opportunity occasioned by social and environmental forces within the countryside' (Cloke and Park, 1980).

REFERENCES

Allison, A. 1975. *Environmental planning - a political and philosophical analysis.* (George Allen and Unwin, London).

Ashworth, G. 1974. Natural resources and the future shape of Britain. *The Planner,* 60, 773-8.

Best, R.H. and Rogers, A.W. 1973. *The urban countryside.* (Faber and Faber, London).

Blacksell, M. and Gilg, A. 1981. *The countryside - planning and change.* (George Allen and Unwin, London).

Boote, R.E. 1976. Only one earth - it's no longer why conserve, only how! *Wildlife,* 18, 443-5.

Chapman, J.D. 1969. Interactions between man and his resources. in *Resources and man - a study and recommendations,* Committee on Resources and Man (ed) (Freeman, San Francisco), 31-42.

Cherry, G.E. (ed.) 1976. *Rural planning problems.* (Leonard Hill, London).

Cloke, P.J. and Park, C.C. 1980. Deprivation, resources and planning - some implications for applied rural geography, *Geoforum,* 11, 57-61.

Cloke, P.J. and Park, C.C. 1984. *Resource management in the countryside: a geographical perspective.* (Croom Helm, London).

Coppock, J.T. 1977. The challenge of change - problems of rural land use in Great Britain. *Geography,* 62, 75-86.

Cottrell, A. 1978. *Environmental economics.* (Arnold, London)

Dasmann, R.F. 1976. *Environmental conservation.* (Wiley, New York).

Hall, J.M. 1980. *The geography of planning decisions.* (Oxford University Press, London).

Harrison, A.J. 1977. *Economics and land use planning.* (Croom Helm, London).

Helliwell, D.R. 1973. Priorities and values in nature conservation. *Journal of Environmental Management,* 1, 85-127.

Hines, L.G. 1973. *Environmental issues - population, pollution, economics.* (Norton, New York).

Holling, C.S. and Chambers, A.D. 1973. Resource science - the nurture of an infant. *Bioscience,* 23, 13-20.

Krutilla, J.V. 1967. Conservation reconsidered. *American Economic Review,* 67, 777-86.

Lefaver, S. 1978. A new framework for rural planning. *Urban Land,* 37, 7-13.

Lichfield, N., Kettle, P. and Whitbread, M. 1975. *Evaluation in the planning process*. (Pergamon, London).

Manners, G. 1969. New resource evaluations. in *Trends in geography*, Cooke, R.U. and Johnson, J.H. (eds), (Pergamon, London).

Moore, T. 1978. Why allow planners to do what they do? A justification from economic theory. *Journal of the American Institute of Planners*, 44, 387-98.

Moseley, M.J. 1980. Is rural deprivation really rural? *The Planner*, 66, 97.

O'Riordan, T. 1971. *Perspectives on resource management*. (Pion, London).

Selman, P.H. 1976. Environmental conservation or countryside cosmetics? *The Ecologist*, 6, 334.

Sewell, W.R.D. 1973. Broadening the approach to evaluation in resources management decision making. *Journal of Environmental Management*, 1, 33-60.

Statham, D. 1972. Natural resources in the uplands. *Journal of the Town Planning Institute*, 58, 468-78.

Whitby, M.C. and Willis, K.G. 1978. *Rural resource development: an economic approach*.(Methuen, London).

Zimmermann, E.S. 1951. *World resources and industries*. (Harper, New York).

Chapter 8

Towards a policy-oriented analysis in rural areas: a reflection on living conditions

Tom van der Meulen

INTRODUCTION

In this chapter, I want to indicate the nature of applied rural geography. Here, only one thing counts: the approach to real problems has to make sense; and this should be judged by outsiders. A social science can only be called empirical when its findings have some impact in reality - otherwise it is only academic. First I shall briefly mention some shortcomings in the normal scientific approach as it is practised in rural geography. I propose instead another paradigm, which explicitly admits the fact that social research is communication in a 'forum' (a community of experts). This option is elaborated at the end of the paper with respect to some practical rural problems.

GEOGRAPHY AS A BEHAVIOURAL SCIENCE

The so-called scientific approach in social science produces a number of problems. It is not possible here to present an exhaustive description of all of them. Being a social science, rural geography aims to explain behaviour in rural areas. This explanation is performed by trying to reconstruct the conditions which were responsible for the behaviour that is studied. This holds for causal as well as rational explanations. In the former, the researcher is looking for causes or factors, while in the latter interest is focussed on reasons which people give to explain their behaviour. Some authors have tried to reconcile the two (Stegmüller, 1974). Obviously, the rational explanations have more advantages. Nevertheless, the two types of explanation share some shortcomings, three of which I shall mention.

First, since the aims is to reconstruct antecedent conditions, a determinism has to be invoked. This overlooks the fact that behaviour is most fruitfully studied when it is seen as a decision by the actor who is being studied, and not as the outcome of a larger 'mechanical' system which is wholly defined by the researcher. As rational explanation is directed also at reconstruction (Strik, 1981), this holds for rational explanation too; but it has to be admitted that the rational explanation in theory does not believe in some 'disposition' of man. Second, only the behaviour of 'rational' actors can be explained. Explaining in a *causal* way forces the introduction of some disposition to act in well-defined circumstances to make the behaviour intelligible. Within *rational* explanation, one must assume that the decisions taken by the studied actor are necessarily the same as they would have been if the researcher himself had acted in that situation. In spite of

Dray's 'ladder of rationality' (1970), it should be noted that in
both cases the rational actor is a (re)construction of the
researcher. This means that the number of explanations potentially
equals the number of researchers. It is not possible to separate
valid explanations from non-valid ones. In fact, all explanation
is rationalisation. For example, one can cite the rural geographer's habit of using such holistic concepts as the 'rural-urban
continuum', which camouflages a reality which is really multi-
dimensional (Barth, 1972; Beekman, 1973; and, for an application in
geography, Van der Meulen, 1979). Third, social theories can lose
their validity when they are applied by the subject under study.
This is the case in 'reflective' behaviour (Hofstee, 1980). Since
reflective behaviour is the object of the social sciences, this is
a serious difficulty. Reflective behaviour is the application of
knowledge whether it is true or not. In the normal scientific
approach, it is not supposed that actors learn from their mistakes.
However, they do. In what way, then, should their behaviour be
studied?

PROSPECTIVE AND RETROSPECTIVE APPROACHES

I propose to consider behaviour by its effects instead of its
causes. The orientation is no longer retrospective; it is prospective. Behaviour (in the case of institutions, policy) is considered as a test and an application of scientific knowledge. The
effects of policy can be classified as follows:

Effects of behaviour (policy)

		application of knowledge	
		desired	not desired
test of knowledge	expected	1	2
	not expected	3	4

   Growth of knowledge (by testing hypotheses) takes place through
discussion in the 'forum', as is the case in retrospective research.
The forum is the community of experts (de Groot, 1971). Testing
occurs by comparing expected and unexpected effects on the basis of
methodological criteria. This means that the testing of knowledge
occurs *after* the event or action. The application of knowledge is
in the same way a matter for the forum, but here the effects are
discussed *before* the action is performed. In this model of analysis,
which can be described as 'action research' or 'rational planning',
it is not necessary to assume or construct a rational agent. There
is no need either for assumptions about behavioural dispositions. A
necessity-relation, either manifest or latent, between initial
conditions and performed behaviour is absent. By reflecting on
behaviour, learning processes continue. There is no longer the
paradox that empirical insights lose their validity when they are
applied by the actor under study; reflective behaviour is an essential element in rational planning. The prospective approach meets
the *need* for discussing choices in behaviour, and it *inevitably*
takes account of the possible influence of the researcher not only
as a stimulus (as is done in causal explanations), but also as a
conjecture that is critically relected by the studied actor.

   Rational planning can be described in short as a discussion
about possible (including inevitable) and desired developments. In
this sense, planning can be regarded as a realisation of social
research; behaviour, or a policy, is discussed in the light of
expected and desired effects.

The concept of rationality ('rational planning') has a dynamic heuristic meaning in the sense of 'contributing to the solution of problems' (Toulmin, 1977). Rationality no longer concerns an agent's disposition or something similar, but rather is the *discussion* about both past and future behaviour or policy. The outsider cannot say whether a discussion is rational. Yet it is possible to partici- pate in it by introducing new arguments. Principally a rational discussion is an open one. This is not to say that such a discuss- ion is not restricted by relevant norms; these constitute the paradigm which, of course, can change. This concept of rationality has consequences for the composition of the forum. Since it is the behaviour of the actors under study which is discussed, the agent himself belongs to the forum - he is seen as susceptible to reason. There also are consequences for the validity concept which is the guiding principle in a rational (empirical) discussion. These consequences will be elaborated in the next section.

THE FORUM AND THE VALIDITY CONCEPT

In rational planning, behaviour is analysed with regard to its effects. For public authorities it is justifiable to consider policy as the agent's behaviour. The effects may be expected or unexpected, desired or not desired. The agent's membership of the forum does not entail, however, that the empirical insights, which are based on their expectations and wishes, are always pronounced or deliberate. Nevertheless, the agents can be held responsible for them by the researcher. Thus the insights become explicit. Knowledge cannot be seen as an attribute of the actors, as is the colour of their hair or clothes. Knowledge is a product of rational discussion. The only aim is to discuss the effects of behaviour asking the question, 'what shall we do and what can we learn?'. The forum, then, is the community of investigator and investigated actors; this is the case in testing knowledge as well as in its application (see the previous section). The knowledge which is discussed is knowledge about the living conditions of the studied actor. There is no difference between the 'scientific' knowledge of the researchers and the common sense of the subject under study. Both are engaged in one rational discussion about living conditions; how to change them and how to use them. Know- ledge is only meaningful when you can do something with it.

This is not the place for a discussion about the concept of truth. Suffice it to say that what can be considered as truth is an outcome of the forum's discussion. Truth conceived as a cor- respondence to facts (facts being considered as extra-linguistic entities) is not fruitful. According to Habermas (see Kunneman, 1979; 1980) truth is a claim to validity which is shared by a group (forum). We can speak of a dialogical truth meaning validity in the context of the forum. As the latter also discusses the appli- cation of knowledge, the 'normal' (epistemological) concept of val- idity has to be embedded in an implementary concept of validity - the validity of applied insights. Hypotheses are tested by appli- cation in a real situation. This is provided for in rational planning by testing and applying empirical statements.

Recently the concept of negotiation has been introduced in planning (Buit, 1982). It is meant to denote a style of planning which is based on negotiations between the actors involved. More- over, the concept is used to distinguish this style of planning from one based on research. It can be admitted that negotiations play a major part in the development of plans. However, it is a mistake to think that this style of planning stands in contrast to one based on research. It should be remembered that when a policy is discussed, the arguments have an empirical content; they express expectations and desired developments. Application of the statements has a testable impact (Popper, 1972). The consequences can include such diverse items as prestige on committees, winning of elections, relative deprivation of groups and a better use of resources. Normally many of these items, especially the first, are not the subject of policy-oriented research. But this is no reason to deny their empirical character. All are testable and play a substantial part in policy negotiations. In fact one can say that in those negotiations the actors involved perform an empirical discussion. Afterwards, one can identify which arguments were more of less valid. Essentially, this discussion cannot be distinguished from one performed by professional researchers or scientists. Since Habermas's 'ideal conversation' situation is absent from political discourse and in scientific circles the situation is parallel, we have to consider it as a retrospective construction. Gaining prestige, the defence of old positions and all those other human weaknesses are present in science as well.

The current position can be summarised as follows:
- from a retrospective to a prospective approach;
- no qualitative difference between scientific knowledge and common sense - there is only one criterion for empirical knowledge; effects are testable when knowledge is applied;
- the approach can be called 'action research' or 'rational planning';
- the term 'rational' has a heuristic meaning, in the sense of 'contributing to the solution of problems';
- the occurrence of reflective behaviour, which destroys validity of knowledge, does not undermine this model of analysis; it is well integrated into the approach;
- epistemological validity is not a goal in itself; it should be contained in a concept of implementary validity, to which the conditions for a concept of epistemological validity can also be applied.

POLICY-ORIENTED RESEARCH IN RURAL AREAS

Normally the relationship between research and policy has been analysed in a restrospective way. It was characterised by looking for those conditions which would promote the influence of the social system of researchers on the social system of policy makers (Van Lohuizen and Daamen, 1976; Van de Vall, 1980). It considers variables such as the position of the researcher in the policy-making organisation, the use of academic traditions in the presentation of results (e.g., footnotes) and the number of pages in the report. The evils of this kind of analysis are obvious; a restricted forum (as the reports of the analysis are only addressed to professional researchers), an implicit determinism (since there is supposed to be a relationship between circumstances and policy), the banal fact that empirical statements are judged by their speaker and not on their validity, and the claim that policy makers are not susceptible to reason or critical reflection. Policy makers cannot but be manipulated by the presentation of thin reports, absence of footnotes etc. This opens up the possibility of falsifying the empirical findings when policy makers demonstrate reflective behaviour.

It is an essential characteristic of policy-oriented research that it aims to discuss actual or potential policy. It is not concerned about the long-term validity of the analysis. This is not to say, of course, that thinking about the long-term *consequences* of a policy should be neglected. With regard to policy, research

70

has only one function and this is a critical one. Policy is criticised on the basis of its expected effects. There is an empirical
discussion between policy makers and researchers. The critical
function of research can be described on at least three dimensions.
1. If a policy is based on empirical insights, it is not useful to
    confirm those insights by research.
2. When looking for the consequences of policy measures, research
    should not restrict itself to those items that sound acceptable
    politically.
3a. Research on topics which are not yet the subject of policy by
    the authorities is an expression of a critical attitude towards
    the existing limitations of policy.
3b. Conversely, policy-oriented research can also aim to identify
    an over-organisation of some policy performance (Van der Meulen,
    1982).

    This means that it is not possible to give a general rule as to
whether research is relevant for policy. The only thing that can
be said is that it must be assumed that there is a choice to be
made. This can be an institutionalised one, while it is also
possible that a new junction in the field of policy is made as a
result of discussions between the investigator and investigated
actor. The extent and the structure of fields of policy are permanently under discussion.

    Groot's investigation in 1972 of small villages illustrates
this. The significance of this study was an implicit critique of
the policy of population redistribution in which it was assumed
that people wanted a more urban scene. Groot showed that many
inhabitants of small villages wanted to stay there. Since then,
small villages have been an accepted feature of regional planning.

    There is some evidence of the problems of small villages being
approached in a holistic way. This entails a woolly definition of
management conditions or the 'living conditions of the municipality'.
It is not specified in what respects a certain policy could be
effective, because holism assumes that certain characteristics are
relevant anyhow. Because of this lack of clarity, it is not
possible to articulate a critical dialogue about possible solutions
to real problems. Some examples of this view can be cited (see Van
der Meulen 1980 and 1981 for a more extensive explanation).
1. There is the contention that house building provides the solution
    to all the problems of small villages. This is derived from the
    idea that this will always have *some* effect, which may be true,
    but due to the smallness of villages (among other things), the
    effects are wholly unpredictable.
2. Natural growth is the best way for a village's population to
    expand. This assumes that villages are 'natural entities', and
    that migration is undesirable. There would be no need for any
    planning if all settlement growth was 'natural'. It should be
    noted that the natural growth of small villages is often nega-
    tive as a result of the age structure of the population. Munici-
    palities pleading for natural growth either forget this or use
    the argument in a selective way. In fact, it must be realised
    that neither the ideal growth nor the ideal size for a settle-
    ment can be unambiguously stipulated; this is a matter for
    negotiation between the various actors (province, municipality,
    housing corporations) with their diverse interests and therefore,
    diverse living conditions.
3. There is the view that the viability of villages (the term 'via-
    bility' assumes some vitalism) can be determined by an enumera-
    tion of the diversity of corporate life in the settlements and
    the existence of local leaders. A viable village is seen then
    as a condition for the allocation of further houses. By this
    view, it is considered undesirable for people to live

independent lives, as is normal in cities. Nevertheless, in the latter case a corporate life is no criterion for the allocation of new houses.

These views, which are often expressed by municipalities, overlook that small villages are multi-dimensional with complex problems. It is an illusion that there will come a time when all problems will be solved; this is as true for small villages as for larger settlements. All that can be hoped for is a continuing evaluation of the effects of policies and also for an understanding of the fact that the solution to real problems requires that living conditions be specified. These are not entities *sui generis* but problem-conditioned conjectures.

AN EXAMPLE FROM ZEELAND

In this traditional rural area, many houses in the countryside were abandoned in the years after the war when workers lost their jobs in agriculture and migrated to the Randstad-Holland and the industrial centres in Noord-Brabant. The growth in incomes and leisure time of many city dwellers and suburbanites in Holland, Belgium and West Germany forced them to look for new possibilities for spending their time and money. One such possibility was to buy a second home in Zeeland.

While this process was restricted in the sixties to the dispersed houses of the countryside it presented no problems for local government. The buildings were conserved, and this could be seen as a contribution to living conditions in the countryside, at least from a visual point of view. However, in the seventies the process of conversion from permanent to second home invaded the villages. In some villages a quarter of the housing stock was converted to a second home. Some of these were sub-standard houses in which none of the local people were interested. Increasingly, however, the older houses in the villages were the prize in a battle between potential second-home owners and the local people. Since the latter generally had less money to spend, they lost the battle. Besides, there were people who were eager to sell to second-home owners, since they were able to build a new house on the proceeds. Therefore, new land-use plans were required.

These processes troubled both provincial authorities and national government. The provincial government was confronted with new land-use plans to accommodate those displaced by the incomers. The new plans were regarded as a waste of space. The Ministry of Housing, which is responsible for the distribution of new houses which are built with a contribution from the national government, was confronted with new demands for houses even though there were enough good houses that were being used as second homes. For most municipalities the use of the housing stock was not an element of their living conditions; it did not affect their policy. The use of the housing stock was only seen as a relevant issue in so far as the municipalities saw their task as meeting every demand for housing. This meant that the municipalities had no interest in managing their housing stock. Conversion to second homes and the displacing of the local population from the housing market were seen as unavoidable.

The provincial and national governments forced municipalities to set up management schemes for their housing stock; rules of preference were established to restrict the conversion of permanent houses to second homes. The higher authorities had a strong weapon - not approving land-use plans and not allocating new houses. At the same time, a survey was carried out by the Provincial Planning Department as to the attitudes of the local population to second houses. The result was that second homes were regarded as a threat to the quality of life (Van der Meulen, 1976, and the papers by Clark and Martin in this book). It can be assumed that it was this result, an implicit critique of the municipal housing policy, which provoked the municipalities into adopting a new housing policy, namely restricting the use of houses as second homes. There was also growing protest from the local population.

How can this development be interpreted in terms of the approach which has been proposed in this paper? In the first place, living conditions are not given, but are recognized as relevant for a certain policy. Second homes were not an element in the living conditions of the municipalities *until they were seen as a problem* which could be solved. In general, there is no indication that a characteristic identified as significant by an academic researcher also has the same meaning for the local people.

In the second place, some light is thrown on the relation between policy and research. Research is the mechanism which provides the feedback between an organisation (or any actor) and its surroundings. Living conditions become manifest when they generate debate. The investigation into the attitude of the villagers to second homes was an expression of some doubt about the adequacy of municipal policy. In fact, the study was set up to change this policy, or at least to provide the municipalities with some data to reflect on their current policy. Of course, it is possible that the result of the investigation might have been that people enjoyed second homes. In this case, the investigation would have been a weapon against the provincial and national authorities, but the function would have been the same - criticising policy. It should be noted that the identification of the people's attitudes (with the help of a Likert scale) reflected the investigator's view on the subject. However, other people did not doubt the results, although they were unavoidably nominalistic. The significance of the observed attitudes is to be found in their contribution to a discussion with policy makers. In contrast, the results are *not* relevant to a 'theory of rural areas'.

Thirdly, the discussion between the national and provincial authorities on the one hand and the municipality on the other was not executed in an 'ideal conversation' situation; the higher authorities had the power to influence municipal policy. Nevertheless, the discussion between the parties involved can be seen as an empirical one and, besides that, the change in policy is an example of reflective behaviour. Notice that in this context it is not important to ask whether the change of policy was due to the pressure of the provincial and national government or to the results of the investigation into attitudes to second homes. This question is not important because it asks for causes and therefore it cannot be answered definitively.

Fourthly, we may ask how the change in policy contributes to scientific aspirations like the growth of knowledge. The change of municipal policy can be seen as an experiment to test the hypothesis that some kinds of management of the housing stock improve the living conditions of the villagers. The change of policy was also favourable for the municipalities as organisations, since they could apply for new houses with some chance of getting them. It is not possible to speak here in terms of the growth of knowledge in the sense of a contribution to the construction of impressive

theories. The terms of reference have to be widened to include
concepts like 'human growth' - the term proposed by Faludi (1976)
as a justification for rational planning.

CONCLUSION

Starting from the premise that behaviour (and, as a consequence, a
policy too) is not *caused* but *considered,* the search for causal
explanations is not a fruitful approach to research in the social
sciences. Therefore, a prospective approach is preferred to a
retrospective one. This also solves the paradox of reflective
behaviour. Behaviour (policy) is regarded as a test and an application of empirical knowledge. This knowledge can be made manifest
by a discussion between a researcher and the agent under study
(which can be an institution). The latter discusses his behaviour
or policy with the researcher who need not be an academic but who
always has a critical attitude towards the behaviour under investigation. Given a dialogical concept of truth, truth is regarded as
an outcome of a forum's discussion, the forum being composed of the
investigated actor and the researcher. Truth defined as correspondence to the facts has to be interpreted as correspondence to statements about reality. This means that the concept of 'living
conditions' as such is an empty one. The question is by what
empirical notions behaviour (policy) can be criticised. So, it is
these notions which constitute living conditions; they are considered when reflecting some behaviour or policy. The epistemological validity (truth), which is the steering principle in
normal scientific discussion, has to be contained in (and not replaced by) a concept of implementary validity; an empirical
statement is tested through application.

In this respect, a distinction can be made between the work of
a professional researcher and the way his results play a role in an
empirical discussion between the researcher and the agent under
study who can be regarded in this sense as the client. The latter
has been the focus of attention in this paper. The researcher's
results are a product of his mind and are contributions to a discussion. At the same time, it cannot be stated definitively that
the results of a professional researcher (or anyone else) reflect
the (real) truth. There is always a nominalistic element in
empirical findings. They reflect the truth of a restricted community of researchers who share some methodological ideas.

The result of the rational discussion about a policy is not a
huge theoretical building, which is the aim of those investigators
who adopt a retrospective approach. The most important point is
that the results are applicable and, perhaps, that it is possible
to detect some 'human growth'.

REFERENCES

Barth, E.M. 1972. *Evaluaties.* (Van Gorcum, Assen).

Beekman, A.J. 1973. *Veelvormige werkelijkheid.* (Boom, Meppel).

Boon, L. 1980. Moet de wetenschap rationeel zijn? *Kennis en
    Methode,* 4e jaargang, 140-151.

Buit, J. 1982. Achtergronden en gevolgen van onderhandelingsge-
    domineerde ruimtelijke ordening; een algemene inleiding. in
    Onderhandelen en ruimelijke planning, Veldhuisen, K.J., Hacfoort,
    E.J.H. and Timmermans, H.J.P. (eds), (Bohn, Scheltema and
    Holkema, Utrecht).

Dray, W. 1970. *Laws and explanation in history.* (Prentice Hall, Englewood Cliffs, New Jersey).

Faludi, A. 1976. *Planning theory.* (Pergamon Press, Oxford).

Groot, A.D. de. 1971. *Een minimale methodologie.* (Mouton, Den Haag).

Groot, J.P. 1972. *Kleine plattelandskernen in de Nederlandse samenleving. Schaalvergroting en dorpsbinding.* (Veenman and Zonen N.V., Wageningen).

Hofstee, W.K.B. 1980. *De empirische discussie. Theorie van het sociaal-wetenschappelijk onderzoek.* (Boom, Meppel).

Lohuizen, C.W.W. van and Daamen, J.C. 1976. *Onderzoek en ruimtelijk beleid.* (PSC/-TNO, Delft).

Kunneman, H. 1979. Cognitieve en normatieve rationaliteit. *Kennis en methode,* III, 173-198.

Kunneman, H. 1980. Rationaliteit en consensus. *Kennis en methode,* IV, 229-245.

Meulen, T.v.d. 1976. *Dorpenonderzoek.* (P.P.D. Zeeland, Middelburg).

Meulen, T.v.d. 1979. Hierarchy of centers: some notes on a concept. *Tijdschrift voor Economische en Sociale Geografie,* LXX, 361-365.

Meulen, T.v.d. 1980. Kleine kernen, gezien vanuit onderzoek en beleid. *Zeeuws Tijdschrift,* XXX, 46-54.

Meulen, T.v.d. 1981. De kleine kernen: een probleem? *Deltasignalen,* IV, 10-15.

Meulen, T.v.d. 1982. Het structuurscheme volkshuisvesting en de kleine kernen. in *Symposium structuurscheme volkshuisvesting,* (KNAG, Amsterdam), 76-94.

Popper, K.R. 1972. *The logic of scientific discovery.* (Hutchinson, London).

Stegmüller, W. 1974. *Wissenschaftlichte Erklarung und Begründung.* (Springer Verlag, Berlin).

Strik, S. 1981. Rationele verklaring en hermeneutiek. *Algemeen Nederlands Tijdschrift voor Wijsbegeerte,* LXXIII, 207-228.

Toulmin, S. 1977. *Human understanding. The collective use and understanding of concepts.* (Princeton University Press, Princeton).

Vall, M.v.d. 1980. *Sociaal beleidsonderzoek.* (Samsom, Alphen aan de Rijn).

SECTION TWO:

RURAL STANDARDS OF LIVING

Chapter 9

Mobile services and the rural accessibility problem

Malcom Moseley and John Packman[1]

This paper is concerned with the 'Mobile Services in Rural Areas'
research project which was sponsored by the Department of the
Environment and the Welsh Office and undertaken at the University
of East Anglia by the authors. The work finished in September 1983
and this paper sets out some interim observations. A final report
is now available (Moseley and Packman, 1983).

THE RURAL ACCESSIBILITY PROBLEM AND THE POLICY OPTIONS

The context for the research is the rural accessibility problem
(see also the papers by Huigen and Gant and Smith in this book).
It is now accepted that a rising level of car ownership is unlikely
to bring enhanced mobility and accessibility to all. Indeed,
rising car ownership in rural areas has been eroding the accessibility of those without a car in two distinct ways. First, as car
ownership rises, public transport is undermined. Second, the
mobile majority (over 70 per cent of rural households now have a
car) increasingly use urban services and undermine the village shop,
post office and other low-level services. The pattern of decline
in one English county, Norfolk, has been charted by Packman and
Terry (1981). Moseley (*et al*., 1977; 1979) has stressed that the
rural accessibility problem has three elements:

Link
People ⟶ Activity

A change in any of these elements (for example in the social composition or location of the population, the availability of public
or private transport, or the location and opening hours of shops
and factories) produces a change in accessibility. Although in-
herently a geographical problem in that it derives from the difficulty in traversing space, inaccessibility is suffered ultimately
by people not by places. Accordingly, the study of accessibility,
and the evaluation of alternative accessibility strategies, should
have a human focus and 'score sheet'. It should establish how
different groups (such as the elderly, mothers at home with their
children, or teenagers) fare in different circumstances.

These 'different circumstances' requiring a social evaluation
include, of course, alternative policy options. These options

---

[1] The views expressed in this paper are the authors' and do not necessarily
reflect those of the Department of the Environment or the Welsh Office.

relate principally to the locational and transport dimensions.
More specifically, there are four main groups of policy alternatives, which can (and indeed should) be put together into packages
suiting local circumstances and these are set out below:

i) <u>Passenger transport</u> This could be subsidised, car-sharing
schemes could be organised, or the Post Office persuaded to
carry passengers as well as mail (post buses).

ii) <u>Mobile services</u> These involve taking the activity to the
people rather than *vice versa*. It is worth mentioning at this
point that telecommunications constitute a special sort of
mobile service. British students can study at home for a
university degree by tuning into educational programmes on
television and by consulting their tutors over the telephone.

iii) <u>Key villages</u> This is a generic term for those settlements in
which land-use planners deliberately concentrate service provision and population growth. The need for transport is
thereby reduced.

iv) <u>Fragmented services</u> This option is the converse of the previous one. It involves accepting a loss of scale economies and
a deliberate fostering of small rural outlets in remote locations. The small village school is retained; the village shop
is subsidised.

Of these four options, mobile services have been least re-
searched in the British context. Whereas individual agencies may
be well aware of the characteristics and problems of their own services (for example mobile libraries, domiciliary social services
and postal delivery) no-one seems yet to have attempted an overall
appraisal.

OBJECTIVES, TIMETABLE AND SCOPE

The research was designed to clarify:

i) the range and nature of mobile and delivery services in rural
areas;

ii) their ability to alleviate the social problems of such areas;

iii) the scope for increasing their potential in this respect, by
relaxing identified constraints.

The work fell into three stages of roughly equal duration.

<u>Stage 1</u> involved identifying and cataloguing a wide range of
relevant services in Britain, and appraising from secondary sources
those problems and needs of rural residents to which such services
might conceivably be relevant.

<u>Stage 2</u> comprised in-depth case studies of selected services,
concentrating on service suppliers or operators, consumers and
potential consumers. It also involved studies of small areas in
the north of Scotland, East Anglia, and mid-Wales.

<u>Stage 3</u> involved devising and evaluating alternative policy
proposals, incorporating return interviews with selected service
providers and potential consumers.

Our working definition of mobile services was deliberately broad.
They may tentatively be held to embrace 'services provided by
people to rural households or individuals, discontinuously, at or
close to their home'. Three types of mobile services thus emerged,
and we gathered information on all three:-

i) '<u>Roadside</u>' services, in which the service is provided directly
from a vehicle which stops at a sequence of service points
(e.g. a mobile shop).

ii) 'Home' services, including delivery, collection and domiciliary
    services involving a home visit (e.g. refuse collection, a
    visit by a social worker).
iii) 'Rotating' services in which a sequence of fixed-location out-
     lets is serviced for short periods by a peripatetic service
     provider (e.g. branch surgeries).
Table 9.1 sets out a simple categorisation based on this distinction.

It may be instructive to help define the boundaries of the study
by listing certain items which were excluded:

mobile homes (not a mobile service under any definition);
telecommunications (not a person-to-person service and much
    studied elsewhere);
electricity supply, sewerage, television (services delivered to
    the home but not by a person);
wholesalers and distributors servicing village shops (the re-
    cipient is a 'firm' not a 'household').

Despite these exclusions the list of candidates for further
study was a long one (Table 9.1). Some warranted study because of
their ubiquity: mail is delivered daily to virtually every dwelling
in Britain and daily doorstep deliveries of milk reach 90 per cent
of rural households. So we asked whether these services could
expand their scope to include a wider range of functions. Stamps
could be bought from the postman, newspapers from the milkman.
Others attracted attention because of their novelty. 'Play buses'
take the experience of pre-school playgroups to the children and
mothers of isolated areas. A 'technology bus' takes micro-computers
from school to school to enlarge the pool of children able to grasp
the essentials of this new field of knowledge. Still others needed
consideration because in Britain at least they do not exist; mobile
post offices (which are numerous in the Netherlands) and mobile
pharmacies, for example. Stage 1 tried to appraise the whole range
of mobile services in rural areas as a prelude to more detailed
work. This broad appraisal provides the focus of the rest of the
paper.

DEVELOPMENTS 1950-1980

First, it is useful to seek a time perspective. Work by Haggett
and Mills at Bristol University has charted the post-war rise and
fall of village services in the largely rural county of Somerset
in south-west England. The original survey evidence from H.E.
Bracey related to 1947-50 and has been put alongside the 1980
picture. The evidence specifically on mobile or delivery services
is fragmentary, but the following points emerge.

  i) In 1950, 68 per cent of Somerset's 400 parishes had between 6
     and 15 tradesmen delivering commodities to people's homes. A
     further 25 per cent had more than 15. The most common, in
     descending order, were milkman, grocer, baker, butcher, fish-
     monger, coal merchant, laundry, newsagent and greengrocer.
     By 1980, the proportion of parishes with over 6˙tradesmen
     delivering had fallen from 93 per cent to less than 30 per cent.
 ii) The most dramatic decline has affected laundry delivery (from
     217 parishes in 1950 to 11 in 1980) and 'visiting cinemas'.
     But the mobile library service has greatly expanded at the
     expense of branch libraries, as has the delivery of paraffin
     and bottled gas.
iii) The data do not provide a clear picture of the changing for-
     tunes of mobile shops as such. But Mills observed 'the tend-
     ency in 1950 to find town-based tradesmen delivering to

Table 9.1 Examples of mobile services in contemporary Britain

FUNCTION \ LOCATION	ROADSIDE	HOME	ROTATION
COMMODITY	General store Wet fish Fish and chips Greengrocers Ice cream	Milk Newspapers Bread Other dairy products Meat Greengroceries Groceries Fish Paraffin Coal Oil Prescriptions Pensions Durable goods Tupperware Mail order Door-to-door sales Other domestic retailing	'Periodic markets' Mobile craft exhibitions
SERVICE	Dentistry for schools Bank/Building society Citizens' Advice Bureau Benefits bus Technology bus Training centre Play bus Optician Art gallery Car repairs Library	House-bound reader services Home-helps Chiropodists Doctors Midwives and other para-medics Social workers Meals-on-wheels Womens clubs The vicar Mail delivery House repairs Refuse collection Hairdressers Gas/Electricity meter readers	Part-time bank/building society Mail collection Peripatetic teachers Doctor's surgery Dentists Child Welfare Clinics Chiropodists Cinema Performing arts Blood transfusion service Vet

parishes in response to orders, while in 1980 the mobile store
is the most visible aspect of the service'.

SOME MOBILE SERVICES IN THE EARLY 1980s

A major part of Stage 1 of our rsearch involved an eclectic survey
of mobile and delivery services in contemporary rural Britain,
relying at this stage on widely scattered and largely unpublished
written material. Below we describe some points that emerged in
relation to three such services.

Retailing

Mobile shops are generally in decline. The main causes seem to be
the increased mobility of consumers, the depopulation of remoter
areas and a widening price differential between mobile shops and
supermarkets as the latter grow in efficiency. Also important are
a more 'choosy' clientele and the rising cost of vehicle operation.
The new generation of retailers are unwilling to tolerate long
hours of work and the time spent travelling unproductively between
stops.

In contrast, mail order seems to be growing in importance. It
accounts for about 8 per cent on non-food retail sales in Britain -
more in rural areas. The rural consumer clearly benefits from not
having to visit the shop and from a Post Office tariff which is not
differentiated geographically.

The 'doorstep delivery' of milk warrants close study on four
grounds: it is virtually ubiquitous even in rural areas; it is
generally a daily service (over 5 times per week); it brings other
things to rural consumers including non-milk products and social
benefits; and it is under threat. The threat arises from the
growing share of the milk market held by supermarkets.

Information and advice services

There are about 35 mobile advice and information bureaux operating
in Britain, the vast majority in urban areas. The particular need
for such services in rural areas reflects the inadequacy of public
transport and the closure of small offices by the local authorities
and the Department of Health and Social Security, Also important
are the high proportion of elderly people, and the loss of informal
advisers such as doctors, clergymen and policemen.

Those mobile schemes which are, or have been, operational in
rural areas suggest that it is grossly uneconomic to try to serve
communities with fewer than 4000 inhabitatns on a weekly basis.
Among the more successful urban schemes have often been those which
combine otherwise disparate services. For example, the Bradford
Books and Information Library Bus combines in one double-decker bus
a library and play-scheme for children under five, an adult liter-
acy service and a citizens' advice bureau. The spreading of over-
heads which this achieves would seem to be even more valuable in
sparsely populated areas.

The play bus

The original concept of the play bus, a double-decker bus converted
into a mobile play-centre, was first tried out in Liverpool over
twelve years ago. Today there are over 150 such schemes; again
most are in urban areas. The majority demonstrate great

flexibility and try to service, at appropriate times, children of
varying ages - and often adults with very different needs. For
instance, some provide language classes for Asian women, some serve
as clubs for pensioners, some concern themselves with health education. Because of their flexibility, play buses appear to have
great potential in rural areas since each can provide for the
children of several villages in sequence. Many play buses (and
indeed other mobile services) have also had a clear demonstration
or catalyst effect: after a few months local people often gain the
awareness or confidence to provide their own facility along more
conventional lines.

The authors' first interim report concerned itself largely with
a service-by-service review of the current situation. Other
sections of the report relate to library services, the post office
(collection, delivery and counter services), education and training,
health services (particularly pharmacies), personal social services,
mobile banks and building societies, culture, leisure and the arts.
Several conclusions emerged.

CONCLUSIONS: THE ROLE OF MOBILE SERVICES

1. A continuum

Rural Britain demonstrates a continuum of mobile services.

Coverage	Examples
Virtually ubiquitous	mail delivery and collection milk delivery
Widespread	mobile foodshops mobile libraries meals-on-wheels home-helps bread delivery newspaper delivery
Sporadic	mobile advice centres play buses mobile banks
Isolated examples	mobile daycentres mobile bookshops mobile technology units
Non-existent	mobile pharmacies mobile post offices

2. The flexibility of mobile services

The key element which distinguishes mobile services (of the 'road-
side' variety) from fixed services is their flexibility. This takes
various forms.

 i) Location - Clearly many different locations can be serviced.
    New locations can be tried: unsuccessful ones discontinued.
 ii) Time - The timing of 'service delivery' can be managed by care-
    fully scheduling routes. Certain mobile banks, for example,
    visit villages in the mornings and afternoons, and workplaces
    in the lunch hour. Timing can be revised with experience.
iii) Activity - Many mobile services serve several functions,
    either simultaneously or sequentially.
 iv) Clientele - Following on from iii) a variety of different kinds
    of clientele can be reached with the same vehicle. Play buses
    can double up as adult literacy classrooms and craft centres,
    for example.

3. Why are mobiles used?

This flexibility underlies many of the more specific reasons why mobile services are used.

i) For trial purposes. Mobile advice centres have been used to gauge the demand, and locate the best site, for a fixed advice centre.

ii) As a community catalyst. A play bus might be used to generate the formation of a play-group; a social security benefits bus might stimulate a welfare rights group; a mobile daycentre might stimulate better community care for the elderly in a village.

iii) To reach the immobile. Some 'immobile' clients cannot reach fixed location service points (hence meals-on-wheels). Others are unwilling to (for example, the semi-literate users of books or the reticent claimants unearthed by benefits buses).

iv) To share a facility when demand is limited. The absence of an adequate market threshold within the normal range of consumer travel is the classic justification for mobile retailing. Mobile information centres and the use of peripatetic teachers provide other examples.

v) To share a facility where the capacity to supply is limited but demand is considerable. The Bedfordshire technology bus exemplifies this sharing of a facility.

vi) To respond to a changing or emergency situation. Mobile banks temporarily serve the growing estates of new towns, and mobile information units serve factories where large numbers of workers are being made redundant.

vii) To capitalise on novelty and visual impact. A mobile bank or advice centre not only advertises itself, it also advertises the whole service of which it forms a part.

viii) Tradition and idiosyncracy. The vast disparity in the levels of provision encapsulated in Table 9.1 defies a purely logical explanation. The virtual ubiquity of doorstep milk deliveries in Britain and the total absence of mobile post offices are phenomena with essentially cultural explanations.

It is also striking that what all the public and voluntary-sector mobile services share is an innovative concern to promote social justice and reduce inequalities in the supply of services. From what many people consider the first mobile library - a horse-drawn van run by the Mechanics Institute of Warrington for the working classes in 1859 - through to Bradford Books and Library Bus serving today the Asian community, the aim of serving the under-privileged has been the same. The social security benefits buses run by the Strathclyde Regional Council provides a striking example of a mobile service run by a local authority intent on helping the poor and is in marked contrast to the attitudes of many English shires.

In this connection it is interesting that the principal, private-sector mobile service - the mobile shop - is chiefly used by the less mobile and generally poorer members of society. In short, the mobile service as a precarious agent of social justice provides a theme worthy of further study.

4. What are the disadvantages?

The limited spread of many mobile services suggests their disadvantages. Chief amongst these is the true cost of mobility - not just the often considerable vehicle-related costs, but the opportunity cost of time spent by the service supplier behind the driving wheel; mobile services involve a lot of 'dead time'. Moreover, there is the problem of the limited scope of the service on offer: the range of groceries, books or information literature must be limited compared with those of urban supermarkets, libraries and citizens' advice bureaux. There is sometimes the problem of security which worries bank managers, and has served to restrict the postman's role as a one-man, mobile post office.

The evidence suggests that the pattern of provision which has evolved over the years is not simply the outcome of rational appraisals of the mobile mode compared with alternatives. Policy proposals must rest on a careful appraisal of the whole range of quirks and constraints that have underlain this evolution. The final report of the research project makes a number of recommendations with this in mind.

REFERENCES

A full bibliography of the ephemera and reports which comprise the British literature on mobile services is contained in Moseley and Packman (1982 and 1983).

Mills, L. 1982. *Mobile and delivery services in Somerset and South Avon 1950-1980: changes in the space-economy of an English county*. (Unpublished working paper, Department of Geography, University of Bristol).

Moseley, M.J. 1979. *Accessibility: the rural challenge*. (Methuen, London).

Moseley, M.J., Harman, R.G., Coles, O.B. and Spencer, M.B. 1977. *Rural transport and accessibility*. (University of East Anglia, Norwich), 2 volumes.

Moseley, M.J. and Packman, J. 1982. *Mobile services in rural areas: first interim report*. (University of East Anglia, Norwich).

Moseley, M.J. and Packman, J. 1983. *Mobile services in rural areas: final report*. (University of East Anglia, Norwich).

Packman, J. and Terry, M.H.C. 1981. *Services in rural Norfolk 1950-80*. (Planning Department, Norfolk County Council, Norwich).

Chapter 10

Access in a remote rural area

Paulus Huigen

INTRODUCTION

In every geographical milieu access can be a problem for certain
groups in society. In the peripheral rural areas access is a
general problem: it affects all the inhabitants to varying degrees.
Therefore Moseley (1979) gave his well-known book on access the
subtitle 'the rural challenge'.

This paper is a first report on part of a study on access in
Zuidwest-Friesland. Zuidwest-Friesland is a peripheral rural area
in the north of the Netherlands (see Figure 1.1). This study was
carried out within the framework of the Zuidwest-Friesland project
in the Department of Geography of the State University at Utrecht.

The paper is divided into three parts:
- the context of the study, the research environment;
- a few remarks on the more theoretical aspects of the concept of
access;
- some provisional results of the study.

THE RESEARCH ENVIRONMENT

Peripheral rural areas are characterised by area-bound economic
activities (see Veldman's chapter in this book). The land use is
dominated by agriculture (Symes, 1981). The direct consequence of
this is a settlement pattern characterised by a lot of small villages, oriented to small market centres. The services and facilities are organised on a regional scale because of a lack of sufficient local demand. Employment outside agriculture and related
businesses is scarce and in general concentrated in the bigger
settlements. As a consequence the inhabitants of the peripheral
rural regions will have to travel over relatively large distances
to reach the locations of services and non-agricultural employment.

An economic activity can only be established if a certain minimum volume of activity can be carried out. This threshold applies
especially to the kind of activities that offer so-called central
goods and services. These central goods and services demand
that a person is physically present when accepting them. The

threshold is reached through the level of demand with purchasing power. This consists of two elements, namely the density of population as an indication of the number of potential customers and the mean income of the potential cumstomers that is available for expenditure. Consequently, peripheral rural areas are at a disadvantage compared with cities and urban regions. Firstly, there are higher costs in time and money to traverse the greater distances in order to reach the locations which offer central goods and services. Secondly, the mean income in peripheral rural areas is lower than the national mean and, in any case, lower than the mean income in urban regions.

The disadvantages in income and expenditure go together with relatively less choice between activities that offer the same kind of goods and services. Inhabitants of the peripheral rural areas are not only at a disadvantage concerning central goods and services, but also in relation to those public services which are delivered directly to the house, such as gas, water, electricity, rubbish collection and, to a certain degree also, public transport. Compared to the urban regions, the economic viability of these services in peripheral rural areas is endangered with very high costs of construction and operation and a relatively small turnover. In order to diminish the costs of running such services, spatial policy has in general been directed at the concentration of inhabitants in a few bigger settlements (key-village policy, Cloke, 1979; also Van Bemmel's and Lockhart's papers in this book).

Key-village policy and the process of scale enlargement in the activities that offer central goods and services both lead to a concentration in space of housing and employment for the inhabitants of peripheral rural areas. This development is stimulated by the present economic recession. For the inhabitants of the settlements that are not designated as key villages, participation in activities outside the home will become more difficult.

Will the trouble they have in participating at will in the social, economic and cultural activities of the broader society become so great that one could speak of rural deprivation? Will a situation arise in which the regional differences in this respect are too big to be acceptable within the context of present society? These questions can only be answered by empirical research on the spatial situation in which the inhabitants of peripheral rural areas live. Access, which is according to Moseley (1979) a characteristic of people, is an important concept to be used in this context. Impact studies will make it possible to evaluate the policy that has been pursued and that will be pursued in the future.

THE CONCEPTS OF ACCESS AND ACCESSIBILITY

The concept of accessibility is often used in human geography. In much traditional geographical research it was usually treated as a purely geometrical concept. Nowadays we find the concept being used in a more social context (Harvey, 1973; and Gant and Smith's paper in this book). This arises from the fact that the possibilities for participating in various activities outside the home are not evenly distributed among the members of society (Lenntorp, 1980).

In operationalising the concepts of access and accessibility we link up with the idea that Moseley (1977, 1979) describes as the access and accessibility problem. He has suggested a general verbal model; people - link - activity. The access and accessibility problem has to do with people and their characteristics (age, sex, income, societal role etc.), with links and their

characteristics (road network, timetable of public transport) and
with activities and their characteristics (location, kind, opening
hours). Viewed in spatial terms there are three dimensions to this
model; the source of the need (location of the groups of individuals that want access), the link (in the case of physical access the
distance to be covered), and the destination of the supply (the
location of the activities to which access has to be gained). In
this connection activities are conceived as places or establishments
in which activities are carried out. On the basis of Moseley's
model, a distinction can be made between the concepts of access and
accessibility.

Accessibility originates from the destination dimension. It is
in the first instance a characteristic of a place or an establishment.
It indicates the difficulty (or ease) with which a certain place
can be reached from other locations. Accessibility, expressed in
the number of people that can reach an establishment with a certain
amount of difficulty, can be seen as the carrying capacity of this
establishment. This capacity is decisive for the presence of an
establishment at a certain location. The concept of accessibility
is of particular importance in retail planning.

Access is a characteristic of people. It can be described as
the ability of individuals or groups to participate in various
activities outside the home. Because physical access is the central point here, it can be defined as the ability of (groups of)
individuals to move so as to participate in the activities that
they want, within the constraints of their time- and money-budgets.
The quality of physical access is determined by the number of
choices that individuals have for participating in both different
kinds of activities and alternative activities of the same kind.
With regard to researching access and accessibility three general
remarks can be made.

Access has to do with the possibilities that (groups of) individuals have for participating in various activities. Because manifested behaviour is adjusted behaviour (Hägerstrand, 1974), possible behaviour is considered to be of more importance than the manifested behaviour of individuals. The manifested behaviour is adjusted to circumstances. Generally speaking these circumstances
can be described as the ruling values, norms and organisation of
society in time and space. This means that one can not suppose
that manifested behaviour is an expression of the needs and wishes
of individuals. In this context Moseley (1979) stated in his
criticism of conventional transport planning that extrapolation of
manifested demand is to assume the rural transport problem out of
existence. Lenntorp (1980) too warns of this trap. Yet in the
Netherlands the system for standardising the service level of
buses is mainly based on a certain ratio between offered bus capacity and the actual demand per hour and per section of a route (De
Kogel, 1978).

A second point with regard to researching the access and accessibility problems concerns the level of aggregation. Several authors
on access and accessibility (Moseley, 1979, 1977; Breheny, 1978;
Wachs and Kumagai, 1973; De Boer, 1977; Lenntorp, 1976, 1980) have
favoured disaggregation in the field of access and accessibility.
Not enough attention has been paid, they argue, to what comes into
access, in which way and for whom. The consequence of this should
be that within the access and accessibility problem as described by
Moseley, one should disaggregate the three elements: people (social-
economic groups, age groups, sex), link (private and public transport) and activities (kind, location and scale).

A final point arises from the time-space approach to reality developed by Hägerstrand and his followers in Lund (Carlstein et al. 1978). The most important merit of this approach is that time is explicitly taken into consideration. In a lot of traditional research concerning access and accessibility hardly any attention has been paid to when activities are accessible. One should make clear which points in time are under consideration. Seen against the background of developments in technology which make it possible to cover greater distances in a shorter time, the question is whether problems of access and accessibility originate more from synchronisation than from synchorisation (Parkes and Thrift, 1980). Synchronisation means that the needs of individuals and the supply of activities occur at the same points in time. The physical togetherness of individuals and activities in space constitutes synchorisation. Participating in activities in physical terms is only possible if there is both synchronisation and synchorisation of individuals and activities.

THE AIMS OF THE ACCESS STUDY IN ZUIDWEST-FRIESLAND

The focus of the Zuidwest-Friesland project at Utrecht is the question of 'the interaction between physical-spatial structure and the social-spatial system in rural areas' (Van Bemmel et al.,1979). The access study tried to contribute to the development of a method for evaluating the intention, arrangement and management of the physical-spatial structure especially concerning the location of residences, employment, services and the travel possibilities between these activities.

The study starts more from the social aspect of individual needs, access, than from the economic aspect of the supply, accessibility. By developing such a method we try to link up with what Lichfield (1974) calls planning balance-sheet analysis. This technique is related to types of analyses such as cost-benefit analysis, cost-effectivity analysis and goals-achievement matrices (Stijnenbosch, 1978). The aim is to identify who gains and loses in terms of physical access when a decision on location is being taken. For whom will it be easier, and for whom will it give more trouble, to gain access to a certain activity at a certain location if this activity is established or removed?

The access study focussed on five topics.
i) How can the concept of access be made operational?
ii) What is the current situation with respect to access in peripheral rural areas?
iii) Which are the constraints which influence the actual physical access of groups of inhabitants in the peripheral rural areas?
iv) What are the consequences of attaining (or not attaining) certain facilities for the daily life of groups of inhabitants in peripheral rural areas?
v) What conclusions can be drawn from the results of the analysis concerning the improvement of access in peripheral rural areas?

The research programme consisted of three parts. The first involved an exploration of access in Zuidwest-Friesland. This was done using location profiles. In the second section a time-space survey was carried out to get information on the daily behaviour in time and space of the inhabitants of Zuidwest-Friesland. The third part of the study was directed at the development and application of a micro-simulation model. Zuidwest-Friesland serves as an example of the peripheral rural areas of the Netherlands (see Hauer's chapter in this book and Hauer and Veldman, 1980).

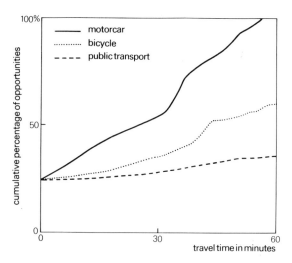

Figure 10.1   Locational profile of a settlement

LOCATION PROFILES IN ZUIDWEST-FRIESLAND

An initial exploration into access in Zuidwest-Friesland was made
by constructing location profiles for the 124 settlements in the
region. These location profiles are what Moseley calls comparative
measures of access. Within our framework we see them as indications of the number of spatial possibilities for the people living
in a specific settlement. They describe the level of access of
individuals living at a certain location. The profiles indicate
the number of opportunities which come into reach as more time is
spent on travel (Figure 10.1). Examples of the same kind of operationalising of the access concept can be found in Breheny (1978)
and Wachs and Kumagai (1973).

In the location profiles the 'opportunities' were 60 different
kinds of establishment offering different services; shops, schools,
establishments offering medical and social services, and places of
entertainment. The travel times were calculated for three different modes of transport. For the motorcar, travel time was calculated from a mean velocity for different kinds of roads and the
shortest road distance between settlements. For the bicycle we
took the same mean velocity for all kinds of roads. Measuring
travel time on public transport presented some difficulties.
Public transport differs from privately owned modes of transport in
its availability during the day. It is possible to take into
account the availability of public transport by including waiting
times. If this is done, travel time by public transport depends
heavily on the time of departure (Figure 10.2). We calculated the
travel time by public transport as follows. For every settlement
13 travel times (including waiting times) to every other settlement
were calculated, namely every quarter of the hour starting at 1 pm
and ending at 4 pm. This period was chosen because it could be
seen as a representative period for the whole day. The mean of
these 13 travel times was used as the travel time for public transport in the analysis.

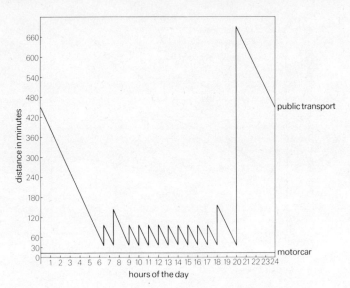

Figure 10.2   Distance in time from Gaastmeer to Sneek
(Zuidwest-Friesland)

The profiles were analysed by means of the sizes of the surfaces beneath the three curves. These curves represent the three modes of transport. To avoid the problem that differently shaped curves can have the same size of surface beneath them, surfaces for every quarter of an hour were analysed.

With regard to the mode of transport we found, of course, that travelling by car offered the most spatial possibilities. The bicycle came second and public transport seems to offer the least possibilities to participate in activities in settlements other than the place of residence.

On the basis of a cluster analysis, the location profiles were aggregated into five groups. The first group comprised settlements with a large number of establishments offering different kinds of services within their boundaries. The location profiles of the settlements in the second group can be characterised by rapidly widening spatial possibilities with a relatively small increase in travel time. Group three contained the settlements with an intermediate quality of access. In group four were the settlements which are at best very infrequently served by public transport. The settlements in group five suffered a relatively high degree of isolation. This division of the settlements in Zuidwest-Friesland into five groups, which were relatively homogeneous with regard to their location profiles, was used to determine in which settlements the time-space survey was to be carried out. One settlement was chosen from every group. In this way it was possible to get a spatially representative picture of the access problem in Zuidwest-Friesland.

THE TIME-SPACE SURVEY IN ZUIDWEST-FRIESLAND

Method and design

In five settlements with different location profiles a time-space
survey was carried out. This survey measured the actual access of
individuals, the way they spent their time, the availability and
use of modes of transport and background variables such as age,
profession, education and income. Actual access and the use of
time, together with behaviour in time and space, are measured by
the 'fresh diary appraoch'. This method involves asking the
respondents to describe how they passed the day before the survey
date. The daily behaviour in time and space is recorded by taking
down, among other things, the starting and ending time of every
activity, the kind of activity carried out, the location of the
activity and, when the activity involved travel, the mode of
transport used.

This time-space survey has three interwoven aims. Firstly,
it is our intention to describe daily life in a peripheral rural
area based on empirical evidence. This emphasis on daily life
arises from the consideration that the experiences of individuals
in their daily lives shape decisions which have consequences over
a longer period, such as moving to another place of residence or
changing job (Cullen, 1978). Secondly, the analysis of daily
behaviour in time and space gives an insight into some of the constraints that influence daily behaviour. Lastly, this insight
together with the empirical facts of daily life can be used for an
analysis directed at the evaluation of decisions with respect to
the geographical milieu in Zuidwest-Friesland in time and space.

Some provisional results

One of the topics in the analysis is the influence of the physical-
spatial structure of Zuidwest-Friesland on the daily life of the
inhabitants. This has been operationalised by looking at the differences between the five settlements in the space-time behaviour
of their inhabitants. At the same time we can find out if the differences in spatial possibilities indicated by the location profile
can be traced in daily life.

Analysing the data on the scale of the total population of the
settlements shows differences in the daily behaviour in time and
space between them. The time spent on activities outside the house
and the movement behaviour (length, duration, use of modes of
transport and direction of the trips) result in a picture that is
reasonably comparable with the impression obtained from analysing
the location profiles of the settlements. In general, the orientation for activities outside the home towards the place of residence
increases when there are more facilities and services in the place
of residence and also when the location is more isolated. In these
analyses some indications of differences in access between peripheral rural areas and urban regions could be found. Vidakovic
(1979) reports from a time-space survey carried out in Amsterdam
that about fifty per cent of the total number of trips were multi-
purpose trips. In Zuidwest-Friesland only about ten per cent were
multi-purpose. Because the definition of a multi-purpose trip
differed between the two surveys, we have tried to make the Zuidwest-Friesland data reasonably comparable. Even when one assumes
that every trip to a shop, private business facility (post-office,

bank etc.) or to a social or medical facility is a multi-purpose
trip during which five different establishments are visited, the
number of multi-purpose trips in Zuidwest-Friesland is still only
about 35 per cent. That is still very different from the 50 per
cent in Amsterdam.

Vidakovic (1980) also looked at the time spent on activities
outside the home in Amsterdam. He regarded time spent in relation
to distance from the home as a measure of the population's access-
ability. If more minutes per kilometre are spent, or a larger
percentage of time spent is closer to the home, then one can speak
of better access.

In Amsterdam about 90 per cent of the time spent outside the
home is within a distance of 9 to 10 kilometres as the crow flies.
In Zuidwest-Friesland in the most favourable conditions 90 per cent
of the time spent outside the home is reached within a road distance of about 25 kilometres. About 30 kilometres was the distance
for the most isolated settlement in the Zuidwest-Friesland survey.
Looking in more detail at these analyses, we see that the differences between the settlements are mainly caused by a few activities.
Disaggregation of the time spent outside the home by the kind of
activity shows that the observed differences are mainly to be found
in travelling to work and attending secondary education. No big
differences were found with regard to visiting facilities.
Although we expected to find some differences in this activity, this
can be explained in two ways. Firstly, the activity is a catch-all
category. It includes visiting a supermarket as well as a hospital.
Secondly, on a normal weekday there are few visits to facilities
with a relatively high level of specialisation. These results
strengthen the impression that poor access is mainly associated
with visiting facilities with a relatively high level of specialisation and places of work.

We also looked at some aspects of the behaviour in time and
space of groups of inhabitants. For this analysis we divided the
inhabitants into seven groups according to their role in society.
These groups are the working males, the working females, women with
children younger than 13 years of age, the elderly (over 64),
scholars (mainly attending secondary education), remaining females
(mainly housewives) and remaining males (mainly those receiving
some kind of social security benefit). A discriminant analysis for
these groups on the time spent on seven categories of activities
(personal care, household care, looking after children, education,
social activities, family activities and leisure) resulted in a
good score of 67 per cent. This means that 67 per cent of the
resp.ndents were classified in the right group on the basis of
their time spent. The time spent on working, household care and
education were the most discriminating activities. The biggest
differences in time spent are found between working males and the
elderly, the smallest differences between housewives and remaining
males. The elderly are the most homogeneous group, with the housewives second. The working females show the most internal differences in time spent.

To determine the influence of the location of the place of
residence on time spent, we looked at the differences between the
different groups in each settlement (constant location) and the
differences between the same groups in different settlements (variable location). Partly because the activity categories are broad,

the differences between the groups in each settlement are, in
nearly all the cases, bigger than the differences between the same
groups in different settlements. This observation could also be
made by putting the daily behaviour in time-space terms. In the
first instance we constructed 18 variables, which described the
daily behaviour in time-space terms. Following both a principal
components analysis of these eighteen variables and theoretical
considerations, four variables were chosen for further analysis.
These are the surface and length of the daily paths of individuals
in the time and space, the number of daily trips and the daily
amount of time which is spent outside the home. With regard to
each of these variables, the general conclusion can be drawn that
the differences between the groups in each settlement are bigger
than the differences between the same groups in different settlements. Although one can trace the differences in spatial possibilities as indicated by the location profiles, the provisional
results of this part of the study lead to the conclusion that
variations in access in Zuidwest-Friesland depend more on the socio-
economic characteristics of groups of individuals than on spatial
characteristics such as the location of the residence within the
research area.

This links up with the statement by Moseley *et al.*(1977, 63)
that, 'space has very different implications for different people.'
To make this operational within the study of Zuidwest-Friesland, we
are trying to develop a simulation model which operates at the
level of the individual.

TOWARDS AN INDIVIDUAL-ACCESS SIMULATION MODEL

The first aim of the model is to create the possibility of discover-
ing the frictions in individual physical access. The model has to
answer the question of whether it is possible for an individual,
given the time and the modes of transport available (based on the
empirical data from the time-space survey) to participate in a
certain activity. Participating in an activity is made operational
by looking at the time spent in the establishment in which the
activity is carried out. This takes place within the time-space
structure of the research area. The elements of the time-space
structure are the establishments to be visited and the travel
possibilities between the location of the individual and the location of the establishment. These travel possibilities are made
operational by taking into account the road system of Zuidwest-
Friesland and the timetable of public transport. Together with the
kind, location and opening hours of the establishments to be
visited, the time-space structure of the research area can be
constructed.

The analysis of possible frictions in access is designed as
follows. By means of the analysis of actual access and the daily
spending of time, activity programmes are constructed. An activity
programme can be described as a series of activities to be carried
out. Because physical access is the core of this study, only those
activities linked with movement are considered. An activity programme therefore is essentially a series of visits. It is possible to take into account a number of constraints that result from
analysing the data from the time-space survey. For example,
activities that are fixed in time and/or space (working, attending
secondary education, taking care of children etc.) can be inserted.
Account can also be taken of when each mode of transport is available to each individual. Possible frictions in access can be discovered by looking at the possibilities for carrying out the
activity programme and, if this is possible, at the difficulties
and costs involved.

A second reason for developing an individual-access simulation model is to analyse the effects of changes within the three elements that constitute the access problem, namely changes with respect to:
- the time-space characteristics of the facility to be reached location, opening hours, etc.);
- the time-space characteristics of the transport possibilities between the residence and facilities to be reached (a new road, altered frequency or route for public transport, etc.);
- the time-space characteristics of groups of individuals in the settlements (area of residence, shortening working hours, higher costs of using privately owned modes of transport, etc.).

Any change in one of these three elements results in a different situation with regard to the access of individuals. By simulating the carrying out of the activity programmes, insights can be gained into the effects of these changes.

Taking into consideration the aim of the model to be developed, we can state some general conditions which the model has to fulfil.
- The model has to simulate the carrying out of activity programmes in such a way that the resemblance with reality is as close as possible.
- The model has to give results on:
  - the number of alternatives for carrying out an activity programme;
  - how the programme might be accomplished;
  - the difficulties involved in carrying out the activity programme.

In the first instance the model is not explanatory in character. However, this does not mean that application of the model cannot contribute to explanations. If a certain activity programme cannot be carried out, a first explanation may be found in the constraints on the activity programme.

Initially, the model resembles the PESASP model developed and applied by Lenntorp (1976). Differences between the two models can be found in the operationalisation of activity programmes and the time-space structure. With regard to the application of the model, Openshaw (1978) sees a bright future for these micro-simulation, time-space models. Especially because of their simplicity, they should be useful tools for analysing the impact of policies.

REFERENCES

Bemmel, A.A.B. van, Huigen, P. and Veldman, J. 1979. *Projektkader, de stand van zaken per 1 september 1979, Projkt Zuidwest Friesland*. (Geografisch Instituut), nota nr. 1

Boer, E. de, 1977. Bereikbaarheidsbeleid. in *Colloquium vervoers-planologisch speurwerk 1977*. Jansen, G.R.M. *et al*. (eds), 799-819.

Breheny, M.J. 1978. The measurement of spatial opportunity in strategic planning. *Regional Studies,* 2, 463-479.

Carlstein, T., Parkes, D. and Thrift, N. (eds). 1978. *Timing space and spacing time*. (London), Volumes 1, 2, 3.

Cloke, P. 1979. *Key settlements in rural areas*. (Methuen, London).

Cullen, I.G. 1978. The treatment of time in the explanation of spatial behaviour. in *Timing space and spacing time,* Carlstein, T. *et al.* (eds), 2, 27-38.

Hägerstrand, T. 1974. The impact of transport on the quality of life. Introductory reports on the fifth international symposium on the theory and practice in transport economics, Athens, 22-23 October 1973. *Transport in the 1980-1990 decade,* 1.

Harvey, D. 1973. *Social justice and the city.* (Edward Arnold, London).

Hauer, J. and Veldman, J. 1980. *Kenmerken van landelijke gebieden op COROP-niveay, Projekt Zuidwest Friesland.* (Geografisch Instituut, Utrecht), nota nr. 2.

Kogel, G.G. de, 1978. Rural transport policy in the Netherlands. in *Rural transport and country planning,* Cresswell, R. (ed), (Leonard Hill, Glasgow).

Lenntorp, B. 1976. *Paths in space-time environments: a time geographic study of movement possibilities of individuals.* (Lund).

Lentorp, B. 1980. On behaviour, accessibility and production. *Rapporter och notiser.* (Lund), 58.

Lichfield, N. 1974. Costs and benefits in residential accessibility. *Planning and Administration,* 1, 88-101.

Moseley, M.J., Harman, R.G., Coles, O.B. and Spencer, M.B. 1977. *Rural transport and accessibility.* (Norwich), 1, 2.

Moseley, M.J. 1979. *Accessibility: the rural challenge.* (Methuen, London).

Parkes, D.N. and Thrift, N.J. 1980. *Times, spaces and places, a chronogeographic perspective.* (John Wiley, Chichester).

Openshaw, S. 1978. *Using models in planning: a practical guide.* (Newcastle upon Tyne).

Stijnenbosch, M.H. 1978. *Een geografische visie op de non-profit sektor.* Utrechtse geografische studies, 9, (Geografisch Instituut, Utrecht).

Symes, D.G. 1981. *Settlement and infrastructural development.* (a discussion paper), Paper for the XI Congress for European Rural Sociology, (Helsinki).

Vidakovic, V. 1979. De ontwikkeling van de paden-theorie. *Verkeerskunde,* 6, 271-276.

Vidakovic, V. 1980. *Mens-tijd-ruimte. Uit de dagboeken van 1400 Amsterdammers. Een essay gebaseerd op onderzoek naar activiteiten en verplaatsingen van inwoners uit drie stadsdelen.* (Dienst Ruimtelijke Ordening Amsterdam, Amsterdam).

Wachs, M. and Kumagai, T.G. 1973. Physical accessibility as a social indicator. *Socio-economic Planning,* 5, 437-456.

Chapter 11

The downward development of small service centres in rural areas: a theoretical exploration

Henk de Haard

INTRODUCTION

At the beginning of the 1970s we were confronted with some remarkable consequences of a not uncommon feature of villages and hamlets
- the closure of a service. We were conducting a survey in
Schnackenburg, a small village on the Elbe in a remote corner of
Lower Saxony, West Germany (see Figure 11.2). The village still
had ten services but some months before, the blacksmith, a jack of
all trades, had died. As in so many cases none of the family
wanted to take over the business. After the forge had closed, the
the other shopkeepers complained that fewer people came from surround
ing hamlets to do their shopping.

Following this, we suspected that perhaps people do not find it
worthwhile to go to a small centre when a service has disappeared
and the other shop and services cannot be visited during the one
shopping trip as before. Probably a trip to such a service centre
set up a conflict in the time budget of those people since the
opportunities to do whatever one wishes diminish considerably (Van
Dijk et al., 1982.

Essential to the analysis of this phenomenon is the concept of
the multi-purpose trip. One is not considering here the purchase
of all kinds of goods at once, irrespective of the number of shops
involved (Timmermans, 1980), but rather one is focussing on the
number of services which a person can use when he visits a service
centre. Thus a multi-purpose trip is defined as where one visits
at least two service elements (e.g. shops) to obtain one or more
goods or services. If one calls at only one service element, then
one is making a single-purpose trip.

In the next section we shall attempt to trace the mechanism
which stimulates or constrains the combined visits of people to
services and show what are the repercussions of this mechanism with
regard to both the development of small centre as such and the
functional hierarchy of service centres in a rural region.

THE CONCEPT OF CALL FREQUENCY

Consumers' spatial behaviour with respect to service centres consists of two essential components - the trip frequency and call
frequency. Trip frequency is the number of times someone pays a
visit to a centre within a given period. It gives an impression of

someone's spatial orientation and his attachment to different centres. Much attention has been paid to this concept in central-place theories

Less theoretically developed is the concept of the call frequency of a trip; that is, the number of successive contacts which someone has with service elements during a single visit to a service centre to obtain one or more goods. This phenomenon also affects people's spatial orientation and their attachment to centres. It matters greatly whether someone visits a centre once a week and calls at five shops or whether he visits the centre three times and visits only one shop each time. The attachment to that centre is probably stronger in the first case than in the second.

How many successive contacts can people have with shops and services when they are in a centre? This is difficult to indicate for the present. We can say that consumer behaviour is subject to constraints of a physical, social or economic character. For that reason people have limited time available to spend in town. Therefore they visit only a few elements every time they are there. In other words, there is an ultimate call frequency which differentiates between individuals and groups. In this study, we assume that people normally do not make a trip with a call frequency higher than five.

THE CONCEPT OF SERVICE SEQUENCE

If we suppose that people call at various shops and services in succession during a visit to a centre, then this implies that we also have to approach the supply side of centres in a particular way. We should not represent the supply of services by numbers of service elements, as investigators have done in many theoretical and empirical central-place studies, but rather by using the numbers of service sequences.

Let us assume a very small centre p consisting of four service elements: p = (A, B, C, D). A may represent a baker's shop, B an infant school, C a post office and D a grocery. Suppose that some-one makes a trip to that centre and that he is visiting three ser-vices in succession. Then our centre offers four combinations of service elements ABC, ABD, BDC, ADC; in a formula -

$$S_{(nx)} = \frac{N!}{X!(N-X)!} \tag{1}$$

X = value for the call frequency
N = number of service elements in a centre
$S_{(nx)}$ = number of service sequences in a centre, given X and N

Such a combination of service elements we shall call a service sequence. Provided X is less than or equal to N, each sequence has a size which is similar to the value for some specific call frequency. Service sequences are binomial coefficients which may be derived from Pascal's triangle.

THE CONCEPT OF 'MARGINAL' AND 'FULL' CENTRES:
SOME PROPOSITIONS

In Figure 11.1 some polygons of service sequences belonging to centres of different size have been shown. Maximum consumer be-haviour is represented by a perpendicular line from a point on the base line. This particular point is the assumed ultimate call frequency ($C_u = 5$). Left of this line lie the service sequences

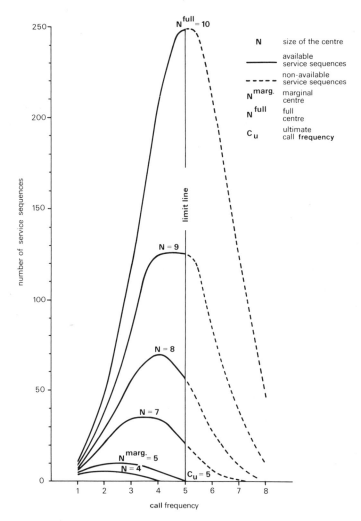

Figure 11.1   Number of service sequences per value of call
frequency, given the size of centre

which are available while to the right of it are those which are not
available.

In fact we are dealing with a number of distribution curves,
some being cut by the vertical limit line. When service centres
become larger, (a higher value for N) then the shape of their
curves changes gradually. First the curves become broader, then
steeper and more pointed and at last their tops slowly move to the
right.

Paying attention only to the area left of the limit line, we
see that a decreasing part of the distribution curves remains as
centres become bigger. We may also note that these parts quickly
transform into steep curves. Only those belonging to centres which
have five or fewer service elements lie completely on the left.

If a service centre has that quantity of service elements which
allows anybody with any call frequency at least one single service
sequence, then this centre is 'marginal' ($N^{marg.}$). The marginal
centre supplies a quantity of service elements which is exactly
equal to the value for the ultimate call frequency. In our study
we assume that this is five, so the size of this centre is five.
For the first time people with the highest or ultimate call frequency have at their disposal the highest supply of service
sequences in centres which are twice as big as the marginal one.
Such a centre is called a 'full' centre (N^{full}).

After these initial considerations, we may state the following
propositions:
1. The law of the increasing supply of service sequences. Given
   some call frequency, the supply of service sequences will increase as centres become bigger.
2. The law of the increasing divergence in the supply of service
   sequences. The quantity of supplied service sequences diverges
   enormously between big and small centres as call frequencies
   become higher.
3. The law concerning the supply of service sequences in small
   centres. In the case that centres are smaller than the 'full'
   centre, we can say that the quantity of supplied service sequences will always diminish, after an initial increase, as call
   frequencies become higher.
4. The law concerning the supply of service sequences in big
   centres. In the case that centres are greater than or equal to
   a 'full' centre, we can say that the quantity of supplied service
   sequences will always increase as call frequencies become higher.

Later we will demonstrate how these propositions, together with some
others, can be applied to a theoretical framework for the decline
of centres.

THE ELASTICITY OF THE QUANTITY OF SUPPLIED SERVICE SEQUENCES:

SOME PROPOSITIONS

We can now start to tackle the Schnackenburg problem; the loss of
service elements in small centres accompanied by a reduction in the
quantity of service sequences, given any call frequency.

Let us assume that in a small village with only nine service
elements a grocery has closed. Soon after, the post office disappears too (Table 11.1). The decrease in the quantity of service
sequences, absolute and relative, is astonishing! The attractiveness of this small centre immediately diminishes.

However, the most striking thing is that a relative decrease in
the supply of service sequences is always greater than or equal to
a relative decrease in the supply of service elements. The meaning
of this relationship is that a change in the quantity of supplied
service elements causes an equally strong or even stronger change
in the quantity of supplied service sequences, given any call
frequency. More formally:

$$\eta = \frac{\text{percentage change in the quantity of supplied}}{\text{percentage change in the quantity of supplied}} \quad (2)$$

percentage change in the quantity of supplied service sequences, given any call frequency — percentage change in the quantity of supplied service elements

η = service sequence elasticity.

Table 11.1   The absolute and relative reduction in the quantity of supplied service sequences in a centre with nine service elements, given the loss of two service elements

absolute reduction in the quantity of supplied service elements 2

relative reduction in the quantity of supplied service elements 22%

		quantity of service sequences, given the supply of service elements (N)		decrease in the quantity of service sequences	
		9	7	absolute	relative
quantity of	1	9	7	2	22%
service sequences,	2	36	21	15	42%
given some call	3	84	35	49	58%
frequency (X)	4	126	35	91	72%
	5	126	21	105	83%

In Table 11.2 we present some results of these elasticity calculations. Again we can derive some propositions from these.

Table 11.2   The elasticity in the quantity of supplied service sequences with respect to some service centres given any call frequency

		call frequency (X)				
		1	2	3	4	5
service centres	9	1.00	2.11	3.29	4.52	5.71
of different	19	1.00	2.05	3.15	4.30	5.50
size (N)	29	1.00	2.03	3.10	4.20	5.34

5. The law of the proportional or more than proportional increase or decrease in the quantity of supplied service sequences. Every increase or decrease in the quantity of supplied service elements in a centre causes a proportional or more than proportional increase or decrease in the quantity of supplied service sequences, given any call frequency.
6. The law of the increasing elasticity in the quantity of supplied service sequences. The quantity of supplied service sequences becomes more elastic as call frequencies become higher.

In the next section some of our propositions play a part in a theory concerning the decline of centres.

The process of decline takes the following course. It starts with a number of consumers ignoring more or less systematically some service element, for instance a grocery. The reason why they do so may be because they are visiting a supermarket in another town. Anyhow, these people act as though this particular service element does not exist any longer. Initially, the call and trip frequency of these consumers will probably stay the same (stage one).

Consciously bypassing a service element means that the service sequences based on this service element together with other ones also disappear. So these consumers accept voluntarily a proportional or more than proportional decrease in service sequences (5th law). People with a high call frequency give up more service sequences than those with a low one (6th law) (stage two).

Now it may be that, because of poor quality, the group no longer demands any service sequence consisting of the remaining service elements and the original call frequency of the group. Some people may even start to ignore another service element which they used when they had to visit the grocery anyway. If that occurs, then those consumers return to the first stage of the process. However, most of them will probably reduce their call frequency (stage three).

By reducing their call frequency, changes again take place in the quantity of supplied service sequences. These changes could be a slight increase or decrease in that supply (3rd and 4th law). The most important thing is that, given the same trip frequency and a reduced call frequency, our consumer group will visit the other service elements less often than before (stage four).

In consequence of this adaptation, one or more of the already struggling service elements may also disappear, although the grocery need not be one of them. Suppose that another service element is indeed closed, then all consumers, whether they like it or not, are faced with a proportional or more than proportional decrease of service sequences (5th law). Those with a high call frequency are forced to give up more service sequences than those with a low one (6th law). Moreover, that loss will be considerable in very small centres (stage five). This may disappoint other visitors so that they too start to disregard one or more service elements (stage one).

THE CONCEPT OF A FUNCTION COMBINATION

The first two propositions affect enormously the functional hierarchy of service centres in each region. They imply that the quantity of supplied service sequences will increase as centres become bigger and that considerable divergences in this supplied quantity will occur as call frequencies become higher. If that is true, then the classical arrangement of centres according to their service level will be disturbed.

In order to construct a functional hierarchy of centres, based on the service level or scarcity of service sequences, it is necessary to group those sequences. However, such a grouping is rather complicated. We have to keep in mind the following items:
  i) the type of function;
 ii) the quantity of service elements per type of function;
iii) the quantity of service sequences per type of function;

iv) the type of function combination;
v) the quantity of service elements per type of function combination;
vi) the quantity of service sequences per type of function combination.

i) and ii) Let us assume a centre g with seven service elements. These elements are: a dentist, general practitioner, infant school, primary school, grocery, baker's shop and chemist. It is possible to group these services into various categories. The infant and primary school belong to the category of schools, symbol (1), the dentist and general practitioner are medical services (2) and the other services can be placed in the category of shops (3). These categories delimit service functions.

iii) Functions are subsets of set g and our propositions are valid for subsets as well as sets. With the help of formula 1 we can determine the number of service sequences per function. Functions are also useful for grouping service sequences. An example is given in Table 11.3. It concerns the opportunities which someone has when he visits at least two service elements in centre g.

Table 11.3   The number and kind of service sequences per type of function in centre g given a call frequency of two

type of function	number of service sequences	kind of service sequence
(1)	1	(infant school, primary school)
(2)	1	(dentist, general practitioner)
(3)	3	(baker's shop, grocery)
		(baker's shop, chemist)
		(chemist, grocery)

iv) However, many service sequences have not been grouped by functions, given a call frequency of two, for instance the sequence (dentist, grocery). That particular sequence involves a combination of medical functions and shopping functions and is symbolised by ($\underline{23}$). If we want to classify each service sequence for any centre, then we have to take as a starting point the power set of the original functions. The elements of the power set (that is, the function combinations) in our example are:
($\underline{1}$), ($\underline{2}$), ($\underline{3}$), ($\underline{12}$), ($\underline{13}$), ($\underline{23}$), ($\underline{123}$). ($\underline{\phi}$)

The ϕ set is excluded. The original functions (1), (2), (3) are identical to the function combinations ($\underline{1}$), ($\underline{2}$), ($\underline{3}$).

v) The quantity of supplied service elements belonging to each function combination is the sum of the supplied service elements of all separate functions included in any function combination. For instance, when in our centre g ($\underline{123}$) is a combination composed of the functions (1), (2) and (3), then the quantity of supplied service elements which belongs to that combination is:
$$n_{123} = m_1 + m_2 + m_3 = 7.$$
n_{123} = the quantity of supplied service elements of the function combination ($\underline{123}$).
m_2 = the quantity of supplied service elements of function (2).

More formally:

$$n_{abc...z} = m_a + m_b + m_c + ...m_z = \sum_{A=a}^{z} m_A \qquad (3)$$

vi) The determination of the quantity of supplied service sequences belonging to one particular function combination is more complicated. It is only possible to calculate this after correcting the results by formula 1 (see appendix 11.1).

THE CENTRALITY OF SERVICE CENTRES, BASED ON SERVICE SEQUENCES

To illustrate how the hierarchical arrangement of centres is disturbed, we may consider as a case study the district of Lüchow-Dannenberg in Lower Saxony (appendix 11.2). Settlements are quite small in this area; Lüchow is the biggest with about 5000 inhabitants. Schnackenburg has only 500 inhabitants.

The centrality of each settlement is expressed as a functional index. These indices, based on supplied service sequences for a given call frequency, were determined with the aid of a slightly modified version of Davies's location coefficient (Davies, 1967).

$$c_x = \frac{t_x}{T_x} . 100 \qquad (4)$$

c_x = location coefficient for any type of function combination, t, given a call frequency of x.

t_x = 1; a service sequence of any type of function combination, t, given a call frequency of x.

T_x = total number of service sequences of any type of function combination, t, in the region, given a call frequency of x.

The centrality value is a product of the relevant location coefficient, c_x, and the quantity of supplie service sequences of each type of function combination in a settlement, given a call frequency of x. The centrality of each settlement, the functional index, is the sum of all centrality values for a settlement, given a call frequency of x.

The results are set out in Table 11.4. In fact, the functional hierarchy of centres, which exists when all consumers have a call frequency of one, is of the classical type. Until now, all hierarchical arrangements had this appearance. When everybody has a call frequency of two, however, we see an enormous increase in Lüchow's centrality. Dannenberg, Hitzacker and Clenze gain too, but the other ones lose centrality, especially Kuesten and Prisser. Striking also is a levelling down as differences in centrality for small centres disappear gradually. Sometimes they even alter their rank. When the value for the call frequency is three, the same changes occur in the hierarchy of centres except that now Dannenberg, Hitzacker and Clenze also lose some of their centrality, and there are more centres of no importance. Finally, when all consumers visit four service elements, then only two centres, Lüchow and Dannenberg, are left with considerable centrality, followed at a lower level by Hitzacker and Clenze (Figures 11.2 and 11.3).

Multi-purpose trips cause a dichotomy among service centres; one or two acquire high centrality and the others lose out. Probably the picture of developments occurring in this hierarchical arrangement is somewhat exaggerated, since multi-purpose trips are not relevant to all consumers, service elements and functions. Moreover, service centres in their turn can stimulate consumers' call frequency to some extent by the quality of their function

Table 11.4   The centrality of service centres in the district of
              Lüchow-Dannenberg, based on service sequences, given
              various values for the call frequency.
              The numerical values are functional indices.

service centres	functional indices per call frequency			
	1	2	3	4
Lüchow	127.16	618.41	1116.47	1320.90
Dannenberg	73.40	202.66	201.96	149.15
Hitzacker	45.80	79.21	53.22	24.51
Clenze	30.27	34.66	14.39	3.16
Wustrow	22.08	18.37	5.18	0.85
Bergen	20.59	14.84	3.81	0.60
Gartow	17.96	11.07	2.45	0.34
Schnega	14.68	8.33	1.46	0.13
Zernien	12.30	5.03	0.70	0.06
Gorleben	7.67	2.04	0.13	0.01
Woltersdorf	7.14	1.56	0.16	0.01
Neu Darchau	6.59	1.21	0.07	0.00
Schnackenburg	5.42	1.58	0.09	0.01
Kuesten	4.63	0.60	0.04	0.00
Prisser	4.36	0.50	0.08	0.00

combinations and service sequences. Therefore call frequencies may
even diverge between service centres. However, for the time being,
we suggest that processes of levelling, dichotomising and perhaps
also variations in rank order are repercussions of multi-purpose
trips.

SUGGESTIONS FOR FURTHER RESEARCH

This paper has displayed a few new insights into consumer behaviour
and the supply side of service centres. However, much work needs to
be done on the following issues:
- the derivation of hypotheses from the theory, especially with respect to the influence of the downward spiral of attractiveness;
- a geographical and not a longitudinal approach to testing the
  hypotheses, e.g. a comparative study of consumer behaviour in
  centres of different sizes in several areas;
- a study of the sociological and geographical determinants of multi-
  purpose trips, for instance the influence of sex, age and family
  composition, the role of private-car ownership and public transport, the amount and quality of services in the centres themselves and the distance between centres;
- the construction of a more integrated hierarchy based on the sum
  of all available service sequences which people visit in their
  daily lives;
- the way in which processes like levelling and dichotomising affect
  the spatial organisation of a region. Which service centres are
  people still visiting and to what extent? Will there be a change
  in the configuration of market areas? Is it true that regional
  service centres are extremely dominant? Investigations

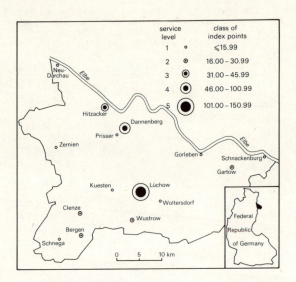

Figure 11.2   The functional hierarchy of centres in Lüchow-Dannenberg based on service sequences, given a call frequency of one

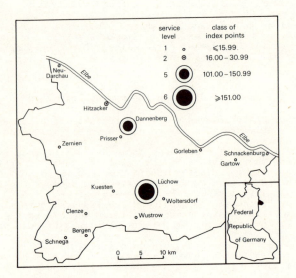

Figure 11.3   The functional hierarchy of centres in Lüchow-Dannenberg based on service sequences, given a call frequency of four

108

demonstrate that 'geographical hierarchy in the sense of functioning systems of centres and nesting areas, held together by
centralistic patterns of consumer trips, is only recognized at
the scale of regional service centres. Below that scale, hierarchy is not apparent' (Buursink, 1975). There is perhaps a link
between these empirical findings and our theoretical foundations.

APPENDIX 11.1 Correction formula for supplied service sequences
belonging to any type of function combination

When we attempt to determine the amount of supplied service sequences belonging to any type of function combination (given any
value for the call frequency) then we are faced with the difficulty
that the outcomes of formula (1) are no longer logically exclusive
sets. This deviation can be corrected by formula (4). Since the
length of this algorithm increases as we use more types of function,
we limit ourselves to three types.

In formula:

$$q_1 = k_1$$
$$q_2 = k_2$$
$$q_3 = k_3 \qquad\qquad (4)$$
$$q_{12} = k_{12} - (q_1 + q_2)$$
$$q_{13} = k_{13} - (q_1 + q_3)$$
$$q_{23} = k_{23} - (q_2 + q_3)$$
$$q_{123} = k_{123} - (q_1 + q_2 + q_3 + q_{12} + q_{13} + q_{23})$$

q_{23} = net quantity of supplied service sequences of function
combination 23

k_{23} = gross quantity of supplied service sequences of function
combination 23, after applying formula 1.

APPENDIX 11.2 Selected types of functions in the case-study area

The statistical material stems from the census of the Federal
Republic of Germany in 1970. We assumed that the service elements,
like retail trade, health and veterinary services, bank and
insurance companies, laundry, cleaning and the like, are relevant
to consumer behaviour. The types of function are 'Wirtschaftsunter-
abteilungen' or economic subsectors. Those chosen are: 'Einzel-
handel', 'Wascherei, Reinigung und Körperpflege', 'Gesundheits -
und Veterinärwesen' and 'Kreditinstitute und Versicherungsgewerbe'.

REFERENCES

Davies, W.K.D. 1967. Centrality and the central place hierarchy.
Urban Studies, 4.

Dijk, H. van, and Trimp, H.C. 1982. Het tijd-geografie-model en
haar toepasbaarheid in de schoolaardrijkskunde. Geografisch
Tijdschrift, 16(2).

Buursink, J. 1975. Hierarchy, a concept between theoretical and
applied geography. Tijdschrift voor Economische en Sociale
Geografie, 66(4).

Timmermans, H.J.P. 1980. Centrale plaatsen theorieën en ruimtelijk
koopgedrag. (Ergon Bedrijven, Eindhoven).

Chapter 12

Rural education services:

the social effects of reorganisation

Michael Tricker

INTRODUCTION

The major problems facing education authorities in rural areas are the declining number of pupils and shrinking resources. Recent cuts in local authority spending, coupled with the dramatic fall in the birth rate in England and Wales have prompted many local educa- tion authorities to review the educational and economic viability of small rural schools. The classic local education authority (LEA) response to falling rolls in rural schools has been to advo- cate the amalgamation of schools - the closure of the smallest and the utilisation of spare capacity elsewhere. In addition, several authorities have implemented consolidation schemes - involving the construction of new 'area' schools (often in key villages) to re- place groups of small schools. The quickening pace of closures and the extent of current reorganisation proposals has led to increas- ing concern over the possible effects of longer school journeys on the children themselves; the effect of increasing remoteness of the school on parental and community involvement; and the potential long-term effects of closures on the viability of rural communities.

Obviously education facilities are only one of a range of services and facilities which are often regarded as important to the functioning of rural communities. However, they are perhaps worthy of special attention since it has often been suggested that a village school may act as a focus for social and community activities. Thus, the possibility of village school closures raises not only issues concerned with the educational welfare of the pupils, but may also have a wide range of indirect social conse- quences. Central Government guidance to LEAs stresses the need to weigh such potential social consequences against the possible educational and economic costs and benefits associated with reorganisation proposals (Department of Education and Science, 1977).

THE RESEARCH PROGRAMME

A great deal has been written about the social role which rural schools are thought to play within the communities which they serve and the possible effects of closure (House of Lords, 1980; Jones, 1980; Cooper, 1979; Plowden, 1978; Rogers, 1979), but there has been a surprising lack of systematic research aimed at testing these hypotheses.

The principal aims of the study discussed in this chapter were therefore:
a) to provide an insight into the role rural primary schools play in helping to sustain and maintain viable local communities, and
b) to assess the implications of alternative patterns of reorganisation in a series of situations ranging from remote, sparsely populated areas with static or declining population levels, through to areas close to the West Midlands conurbation which are subject to pressures of growth.

The research programme included a series of detailed case studies, including 'before and after' studies of the effects of closure and reorganisation on different sections of the population. These involved participant observation in the schools themselves, as well as the collection of detailed information on the attitudes and involvement of large samples of parents and members of the wider community living within the catchment areas of the schools. In all, 20 case studies were carried out comprising 10 villages which had lost schools ('satellite' villages) and 10 villages associated with the schools to which the children had been transferred ('host' villages). The surveys conducted within these case study villages were designed to identify attitudes towards local schools and the perceived or anticipated effects of closure (JURUE et al.,1981). The locations of the villages within which these detailed case studies were carried out are shown in Figure 12.1.

PERCEIVED SOCIAL AND COMMUNITY ROLES OF RURAL SCHOOLS

Several writers have argued that schools, and the range of activities associated with them, play an important social role in the life of communities by providing a 'neutral' centre for contact between different generations and between people from different social classes (SCRCC, 1978; Garvey, 1976). In addition, the importance of the rural school as a social and community asset has been linked to its hypothesised role in stemming depopulation and in maintaining a desired social balance in rural communities. It is argued, for example, that a school helps to retain or attract families who have, or intend to have, children and thereby to maintain a balance in the age structure of the community (ADC, 1978; Cooper, 1979; Lee, 1960).

Analysis of survey data highlighted several factors which appear to have influenced the formal and informal role that individual schools have developed for the communities within their catchment areas and which in turn condition the anticipated and observed effects of closure and reorganisation. These factors include:
a) presence of other facilities
b) location of the school itself
c) the nature of the settlement pattern
d) the age and social structure of the community
e) the attitudes and personal qualities of the staff.
These factors combine to produce the relatively unique relationship between a school and the community it serves which forms the base-line against which changes consequent upon closure and reorganisation must be assessed. In most instances, however, it was clear that the popularly held image of a village school, whose children live close by and where the head teacher lives in the same community, was no longer valid for a variety of reasons.

Firstly, most of the teachers at the schools studied did not live in the village in which they worked and many lived in nearby

Figure 12.1   Location of case study villages

towns. Only one lived in the school house. The reasons for this
were partly attributable to the local housing market, but in
several instances they also reflected a deliberate choice stemming
from a desire to escape from work at the end of the school day.

Secondly, even where the school has survived, its formal role
as a venue for social events has often been taken over by village

113

halls or community centres, and the vast majority of people interviewed in such villages considered the school was not an important
source of social contact. Not surprisingly, participation in
school-related activities was closely related to the family life-
cycle with involvement often ceasing when children or grandchildren
left the school. The importance of schools for those who are not
parents seems to have diminished considerably in the face of
alternative entertainment, increased sophistication of tastes and
greater affluence and mobility to meet these tastes. More than a
third of those interviewed had never attended a function at their
local school and almost 60 per cent had not attended one in the
previous twelve months.

Levels of involvement in school-based activities and the perceived importance of the school as a social focus did, however,
vary appreciably with distance from major urban centres. Thus,
rates of participation and perceived importance were generally
lowest in those settlements which were closest to the West Midlands
conurbation and other large urban centres, and highest in more
remote areas. In several instances this variation seems to be
associated with changes in the social structure of villages arising
from an influx of commuters and householders of retirement age.

By far the highest levels of community involvement were found
in a school where an effort had been made to develop its community
role by incorporating dual-use facilities. In this case the school
now acts as a venue for an extensive and varied programme of social
activities. In most of the cases studied, however, the local
school's importance as a focus for village community life appears
to have diminished markedly and it is against this base-line of
reduced involvement and diminished perceived importance that the
likely effects of closure and reorganisation must be judged.

EFFECTS OF CLOSURE AND REORGANISATION

Arguments advanced in the debate on the future of small schools and
surveys conducted in villages which still had schools, led us to
expect that closure and reorganisation would have a range of effects
on children, their parents and the wider community.

Effects on children

The main effects anticipated by parents involved possible effects
of longer travel time and distances on the social and educational
development of their children, and constraints on their involvement
in after-school activities. In most instances, however, the
perceived effects did not appear to have been as serious as had been
anticipated prior to closure and, in some cases, a striking change
in attitudes seems to have taken place after reorganisation. Thus,
despite widespread feelings of regret that their local school had
closed, a substantial proportion of parents seem to have come to
the view after a relatively short period of time that their
children's new schools were better in terms of curriculum, teaching
and facilities than the old ones. Many parents did, however, draw
an important distinction between the merits of small schools for
older and younger primary children. Thus, whilst the vast majority
of parents interviewed considered small village schools were better
than larger ones for children between 5 and 7 years, (a better
family atmosphere especially for small children starting school),
opinions on the balance of advantages and disadvantages for children
between 8 and 11 years were quite different. A large proportion of
parents stressed the superior facilities and the greater degree of

competition in larger schools as significant advantages for these children.

Contrary to expectation, closure and reorganisation did not generally produce drastic changes in children's involvement in after-school activities. Nevertheless, children living in villages which had retained their schools did seem to participate in after-school activities more frequently than those living in the satellite settlements. Several parents (particularly in areas where schools had closed recently) also felt that closure had resulted in local children meeting less often than before and expressed regret that their children had 'grown away from the village' and 'lost interest' in the community in which they lived. Nevertheless, similar proportions of parents (particularly in areas where reorganisation had occurred some years earlier) felt that their children's social development had benefited from contact with a wider range of children and teachers at the larger host schools. Since the children had become more confident, they expected them to have fewer problems in their transition to secondary education. The indications were that the fears of parents were resolved fairly quickly as their children settled down in the host schools and they themselves became more familiar with them.

Effects on parents

Parent's perceptions of the impact of reorgansiation on themselves were closely related to constraints on their mobility. Although free transport for children was generally provided by the LEA after closure, some parents had undoubtedly borne increased costs for transport. Moreover, in at least one of the cases studied, the pattern of transport provision seems to have constrained the choice of alternative schools for those families without day-time access to private transport and exposed income and class divisions which had remained latent whilst the majority of children attended the same school.

More generally, reorganisation also seems to have increased constraints on parents' involvement with their children's school. Whilst this seems to have had little effect on their participation in occasional formal events staged by the school, it quite clearly reduced opportunities for regular informal contact with their children's teachers and other parents. Nevertheless, whilst parents living in the host settlements were generally able to maintain closer links with their children's school and generally knew the teachers better, very few parents in the satellite settlements expressed any anxieties over loss of contact with their children's teachers or control over their education.

In some instances reorganisation had also reduced the amount of contact between parents and children and produced changes in household activity patterns. Children travelling by school transport usually left home earlier and returned slightly later than they had done prior to reorganisation. This was not always regarded as a disadvantage, however, and in a few instances the fact that children now remained at school for the midday meal had enabled parents to take up part-time employment.

Effects on the wider community

The literature suggests that the closure of schools has several implications for the wider community. Supposed impacts include a reduction in community participation in school-related activities, a reduction in the amount of social interaction within the community, and consequent adverse effects on the viability of settlements (Garvey, 1976; Plowden, 1978).

Changes in community involvement

As indicated earlier community involvement had usually started from a relatively low base-line. Despite this, a marked reduction in participation in school-related activities does seem to have occurred after closure and reorganisation. Whereas a gradual process of adjustment seems to have occurred as parents in the satellite villages were drawn into the community of interest focussing on their children's new school, involvement by non-parents seems to have remained well below that of similar groups living closer to school. Transport difficulties clearly play an important part in explaining this fall-off in participation with increasing distance from the school. Nevertheless, 'lack of time' and 'lack of interest' were cited more commonly as reasons for not attending school functions. This attitude clearly reflected antipathy and resentment arising from the closure itself where this had taken place recently.

Reorganisation also seems to have initiated a reduction in informal involvement in the life of the school. Respondents living in satellite settlements were generally far less aware of school activities and of any encouragement the teachers gave to involvement in them.

Changes in community interaction

Not surprisingly, the schools which were closed were all small. Consequently, the range of school-related activities was often limited and only in villages which had no village hall had the old school been extensively used for regular social events. In most other cases many of the community activities which had previously taken place at the school had been transferred to village halls several years before closure. Those which remained were generally confined to activities connected with the school itself and, in the case of voluntary schools, those connected with the church. Not surprisingly, therefore, the importance of the old school for the social life of an area was consistently rated lower in those villages which had village halls than in those which lacked any alternative meeting places. Consequently, whilst roughly half those interviewed felt that closure had affected community life within their village, the sense of loss related to the disappearance of a focus for informal social contact rather than the loss of a venue for any extensive range of community activities.

Such effects were felt most strongly by parents, just over a third of whom experienced a reduction in contact with other parents and the rest of the community. Although parents did still meet at the school bus stop, a substantial number of children were often picked up at a series of points along the bus routes. Their parents were therefore deprived of a certain amount of casual social contact.

Respondents drew a clear distinction between the effects of closure on the community in general and the effects on their own household. Thus, whilst there was a commonly held view that the community had lost a focus for social interaction, the number of people who indicated that the social life of their own household had been affected was very low. For parents at least, any loss of social contact within the village may to some extent be compensated by new links forged with the host school. However, this is

116

clearly less likely to be true of less mobile groups such as the elderly. Nevertheless, in two instances where new educational links had been reinforced by the provision of community facilities in or adjacent to the host schools, there is evidence that the host settlements are emerging as significant foci for the newly enlarged catchment areas. This is indicated not only by the emergence of new activities, but also by the transfer of a number of activities which used to take place in the satellite settlements. It is probably significant that this was occurring in areas with relatively dispersed settlement patterns. In those areas with relatively nucleated settlement patterns, community life tended to be more self-contained and therefore the links developed with the host settlements seem generally to have been far weaker.

Changes in the viability of settlements

Effects of closure on the viability of the settlements themselves are obviously more difficult to isolate, not only because this requires reference to other factors such as housing, planning and transport policies, but also because the time scale over which changes and adjustments occur may vary. Clearly, retrospective surveys conducted some years after the closure only allow examination of the perceptions and opinions of those households which remain. The major limitation of the survey data therefore relates to the inevitable difficulties in assessing the relationship between closure, population movements and subsequent changes in population structure and the social balance of communities.

The surveys carried out immediately prior to closure did identify widespread concern that closure of the local school would make the village less attractive - especially for households which had, or intended to have, young children. In those areas where closure had already occurred, it proved difficult to locate more than a very small proportion of either those households which had moved after closure or those which may have been deterred from coming because the school had closed. However, the number of residents that indicated an intention to move from a village because of school closure was extremely small.

Those parishes which lost schools have generally experienced a decrease in population whilst those which retained schools have experienced an increase. However, there is very little evidence that school closures have triggered extensive movements of existing households and, in many instances, this differential growth may be attributed to a general trend towards smaller household sizes coupled with restraints on the building of new houses in the smaller 'non-key' settlements. This trend, taken along with the marked decreases in population which occurred prior to the closure of some of the case-study schools, tends to support the argument that school closures are a response to population changes rather than a prime cause. In several instances the number of children of primary school age in the satellite settlements now exceeds the number present at the time of closure. It is therefore difficult to detect any clear link between school closure and subsequent changes in age structure. There is, however, some indication that long-term changes may occur in the population structure of settlements which have lost facilities like schools and that, as a result, their residents are increasingly those who are less sensitive to the availability of local services.

Whilst some parents do tend to make occasional use of post offices and shops in the vicinity of their children's school, evidence for a link between closure of the local school and the

117

subsequent closure of such facilities is ambiguous. Many of the
case-study villages had already lost a range of commercial facilities, and whilst in two cases post offices and general stores had
closed shortly after the school, in two other cases there were effectively no commercial facilities to lose by the time the school
had closed. What is clear, however, is that very few residents
seemed to perceive any link between school closure and the loss of
such facilities, and tended to point instead to increases in personal mobility, more attractive prices elsewhere, and changes in
the social structure of settlements which have all tended to erode
support for local shops.

CONCLUSIONS

School closure and reorganisation generally tends to be associated
with reductions in the amount of social contact between parents,
between parents and members of the wider community and between the
community and the school. However, several of the other effects
which the literature and certain pressure groups have argued result
from school closures were not borne out by this study. Although
the social effects of closures are not insignificant, it is probable that they have been overstated. If significant financial
savings and real improvements in educational provision can be
achieved as a result of reorganisation, these benefits are not
likely to be outweighed by the detrimental social impact of closure,
provided alternative community facilities are available.

This generalisation does, however, conceal a range of impacts
on disadvantaged groups, as well as differences between the needs
of relatively remote rural areas and those with better acess to
urban facilities. The social effects of closure must be seen in
the wider context of social and economic changes which are taking
place in rural areas, within which the reorganisation and rationalisation of educational facilities is merely one element. Those
groups which stand to lose most from school closure - the elderly,
the less affluent and the less mobile - are also likely to be
experiencing varying degrees of deprivation as a result of the progressive withdrawal of other services and facilities. It is
important, therefore, that local authorities should consider educational provision as an integral part of rural settlement planning
and policies aimed at rural socio-economic development.

This study has highlighted the extent of the changes which have
occurred in the actual and perceived role of rural schools. As a
result, many schools are now grossly under-utilised community resources. There is, therefore, an urgent need to develop coherent
policies for promoting the wider use of schools aimed at realising
their full potential for community development. Where a small
school already plays a wider social role, this may not be sufficient
to justify its continuation. However, where new 'area' schools are
developed, or existing schools are remodelled to perform a wider
role, fuller consideration should be given to developing joint
educational and community facilities in order to foster links with
communities in their enlarged catchment areas.

FOOTNOTE

The research which is discussed in this chapter was carried out for
the Department of the Environment and Department of Education and
Science by a team of researchers drawn from JURUE and the Department
of Educational Enquiry at the University of Aston in Birmingham.
The views expressed do not necessarily represent those of the
sponsors.

REFERENCES

ADC (Association of District Councils) 1978. *Rural recovery: Strategy for survival*. (ADC, London).

Cooper, G. 1979. The village school. *Town and Country Planning*, 48, 190-1.

Department of Education and Science, 1977. *Falling numbers and school closures*. Circular, 3/77, (DES, London).

Garvey, R. 1976. Closing down the village schools. *Where?*, 119.

House of Lords, 1980. Village schools: Government policy. *Hansard*, Vol.407, No.101.

Jones, P. 1980. Primary school provision in rural areas. *The Planner*, 66, 4-6.

JURUE et al. 1981. *The social effects of rural primary school reorganisation*. (JURUE and the Department of Educational Enquiry, University of Aston in Birmingham).

Lee, T. 1961. A test of the hypothesis that school reorganisation is a cause of rural depopulation. *Durham Research Review*, 3, 64-73.

Plowden, 1978. Policy on village schools. *The Times*, 7th September, 1978.

Rogers, R. 1979. *Schools under threat*. (Advisory Centre for Education, London).

SCRCC (Standing Conference of Rural Community Councils), 1978. *The decline of rural services*. (National Council for Social Services, London).

Chapter 13

Spatial mobility problems of the elderly
and disabled in the Cotswolds

Robert Gant and José Smith

INTRODUCTION

The changes occurring in service provision and transport availability in rural Britain and their impact on local communities are
now well documented (Association of District Councils, 1978;
National Association of Local Councils, 1980). While many groups
in rural society have benefited from increased personal mobility,
others like the elderly and the handicapped have been detrimentally
affected by the processes of rural change. With justification, the
elderly have become a focus of concern as a result of their increasing numbers and proportionate representation in local populations (Allon-Smith, 1982; Hunt, 1978). As a group they are
characterised by low incomes, low levels of car ownership and
limited access to essential services. Moreover, the process of
ageing can lead to disabilities and restrictions on personal
mobility (Borsay, 1982; Buchanan, 1983; Moseley, 1978a). Although
in Britain almost two-thirds of the handicapped are also elderly,
neither group is homogenous and marked variations in personal
circumstances can considerably influence the degree of individual
deprivation (Neate, 1981; Williams, et al., 1980).

In the context of rural change, early geographical studies
identified the common mobility problems of disadvantaged groups
(Moseley et al., 1977; Moseley, 1978b). Later studies, in contrast,
have searched for solutions to the welfare problems induced by
service decline, and discuss the need for a range of policy alternatives related to individual mobility opportunities and constraints
at the local scale (Banister, 1980; Stanley and Farrington, 1981;
Moseley, 1979; Nutley, 1980). Within such a framework, this paper
reports on a project undertaken jointly by Kingston Polytechnic and
Gloucestershire Social Services Department which aimed to identify
the extent of mobility deprivation amongst the elderly and the disabled in the North Cotswolds, and the consequent demands placed on
community support and social service provision.

STUDY AREA: POPULATION AND SERVICES

The study area comprised four parishes situated in 'Patch 4', the
most southerly sub-division of the North Cotswolds Social Services
area (Figure 13.1). In 1980 the Social Services area had a population of 27 115. 'Patch 4' contained 21 per cent of this total,
and 17 per cent were aged 65 years and over.

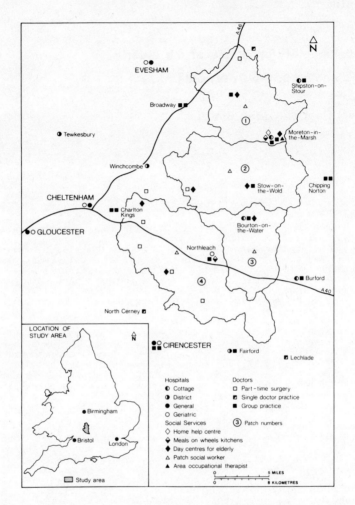

Figure 13.1   North Cotswolds: location of health and social
service facilities used by the survey population

Since 1951 there have been substantial numerical and proportionate increases in the elderly population in Gloucestershire.
For example, between 1971 and 1978 the proportion aged 65 years and
over increased by 17 per cent compared to the national average of
11 per cent; the corresponding statistics for those aged over 75
years are 25 per cent and 15 per cent respectively (Gloucestershire
County Council, 1978).   These trends are likely to continue as a
consequence of structural change and retirement migration.   Furthermore, the forecasts prepared by the County Plannning Department
indicate increases by 1996 of 10 per cent for those aged over 65
years and 26 per cent for those aged 75 years and over.

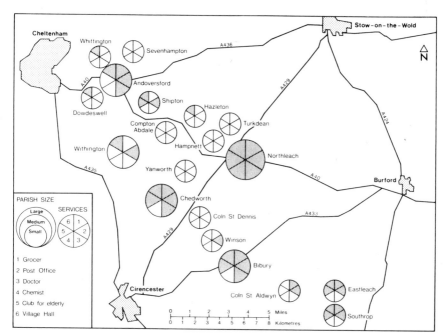

Figure 13.2 Availability of services

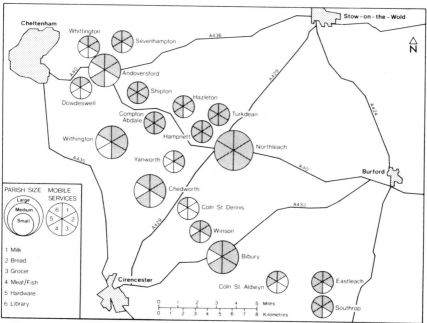

Figure 13.3 Availability of mobile services

123

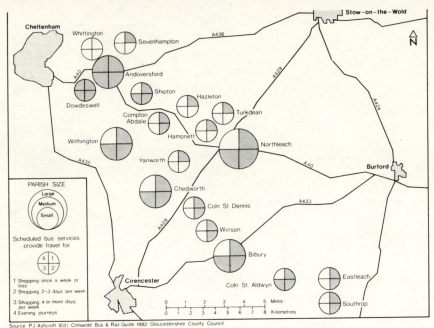

Source: P.J Ashcroft (Ed.) Cotswold Bus & Rail Guide 1982 Gloucestershire County Council

Figure 13.4   Availability of public transport

Throughout the Cotswold region the proportion of the elderly has consistently exceeded the county average. There are, however, marked variations in the percentage distribution of those aged over 65 in the 19 parishes of 'Patch 4', the median being 16 per cent with values ranging from 8 per cent at Winson to 29 per cent at Hampnett.

This can be explained by changes in population structure and the economic base of the region (Gant and Smith, 1980). Between 1921 and 1981 only the four largest parishes recorded increases in population. Without exception, the decline in agricultural employment in the small parishes and selective outward migration have produced an advanced age structure in the residual population (Dunn *et al.*, 1981). In contrast, the special provision of council housing for the elderly in larger parishes has encouraged localised movements from smaller settlements (Gant and Smith, 1981). Furthermore, many parishes have sustained an influx of retired households.

There have also been substantial changes in the pattern and level of service provision; for example, in the period 1972-1977 8 per cent of Gloucestershire villages lost their sub-Post Office, 13 per cent a shop, and 3 per cent a doctor's surgery (Standing Conference of Rural Community Councils, 1978). This trend to concentrate facilities in large urban centres has been detrimental to the North Cotswolds, a region lacking its own urban focus. Although primary health care, social services and low-order goods are available in the region, only four centres provide a full selection of services (Figures 13.2 and 13.3). Shops for high-order goods and major hospital facilities exist only in urban centres outside the region (Figure 13.1). Within 'Patch 4' there are low-order services in several parishes but Northleach alone offers the full range (Figure 13.2). As a result of several decades of rural decline, some parishes are without permanent services, although most are covered by a variety of mobile services (Figure 13.3). Transport is therefore a crucial consideration in linking small parishes with basic facilities and is important to all parishes for access to high-order services. Figure 13.4 shows that while public transport is available in all parishes but one, it offers a very limited service. In five parishes a shopping service is time-tabled on a fortnightly or weekly basis while only two parishes have an evening bus service. Not unexpectantly, the destinations of the bus service reflect the demand for travel to major centres outside the area, such as Cheltenham, Cirencester, Oxford and Swindon. Consequently, only two parishes in 'Patch 4' have a direct, daily link with the Social Services area office at Moreton-in-Marsh.

THE SURVEY: ORIGINS, AIMS AND SAMPLE CHARACTERISTICS

On three occasions since 1978 undergraduates from Kingston Polytechnic have studied aspects of rural deprivation in the North Cotswolds (Smith and Gant, 1981; 1982). This work was extended in conjunction with Gloucestershire Social Services Department in 1982. It was guided by the provisions of the Chronically Sick and Disabled Persons Act 1970 which charged:
'every local authority having functions under Section 29 of the National Assistance Act 1948 to inform themselves of the number of persons to whom that section applies within their area and of the need for the making by the authority of arrangements under that section for such persons' (quoted in Taylor, 1977).
The exercise aimed to assess the potential consequences for the Social Services of a substantial case load of elderly and handicapped persons in the North Cotswolds (Table 13.1; Appendix 13.1). The immediate concerns of the Social Services were threefold: firstly, the large increase in the numbers of the elderly, particularly those aged over 75 years; secondly the tendency for the elderly to live alone or as a retired couple, often at some distance from the rest of the family; thirdly the spatial concentration

Table 13.1 Elderly referrals to the Social Services

Population /Referrals	Patch 4	Survey parishes
Population (1978)		
Total population	5922	1335
(a) Male + Female ⩾ 65 years	1028	257
(b) Male + Female ⩾ 75 years	426	106
Persons referred (1974-1982)		
Total referrals*	131	24
(c) Male + Female ⩾ 65 years	105	18
(d) Male + Female ⩾ 75 years	64	13
Ratios of:		
(c) : (a)	1 : 9.8	1 : 14.3
(d) : (b)	1 : 6.6	1 : 8.2

* For reasons of physical disability and handicap

Source: Social Services Case Records

of the elderly in rural communities where retail and transport services are declining. The field survey investigated these problems and aimed to identify the basic mobility needs and personal difficulties of the elderly and handicapped in those areas which lacked or had only minimal public transport services.

In total, the field survey covered 806 households in 24 parishes in the North Cotswolds. These were drawn proportionately from the four Social Services patches to represent different levels in the provision of public transport, the range of local services, the geographical location of settlements relative to major roads and towns, and local demographic trends. In advance of the main survey a short, returnable questionnaire was delivered to every household to identify those with at least one elderly and/or disabled person. Local enquiries then pinpointed additional households in these categories and, subsequently, interviewers visited each address.

This report is based on 148 household interviews from the first four parishes surveyed - Chedworth, Eastleach, Southrop and Yanworth - and represents a combined contact-response rate of 80 per cent for the target households. Within the sample of 253 persons, 80 per cent were aged over 60 years, and 28 per cent over 75 years. The age-sex distribution of the retired favoured females in the ratio of 3:2. Fifty-three per cent of the survey households included at least one pensioner co-residing with other persons, 42 per cent were lone pensioners and the remainder were younger households with a minimum of one disabled member (Table 13.2).

Table 13.2 Household type

Household type	Disablement of member(s)			
	Adult	Children	None	Total
Lone Pensioner				
Male	1	-	14	15
Female	12	-	35	47
Other Pensioner Households				
Husband and wife pensioners (+ ch/gch.)	10	1+	44	55
Other 2-pensioner	-	-	1	1
Husband or wife pensioner (+ ch/gch.)	4*	1	17	22
Non-Pensioner Disabled	5**Δ	3	-	8
Total	32	5	111	148

*   2 disabled adults in same household

** 2 disabled adults in each of two households

\+   2 disabled children in same household

Δ   1 disabled adult + 1 disabled child in same household

126

While disabled adults were found in 22 per cent of the households, the incidence of age-induced disability was greater where females lived alone. In addition, 4 per cent of the households contained a handicapped child. The distribution of the sample households reflects the nucleated pattern of Cotswold settlement: 86 per cent lived in the main clusters of settlement while only 9 per cent were found more than 1 km from a village. Few households, however, were totally isolated and only 4 per cent were situated more than 100 m from a neighbouring house.

In each parish the proportion of long-established households was balanced by more recent arrivals and almost half those enumerated had previously lived in either the same or a neighbouring settlement, while 40 per cent had originated outside the county. Only 13 per cent of those aged over 15 years were gainfully employed, mainly in part-time work within walking distance of home. Overall, 21 per cent of the housing stock had been adapted, in various ways, to meet certain basic needs of the elderly and handicapped residents.

DISABILITY IN THE HOUSEHOLD CONTEXT

With regard to self-reported disability and living arrangements, the survey households can be classified into three groups: 'independent' households (76 per cent of the total) contain lone retired people in reasonable health and those with at least one pensionable member; the disabled in the 'partially independent' households (15 per cent of the total) have the nominal support of co-residents; in contrast, the lone-disabled in 'dependent' households (9 per cent) are potentially in need of the greatest level of external support.

The majority of the retired people were dependent on simple health aids to meet the challenges of daily life, including spectacles, walking frames, sticks and hearing aids. Far more serious, however, was the plight of the disabled, 38 per cent of whom were pensioners and almost one-third of whom suffered multiple disabilities. Arthritis, the most common cause of self-reported disability, affected one-third of the population; moreover, in the case of the elderly it often aggravated situations of poor eyesight, deafness, angina and immobility of limbs. All members of the dependent households were afflicted by at least one serious disability, 40 per cent suffering multiple disabilities. The corresponding proportions in the partially dependent and independent households were 32 per cent and 6 per cent respectively.

The personal restrictions imposed by impairment of movement are summarised in Table 13.3. In general, few members of the independent households experienced difficulty with any of the specified tasks, with the possible exception of caring for their feet. In contrast, the support given to disabled members in the partially dependent households is crucial to their continued life as part of the family and community. Besides problems with personal hygiene, this group which includes young children is severely restricted in its movements, one-third not being able to use a bus or visit a shop even with assistance. Those in dependent households suffered a real but reduced degree of difficulty in performing the range of tasks.

127

Table 13.3 Mobility and personal care

Task	Percentage needing help			Percentage unable to do task		
	Indep.	Part Dep.	Dep.	Indep.	Part Dep.	Dep.
Getting out of house	3	20	-	1	11	7
Using buses	3	6	21	4	34	21
Shopping	3	9	7	4	34	7
Movement inside house	1	6	-	3	17	14
Personal care and toilet	3	-	-	-	6	7
Cutting toe-nails	9	26	14	2	20	14
Getting into bed	2	9	-	-	6	7
Number of persons	159	35	14	159	35	14

ACCESS TO TRANSPORT AND TRAVEL PATTERNS

Given that physical disability prevents only a small proportion of
the survey population from travelling, access to transport is an
essential determinant of personal mobility. In the survey parishes,
as in other areas of rural Britain, the elderly were disadvantaged
by low levels of car ownership. Whereas 75 per cent of all households owned at least one car, only 57 per cent of those with retired or disabled members had private transport (Hillman et al.,
1976; Smith and Gant, 1982). However, even within the survey population there were marked variations in car ownership with the
single-person, independent and dependent households having proportionately fewer cars (47 per cent and 40 per cent respectively)
than the remaining independent (64 per cent) and partially-dependent
households (66 per cent). In addition, the ownership of a vehicle
varied markedly by settlement and social class. Over 81 per cent
of the households in Southrop had a car in contrast to 37 per cent
at Yanworth, while over 80 per cent of the households in Social
Classes I and II had their own transport as against 45 per cent in
Social Classes V and VI. Somewhat surprisingly, only 2 households
used specially adapted vehicles or invalid cars.

The inability to drive further aggravated the problem of
personal immobility since 62 per cent of the survey population and
74 per cent of the disabled had no driving licence. Many of the
non-elderly disabled depended for transport on healthy members of
their households. Mobility problems are particularly severe for
women, the elderly-disabled and those over 75 years of age.
Advanced age and physical disability had forced 25 households to
give up the ownership of a car in recent years (Norman, 1977).
This loss was felt most acutely by the dependent households, one-
third of which had relinquished the sole household car. For most,
the decision to sell was taken for financial and health reasons
(35 per cent and 29 per cent respectively), rather than because of
the death of the only driver. Access to a car and a driving licence
was particularly uncommon among women and those over 75 years old.

The extent to which the survey population used the very limited
public transport system was related to car ownership rather than
the frequency of the bus service. Table 13.4 confirms variations

Table 13.4 Bus journeys made by the survey population

	% individuals using bus service in March'82	% individuals using bus service since Christmas 81
Settlement		
Chedworth	40	44
Eastleach	39	54
Southrop	15	23
Yanworth	33	50
Household type		
Independent	40	46
Partially dependent	12	29
Dependent	53	60
All Households	37	44

(Based on the individuals in 148 households)

between settlements and household types in the use made of the
local bus service. It also demonstrates the impact of transport
deprivation experienced by the dependent households. Contrary to
expectations, this group were not the main beneficiaries of the
bus-token scheme operated by Gloucestershire County Council, the
proportion using bus tokens declining from 53 per cent for the active
elderly in independent households to 33 per cent of those in dependent households and 18 per cent in partially dependent households.

There were also contrasts in travel patterns among the elderly
and handicapped (Hanson, 1977; Markovitz, 1971; Smith and Gant,
1982). Relatively less travel was undertaken by the disabled in
partially-dependent households, presumably because another member
of the household could make a journey on their behalf (Table 13.5).
Despite their disability, those in dependent households were
extremely active, the proportion making journeys being equal to, or
above, that in independent households.

Table 13.5 Journeys made in March 1982 by different household
types

Journey purpose	Household type			
	Independent		Partially dependent	Dependent
	Single	Other		
	(% individuals making journey)			
Grocer	79	75	32	87
Post Office	69	69	26	74
Doctor	15	30	26	47
Chemist	52	55	26	66
Social Club	42	25	17	34

(Based on 148 households)

Table 13.6   Journey destination, frequencies and modes, March 1982

| Destination | Journey purpose (%) | | | | |
	Grocer	Post Office	Doctor	Chemist	Social Club
Same parish	6	53	47	6	85
Neighbouring parish	14	14	18	15	5
Within 15 km	68	26	32	72	7
Over 15 km	12	7	3	7	3
Total	100	100	100	100	100
Frequency					
(No.of times during March)					
1	3	7	57	22	57
2	12	14	17	20	17
3	6	4	10	4	4
4	63	72	14	52	13
5+	16	3	2	2	9
Total	100	100	100	100	100
Mode of travel					
Walk	5	34	27	3	40
Bus	32	15	13	35	12
Household car	48	40	23	44	28
Other car	11	8	25	14	12
Other	4	3	12	4	8
Total	100	100	100	100	100
No. of persons =	146	131	57	107	59

Journey destinations, in contrast, varied little between the groups but reflected the locations of the nearest available services (Table 13.6). Few people travelled over 15 km for any of the 5 journey purposes examined. Where possible, needs were met locally: thus a high percentage of travel to social clubs, Post Offices and doctors occurred within the same settlement. Where a facility was not available locally (chemist) or offered only a limited choice (grocers), all groups met most of their requirements within a 15 km radius, usually in Cheltenham, Cirencester or Fairford (Figure 13.1). Journey frequency likewise varied with the purpose for travel. Thus journeys to a social club or a doctor were most likely to be made once a month or less, whereas journeys for groceries, to a Post Office and, to a lesser extent, to a chemist were usually on a weekly basis.

A high incidence of journeys on foot was recorded in parishes with their own facilities (Table 13.6) and confirms the importance of walking as a form of transport for the elderly and disabled (Hillman and Whalley, 1979). Bus journeys were most common for shopping trips, while a substantial minority of the survey population depended on lifts, especially for journeys to a doctor. Overall, therefore, physical disability appears to have less impact on spatial travel patterns than journey purpose except in the case of partially-dependent households which are characterised by very low levels of travel. The majority of dependent households show considerable ability in adapting to their handicaps and to the limited public transport system on which most of them rely.

COMMUNITY SUPPORT AND PROVISION OF SOCIAL SERVICES

During the past fifteen years in Britain there have been considerable developments in community support through the expansion of the Social Services, the growth of voluntary organisations and availability of financial aid to families caring for an elderly or disabled member (Austin, 1976; Bebbington, 1979). Since such support can compensate for a lack of access to services and thereby enable an individual to remain in his or her own home, the survey examined the level of support received by the elderly and disabled.

Family links remain the most important for the elderly and disabled and many derive great pleasure from seeing their relatives at frequent intervals (Department of Health and Social Security, 1978). There are more tangible benefits like help in times of stress and illness, and assistance with heavy and demanding household tasks (Williams, 1979). Seventy-five per cent of the survey households had close relatives living within 40 km of their homes; half had children and/or grandchildren living independently in the same or a neighbouring village. Frequent contact with these relatives was the norm, usually by reciprocal visits, and only 10 per cent had not been seen in the month preceding the survey.

Care provided by the family is often supplemented by the voluntary organisations which have evolved to serve both the general needs of the elderly and those with specific disabilities (Garden, 1978a). The survey identified 21 voluntary organisations in which a majority of the elderly and disabled seemed to view themselves as participants rather than beneficiaries. Almost a quarter of those interviewed, including several disabled, claimed recent responsibility for organising some form of social activity while only 3 per cent admitted that they had received help.

Although the statutory Social Services have a role to play in community care, many of those eligible under the Chronically Sick and Disabled Persons Act failed to take advantage of its provisions. The problem of under-reported disability and need in rural areas is well documented (Shaw and Stockford, 1979; Stockford and Dorrell, 1978; Taylor, 1977). This situation is said to arise from the traditional independence of country people, lack of information on available services and statutory rights, and the existence of a network of family and community support which provides a basic level of care and assistance in times of need. An analysis of social services received in the home in the month preceding the survey strongly supported this contention. Throughout the four parishes, only thirteen households had the assistance of a home help, one received meals on wheels, whilst five had been visited by a social worker or occupational therapist.

However, it is felt that the survey statistics under-represent the wider contribution of the area Social Services. For example, 80 per cent of those referred to Social Services on grounds of physical disability or handicap in the period March 1974 to April

131

1982 were aged over 65 years (Table 13.1). Of the 264 referrals from 'Patch 4', 34 per cent were raised by officers of the Area Health Authority, 21 per cent were self-reported, while various public authorities were responsible for the rest. In 60 per cent of the cases, the sponsors explicitly requested material assistance for their client; an additional 26 per cent were recommended for full professional assessment of individual needs. As a consequence, almost half the referrals became directly supervised by the Social Services, a further quarter received advice, while continuing assessment and material aids were provided for most of the remainder.

DISCUSSION AND CONCLUSION

Throughout Britain, Social Service Departments have found it diffi-cult to gauge the volume of demand for their services under the provisions of the Chronically Sick and Disabled Persons Act, 1970. This problem is common to many areas of social service work where the expressed demand effectively understates the real pattern of need (Association of County Councils, 1979). Such latent demand poses special and immediate problems for the planning and funding of social service provision in areas like the Cotswolds where a rapidly ageing population structure is associated with declining services. It was for this reason and in recognition of the link between age structure and disability that the Cotswold study was not confined solely to the chronically sick and disabled but inter-preted the provisions of the 1970 Act to include all elderly persons.

The survey population was not homogenous with respect to need or personal resources. The majority of the elderly and some of the disabled were able to lead independent lives and made few demands on the support services. However, this situation will only be maintained if the elderly and disabled retain their access to essential services and facilities and to their social contacts. With the prospect of further declines in transport and local facilities, policies to cater for the needs of the independent elderly will assume greater importance. For the future, the Gloucestershire Structure Plan advocates the dispersal of residen-tial development among the main villages and their supporting settlements to stabilise population levels and to maintain local services (Gloucestershire County Council, 1979a). The County Council envisages the further provision of special units of accom-modation for elderly households and the reinstatement of basic services in the main villages. This settlement strategy will be supported by a transport plan which aims to maintain and, if possible, improve the present levels of public transport provision (Gloucestershire County Council, 1979b). However, the implementa-tion of such forward-looking policies is likely to be limited by financial constraints and short-term alternatives are being sought. As in other rural areas, greater attention is being paid to uncon-ventional forms of transport, to increasing the support for voluntary schemes and to expanding the availability of social services transport (Cresswell, 1977; Garden, 1978b, 1979).

To date, the family, the community and local voluntary organis-ations have been remarkably successful both in giving general sup-port and in meeting many of the specialised needs of the elderly and disabled. In general, the population has relatively low ex-pectations of statutory support and only those with special problems have been directed to the Social Services. Notwithstanding, the Area Social Services have to deal with a wide geographical spread

of varied case-work. Much of this concerns the elderly, and
involves home visits and supervision. The community provides in-
valuable support for this work and successfully underpins the
favoured strategy of domiciliary support as opposed to residential
care.

In the future, however, local support systems will become increasingly stretched as a consequence of two factors: firstly, the
proportionate decline in the relatively young who are prepared,
voluntarily, to undertake the demanding responsibilities of informal
supervision and direct assistance; secondly, the increasing numbers
of the elderly, many of whom retired into the region in the 1960s
and 1970s, who will suffer increasing restrictions on personal
mobility, age-induced disabilities of various kinds and, following
the death of a spouse, isolation. Undoubtedly, at a time of serious
financial restraint it will become increasingly difficult to meet
the greater demands for Social Services. For this reason, and to
compensate for the more general problems of service decline, it is
important to safeguard the independence of the active elderly and
to encourage schemes of community care for those most in need.
Further collaboration between elected representatives, voluntary
agencies and the statutory services can undoubtedly provide an
effective measure of support for both the dependent elderly and the
disabled, but as the survey shows the majority of the elderly and
many of the disabled are still independent and their interests can
best be served by comprehensive policies for improvement in local
accessibility.

ACKNOWLEDGEMENTS

We acknowledge the assistance of the geography students at Kingston
Polytechnic who conducted the interviews, Tony Redpath who helped
with the analysis of survey material, and Mrs P.K. Westley,
Assistant Research Officer, Gloucester Social Services.

APPENDIX 13.1

The term 'chronically sick and disabled' is used to describe a
person who according to the legal definition of disability set out
in Section 29 of the National Assistance Act 1948 is 'substantially
and permanently handicapped by illness, injury or congenital
deformity which can be either physical or mental in character'.
Such a person, whatever the nature of the illness or disability is
unable to 'engage in the activities, participate in the relationships and play roles which are normal for someone of his age and
sex'.

The related term 'impairment' refers to persons suffering from a
disability or illness which may (or may not) reduce their mental
and physical functioning but which does not prevent them from
engaging in most activities, relations and roles which are normal
for their age and sex.

Self-reported disability and impairment are considered together in
this survey. The term handicap is used to refer to the effect on
the person of either disability of impairment (Taylor, 1977).

REFERENCES

Allon-Smith, R.D. 1982. The evolving geography of the elderly in England and Wales. in *Geographical perspectives on the elderly*, Warnes, A.M. (ed), (Wiley, London), 35-52.

Association of District Councils, 1978. *Rural recovery: Strategy for Survival*.(Association of District Councils, London).

Association of County Councils, 1979. *Rural deprivation*. (Association of County Councils, London).

Austin, M.J. 1976. Network of help for England's elderly. *Social Work*, 22, 114-20.

Banister, D.J. 1980. *Transport mobility and deprivation in inter-urban areas*. (Saxon House, Farnborough).

Bebbington, A.C. 1979. Changes in the provision of social services to the elderly in the community over 14 years. *Social Policy and Administration*, 13, 111-23.

Borsay, A. 1982. Equal opportunities? A review of transport and environmental design for people with physical disabilities. *Town Planning Review*, 53, 153-75.

Buchanan, J. 1983. *The mobility of disabled people in a rural environment*. (Royal Association for Disability and Rehabilitation, London).

Campbell, M. 1979. Our ailing rural health service. *The Village*, 34, 3-5.

Cresswell, R. (ed) 1977. *Rural transport and country planning*. (Leonard Hill, London).

Department of Health and Social Security, 1978. *A happier old age*. (H.M.S.O., London).

Dunn, M., Rawson, M. and Rogers, A. 1981. *Rural housing: competition and choice*. (Allen and Unwin, London).

Gant, R.L. and Smith, J.A. 1980. Changes in the occupation structure of six Cotswold villages 1939-1979: a pilot study. *Classroom Geographer*, November, 10-15.

Gant, R.L. and Smith, J.A. 1981. *Key settlement policy and the elderly: a pilot study in the Cotswolds*. (School of Geography, Kingston Polytechnic).

Garden, J. 1978a. The mobility of the elderly and disabled in Great Britain: an overview. in *Mobility for the elderly and handicapped*, Ashford, N. and Bell, W. (eds.), (Loughborough University of Technology, Loughborough), 74-81.

Garden, J. 1978b. *Social services transport and the elderly*. (Beth Johnson Foundation, University of Keele).

Garden, J. 1979. *Community transport*. (Beth Johnson Foundation, University of Keele).

Gloucestershire County Council, 1978. *Draft Report of Survey technical volume no. 11., social aspects*. (Gloucestershire County Council, Gloucester).

Gloucestershire County Council, 1979a. *Structure plan*. (Gloucestershire County Council, Gloucester).

Gloucestershire County Council, 1979b. *Public transport plan*. (Gloucestershire County Council, Gloucester).

Hanson, P. 1977. The activity patterns of elderly households. *Geografiska Annaler*, 59B, 109-24.

Hedley, R. and Norman, A. 1982. *Home help: key issues in service provision*. (Centre for Policy on Ageing, London).

Hillman, M., Henderson, I. and Whalley, A. 1976. *Transport realities and planning policy*. (Political and Economic Planning, London).

Hillman, M. and Whalley,. A. 1979. *Walking is transport*. (Policy Studies Institute, London).

Hunt, A. 1978. *The elderly at home*. (H.M.S.O., London).

Markovitz, J. 1971. Transportation needs of the elderly. *Traffic Quarterly*, 25, 237-53.

Moseley, M.J., Harman, R.G., Coles, O.B. and Spencer, M.B. 1977. *Rural transport and accessibility*. (University of East Anglia, Norwich).

Moseley, M.J. 1978a. The mobility and accessibility problems of the rural elderly: some evidence from Norfolk and possible policies. in *Solving the transport problems of the elderly: the use of resources*, Garden, J. (ed), (Beth Johnson Foundation, Keele), 51-62.

Moseley, M.J. (ed), 1978b. *Social issues in rural Norfolk*. (University of East Anglia, Norwich).

Moseley, M.J. 1979. *Accessibility: the rural challenge*. (Methuen, London).

National Association of Local Councils, 1980. *Rural life: change or decay*. (N.A.L.C., London).

Neate, S. 1981. *Rural deprivation*. (Geo Abstracts, Norwich).

Norman, A. 1977. *Transport and the elderly: problems and possible action*. (National Corporation for the Care of Old People, London).

Nutley, S.D. 1980. Accessibility, mobility and transport related welfare: the case of rural Wales. *Geoforum*, 11, 335-52.

Shaw, M. and Stockford, D. 1979. The role of statutory agencies in rural areas: planning and social services. in *Rural Deprivation*, Shaw, J.M. (ed), (University of East Anglia, Norwich), 117-35.

Smith, J.A. and Gant, R.L. 1981. Transport provision and rural change: a case study from the Cotswolds. in *The spirit and purpose of transport geography*, Whitelegg, J. (ed), (University of Lancaster), 97-114.

Smith, J.A. and Gant, R.L. 1982. The elderly's travel in the Cotswolds. in *Geographical perspectives on the elderly*, Warnes, A.M. (ed), (Wiley, London), 323-36.

Standing Conference of Rural Community Councils, 1978. *The decline of rural services*. (National Council of Social Service, London).

Stanley, P.A. and Farrington, J.H. 1981. The need for rural public transport: a constraints-based case study. *Tijdschrift voor Economische en Sociale Geografie*, 72, 62-80.

Stockford, D. and Dorrell, P. 1978. Social service provision in rural Norfolk. in *Social issues in rural Norfolk*, Moseley, M.J. (ed), (University of East Anglia, Norwich), 59-75.

Taylor, H. 1977. *An evaluation of the effectiveness of social services provision in a rural area*. (University of Birmingham and Worcester County Council).

Williams, I. 1979. *The care of the elderly in the community.* (Croom Helm, London).

Williams, R. *et al.* 1980. Remoteness and disadvantage: findings from a survey of access to health services in the Western Isles. *Scottish Journal of Sociology,* 4, 105-24.

Woollett, S. 1981. *Alternative rural services: a community initiatives manual.* (National Council of Voluntary Organisations, London).

Chapter 14

Incomes of the elderly in rural Norfolk

Roger Gibbins

INTRODUCTION

The current heightened interest in the elderly by social researchers
and policy makers is due to the rapidly increasing numbers of
elderly in the population and the associated political significance
of an enlarged dependent group. Seventeen per cent of the present
population in the United Kingdom are over retirement age compared
with fifteen per cent in 1961 and only six per cent at the turn of
the century. This is happening largely because people who formed
part of the boom in live births around the turn of the century
survived childhood in greater numbers than before because of late-
nineteenth-century improvements in public health, housing and
nutrition.

The proportion of elderly in the population will remain fairly
static from now to the end of the century. However, the balance
between the younger and more active element and the very elderly
will change considerably. The numbers of those 75 and over will
increase by eighteen per cent between now and the year 2000, while
those aged between 65 and 74 will decrease by thirteen per cent.
It is this group of elderly that are by far the largest users of
health and social services. The heightened political significance
of this changing demographic position is thus apparent and can be
measured by the fact that a discussion document and a White Paper
have been produced by the Government in the last four years
(DHSS, 1978; DHSS, 1981).

POVERTY DEFINED

Poverty is increasingly being defined in relative terms. The
subsistence-level approach to the definition of poverty is now
rejected by most academics in favour of a relative concept that
views the plight of the poor in relation to the generally accepted
standards of society as a whole. The state's own definition of
poverty however, based as it is on the supplementary benefit level,
incorporates still the concept of a subsistence income.

Different definitions of 'poverty' will of course alter the
apparent severity of the problem. In 1969, by the British Government's own standard, seven per cent of households and six per cent
of people, representing some 3.3 million individuals, were in
poverty (Townsend, 1979). However, when incomes are considered in

relation to mean incomes, the numbers in poverty are much greater and when a still wider definition of poverty is taken that includes a measure of participation in activities and possession of certain household items widely accepted to be standard within society, then still more are found to be living in poverty (Table 14.1).

Table 14.1   Percentages and numbers in poverty by different definitions

Definition of poverty	Households in poverty		Population in poverty	
	percentage	number (million)	percentage	number (million)
State Standard				
(based on rates paid by the Supplementary Benefits Commission)	7.1	1.34	6.1	3.32
Relative Income Standard				
(a fixed percentage of those on the lowest incomes)	10.6	2.00	9.2	5.00
Deprivation Standard				
(below which people experience deprivation disproportionately to income)	25.2	4.76	22.9	12.46

Source: Townsend, 1979, table 7.1, p. 273

Although this paper is solely concerned with the incomes of the elderly, it is not suggested that incomes adequately describe the numbers in poverty. In fact, by the use of income data the numbers in poverty are underestimated. However, it has been shown that more reflective measures of poverty are closely linked to personal incomes and, by the use of income data as a surrogate for poverty, far more comparisons can be made with published material, especially that from official sources.

INCOMES OF THE ELDERLY

Of the 4.9 million people receiving supplementary benefit in 1976, 1.7 million or 35 per cent were of pensionable age (Government Statistical Office, 1979). Similarly, of the 14 million people in, or on the margins of, poverty (defined as being below 140 per cent of the supplementary benefit level), 37 per cent were elderly (Royal Commission on the distribution of income and wealth, 1978). This group comprises 62 per cent of all pensioners.

The 1980 Family Expenditure Survey (FES) gave details of the incomes of different household types and again showed that in terms of income the elderly are very much worse off than the rest of the population (Department of Employment, 1980). Tables 14.2 and 14.3 indicate that for one-person and two-person households the proportion of elderly households with low incomes is very much greater than the proportion of non-retired households. Whereas 22 per cent of all households earn less than £60 a week, this figure rises to 58 per cent for retired households. Of households with less than £60 a week, 49.4 per cent comprise retired people (Department of Employment, 1980).

Table 14.2   One-adult households with incomes below £40 a week

Household type	percentage with incomes below £40 a week
Retired, mainly dependent on state pensions	91.5
Other retired	30.5
Non-retired	16.9

Table 14.3   One-man, and one-woman households with incomes below £60 a week

Household type	percentage with incomes below £60 a week
Retired, mainly dependent on state pensions	86.4
Other retired	18.0
Non-retired	5.0

Source: Department of Employment, 1980

In general terms, the elderly are living on very much lower incomes than the rest of the population. However, not all the elderly have to rely on such low incomes since twenty per cent of retired households have incomes of over £100 a week. They do however only represent six per cent of households in this income bracket. In terms of income inequality within groups, then, it seems reasonable to concur with Abrams (1980) and suggest that, although inequalities of income do occur between the elderly, the degree of inequality will not be as great as that occurring in the population as a whole.

Inequality in incomes between the elderly is associated with a number of factors, one of which is age. The aged elderly tend to have lower incomes than the younger elderly. Similarly, women are less well off than men in old age as throughout life (Abrams, 1980). Townsend argues that poverty is a function of class and working status (Townsend, 1979). The lower socio-economic groups are likely to be paid less in their working lives, and those who had jobs of low occupational status are more likely than others to be poor in old age. Some old people are poor by virtue of their low life-long class position, others by virtue of society's imposition upon the elderly of 'underclass' status. As they move into old age, people tend to be separated into two groups, one anticipating a comfortable and even early retirement, the other dreading the prospect and depending almost entirely for their livelihood on the resources made available by the state through its social security system.

RURAL INCOMES

Incomes have been shown to vary geographically within the United Kingdom (Coates and Rawstron, 1971). Analysis of the Inland Revenue's Survey of Personal Incomes (SPI) shows that income from employment is much lower in rural than urban areas. In 1964-65 the six counties with the lowest average incomes were all predominantly rural in nature. A more recent study, using the same data source,

139

arrives at the same conclusion - that personal incomes are lower in the countryside than in the cities (Thomas and Winyard, 1979).

The New Earnings Survey (NES), compiled by the Department of Employment, gives a more accurate reflection of personal incomes than the SPI in as much it includes those lower incomes that are not liable to taxation. However, the geographical breakdown is into regions rather than counties. If the South West and East Anglia are taken as characteristically rural regions, then again it can be seen that incomes are substantially less in rural areas than in urban ones. In East Anglia, the average weekly earnings for manual men is 92.6 per cent of the national average for that type of worker, and for non-manual men 93.1 per cent (Gilg, 1976). Women's earnings differ less markedly from the national average than men's; however this reflects the low pay of women in general rather than any regional distinction.

Two processes are occurring in rural areas that partly explain low incomes. One is the concentration of low-paid jobs, such as agriculture, forestry and fishing, and an associated dearth of higher paid manufacturing jobs in the countryside. The other is that within similar industries and types of employment, wage rates are lower in the countryside than the cities. Evidence from the literature suggests that both these processes are occurring (Gilg, 1976).

SURVEY DESIGN

The evidence provided from published sources indicates that the elderly in general are particularly prone to poverty, and that incomes tend to be lower in rural areas than in the cities. The purpose of this paper is to examine the overlap that exists between these two deprived groups, namely, the eldely and those who live in the countryside.

A questionnaire survey of people over the age of 65 was designed to measure all aspects of deprivation and life-style of the elderly in rural Norfolk. This paper concentrates on the income aspect of deprivation. A number of questions were asked to determine the level of personal income, sources of income and satisfaction of the rural elderly with their level of income.

The purpose of the study was to determine the association between different parts of the countryside and deprivation amongst the elderly, one hypothesis being that those elderly living in more rural areas would be deprived relative to those living in the more accessible parts of rural Norfolk. A principal components analysis of census data, information on service provision, measures of remoteness, population levels and population change was carried out for all parishes in Norfolk. It was discovered that indicators such as declining population levels, low provision of services and physical remoteness were all positively correlated. Furthermore, the elderly were disproportionately concentrated in those parishes which performed poorly by the indicators. The next step in the analysis was to determine if the elderly living within these 'worst' parishes suffered from individual deprivation.

Different types of parish can be summarised from the correlations by using measures of remoteness and population size. For sampling purposes six types of parish were identified:

1) close to Norwich, large population;
2) close to Norwich, small population;
3) far from Norwich, close to a service centre, large population;
4) far from Norwich, close to a service centre, small population;
5) far from Norwich, far from a service centre, large population;
6) far from Norwich, far from a service centre, small population;

Two parishes were selected for sampling in the large parish categories, three or four in the small. In the large categories, the population averaged about 600 people, in the small ones, about 200. Those parishes defined as being far from Norwich were all some 30 kilometres distant, and those which were far from a service centre were some 7 or 8 kilometres away. In the small villages a 100 per cent sample of the elderly was achieved, whereas in the large parishes a random sample was taken.

INCOMES OF THE ELDERLY IN RURAL NORFOLK AND THE NATIONAL SITUATION

The respondents were asked to indicate the income group to which they belonged after being shown a card of categories. From the responses the level of income in relation to the supplementary benefit level at that time was calculated for each income unit – e.g., a single person or married couple. Rather than view the income data in absolute terms, it has been considered relative to the supplementary benefit level. This allows much wider scope for comparison with different surveys and data sources. Furthermore, it enables comparison with earlier studies when incomes were usually lower.

At the time of the survey, summer 1981, the supplementary benefit level (which is the state's own standard of poverty) was £27.15 for a single pensioner and £43.45 for a married pensioner couple. The income necessary to raise individuals above the margins of poverty (that is, to more than 140 per cent of the supplementary benefit level) was £38.00 for single people and £60.20 for married couples.

In the rural sample, 44 per cent of the elderly had incomes on or below the state's poverty level, 26 per cent were living on the margins of poverty and 30 per cent had incomes greater than 140 per cent of the supplementary benefit level. Comparable figures for the elderly nationally are given in the Family Expenditure Survey and in Townsend's national survey of poverty (Department of Employment, 1980; Townsend, 1979). These figures are presented in Table 14.4. The data clearly show that there is a higher proportion of elderly people in poverty in the rural sample than nationally and consequently a lower proportion with high incomes.

Account must be taken of differential response rates to the income question by different socio-economic groups since 23 per cent of the professional and managerial group and only 5 per cent of the partly and unskilled group refused to answer the income question. One must also be aware of any variation in the distribution of socio-economic groups between rural Norfolk and the nation as a whole, therefore it is important to consider the income distribution within each socio-economic group.

Table 14.5 indicates the proportion of elderly in different socio-economic groups living on or below the margins of poverty in the rural sample and in Townsend's national sample. First and foremost, the table emphasises the disparity between the incomes of the manual and non-manual classes. Further, it indicates a rural-national difference in the proportions in, or on the margins of,

141

Table 14.4   Percentage of the elderly in poverty by the state's
              own standard, nationally and in rural areas

| | Percentage of elderly who are: | | |
	in poverty	on the margins of poverty	above the margins of poverty
Rural sample	44.3	25.8	29.7
FES, 1980	19.0	39.0	42.0
Townsend's sample	20.0	44.0	36.0

Table 14.5   Percentage of elderly of different socio-economic
              groups on or below the margins of poverty by the
              state's own standard

	non-manual	skilled manual	partly and unskilled manual
Rural sample	41.3	80.5	86.6
Townsend's sample	47.7	73.0	71.6

poverty in each socio-economic group. This difference is significant at the O.05 level using a chi-square test. More importantly
however, the table shows a greater inequality in incomes within the
sample of rural elderly than within the national sample. There are
fewer people below the margins of poverty in the non-manual group
than would be expected but more in the manual groups. Thus, in
rural Norfolk the more wealthy elderly have higher incomes than
would be expected and the less well off are poorer than the national
average for their particular socio-economic group.

     One of the most important factors that determines whether a
person is thrust into poverty on retirement is the possession of an
occupational pension. In Townsend's study 25 per cent of the
elderly were in receipt of an occupational pension (Townsend,
1979); in the rural sample the figure was 28 per cent. Abrams
showed figures similar to these in the rural sample - 28.5 per cent
of his national survey of the elderly had an occupational pension
as a main or secondary source of income (Abrams, 1980). The
proportion of the elderly in rural areas with an occupational pension
seems similar to that nationally and rates everywhere are increasing
with time - Townsend's survey was carried out some eight to ten
years before the other two. When this hypothesis is considered in
relation to the generally similar levels of income found between
the rural sample in total and Townsend's national sample, the
interesting fact emerges that, although occupational pension rates
are increasing with time, the number of elderly being raised above
the poverty level is not.

     In addition, occupational pensions reflect other measures of
income through their association with social class. In the rural
sample, 53 per cent of those in the non-manual occupational group
were in receipt of a pension from a previous job, whereas only 15
per cent of those in the manual group received one.

INCOMES OF THE ELDERLY IN RURAL NORFOLK

The association between rurality and income can be explored further
by consideration of the levels of income within the six parish
categories identified earlier. As socio-economic group is by far
the overriding factor in relation to income, and because of the
biases already noted, it is important to examine income within
socio-economic groups. The only analysis that will allow valid use
of the chi-square test is to consider the incomes of manual groups
in relation to the supplementary benefit level. This involves 211
respondents. However the calculation of chi-square reveals that
there is no significant difference in retired manual workers'
incomes between the different types of rural area. Although the
lower socio-economic groups in rural Norfolk as a whole are poorer
than their national equivalents, there is no evidence to suggest
that degrees of rurality have any effect on income levels.

CONCLUSION

This survey of the elderly in rural Norfolk has shown that in total
terms the rural elderly have lower incomes than the elderly
nationally. The more important finding of the study is, however,
that there exists a greater inequality in income distribution
between the rural elderly than between the elderly nationally.
There exists a polarisation of experience amongst the elderly in
rural Norfolk with the rich better off than would be expected, and
the poor poorer. This situation is replicated when all aspects of
poverty are taken into consideration. Not only is the inequality
experienced with respect to income, but wide differences also exist
in terms of housing conditions, household possessions, health and
access to facilities. This polarisation is closely linked to social
processes involving retirement migration. An affluent elite of
elderly migrants have used their access to resources to enable them
to benefit from the advantages of rural living, while retired farm-
workers are trapped by their poverty in a rural area that presents
to them only hardship and disadvantage.

REFERENCES

Abrams, M. 1980. *Beyond three score and ten, a second report.*
    (Age Concern, London).

Coates, B. and Rawstron, E. 1971. *Regional variations in Britain.*
    (B.T. Batsford, London).

Department of Employment, 1980. *Family expenditure survey.*
    (HMSO, London).

Government Statistical Office, 1979. *Social trends.* (HMSO, London).

DHSS (Department of Health and Social Security), 1978. *A happier
    old age.* (HMSO, London).

DHSS (Department of Health and Social Security), 1981. *Growing
    older.* (HMSO, London).

Gilg, A. 1976. Rural employment. in *Rural planning problems,*
    Cherry, G. (ed), (Leonard Hill, Farnborough).

Royal Commission on the distribution of income and wealth, 1978.
    The causes of poverty. (HMSO, London).

Thomas, C. and Winyard, B. 1979. Rural incomes. in *Rural depriva-
    tion and planning,* Shaw, J.M. (ed), (Geo Abstracts, Norwich).

Townsend, P. 1979. *Poverty in the United Kingdom.* (Penguin,
    Harmondsworth).

Chapter 15

Public spending in rural areas

Ad van Bemmel

INTRODUCTION

Public expenditure in a welfare state is of the utmost importance
but nowadays we are confronted with large reductions in these
expenditures. These savings are reflected in spatial planning and
policies. For example, a report was published in 1981 by the Dutch
Government dealing with regional socio-economic policy for 1981 to
1985 (Ministerie van Economische Zaken, 1981). This report made
clear that the focus of regional policy in the Netherlands had
shifted, at least partially, to the Randstad, the most urbanised
part of the country. The peripheral and most rural regions are now
getting less attention. The high costs of developing the peripheral
regions are an important motive behind this shift.

This shift forms part of the background to this chapter. We
should also note a point made by Johnston (1977) 'Governments are
major spenders within most countries, so that where they allocate
the contents of their budgets can play a crucial role in either
accentuating or reducing spatial patterns of inequality'. These
spatial patterns of inequality are getting increasing attention in
modern human geography. This agrees with Smith's opinion (1977)
that human geography should be the study of 'who gets what, where
and how'.

This development can be seen also in rural geography. According
to Cloke (1980), 'The primary aim within rural geography could
become one of ensuring equity of opportunity, well-being and quality
of life between all sectors of the rural community, coupled with
the creation and maintenance of a productive and self-perpetuating
rural environment'.

These aims are tied directly to the financial resources of the
authorities involved. This chapter tries to give insights into
these financial possibilities because the impression exists that
rural areas, and especially the more peripheral parts, are getting
less financial assistance per head of population than other areas.
This impression is a result of the fact that planning for the
countryside has often adopted an implicitly urban point of view
towards rural issues (Jones, 1973; Bradshaw and Blakely, 1979). As
Warner (1974) said, 'Decisions affecting rural society are being
made more and more in both corporate boardrooms and government
bureaux far distant from rural life'. Was he exaggerating when he
said, that rural society may increasingly take on characteristics

145

like those of an 'inconspicuous' minority group'?

I think it is important for (rural) geographers to study also the spatial effects of government spending. Such an approach is necessary because maintaining the 'liveability' and viability of rural areas will have many financial repercussions.

Bennett (1980) has pioneered such research. He has asked rightly the crucial question, namely, 'Who gets what, where and at what cost?' He states that solutions to the economic, social and political problems created by differences in wealth and income can be found only in methods of public finance which take explicitly geographical factors into account. He puts forward a geography of public finance which involves spatial patterns of revenue-raising, spatial patterns of public expenditure and the spatial balance between revenues and expenditures.

It is a pity, however, that Bennett has chosen a relatively high spatial level in his analysis (e.g. counties, SMSAs) and that the distinction between rural and urban in this respect has been neglected. This chapter attempts to fill this gap to some extent.

This chapter has been written within the context of a study of the advantages and disadvantages of living in rural areas in the Netherlands and especially in Zuidwest-Friesland (Figure 1.1) (Van Bemmel, 1981b). That study aims to make a contribution to the conservation and improvement of the liveability and viability of settlements in peripheral rural areas.

First, we consider some difficulties in interpreting 'raw' financial data. Second, an analysis is made of public expenditure on a regional level; a distinction is made between rural and urban areas. Third, two regions are compared with respect to the budgeted expenditures of the municipalities in these regions. These two regions are a rural area in the North (Zuidwest-Friesland) and Rijnmond, (the urban region of Rotterdam) (Figure 1.1). The choice of Zuidwest-Friesland and Rijnmond is based on a study by Hauer and Veldman (1980) in which a classification is given for the COROP regions in the Netherlands according to their degree of rurality (see Hauer's paper in this book).

DIFFICULTIES OF INTERPRETATION

It seems very easy to say that when, for example, 50 Dutch Guilders (Dfl.) have been spent in rural areas and Dfl.100 in urban areas that people in rural areas get less money. Interpretation is not so easy, however. The interpretation of absolute amounts of money per head is rather difficult. I shall try to explain these difficulties in a general sense (see Boaden, 1971; Johnston, 1979a and 1979b; Bennett, 1980).

First, prices of goods and costs of services differ spatially. In general, costs of services follow a U-shaped curve (Gilder, 1979). Costs of services are higher in regions with a low population density (e.g. a peripheral rural area) and in areas with a very high population density (e.g. an urban area). On the other hand, costs are relatively low in commuter areas with a moderate population density. Second, there are differences in the efficiency of the authorities in spending money. For example, big cities have specialised financial experts for obtaining grants and for monitoring spending. Rural councils often do not have those experts. Third, needs, wants, necessities and possibilities differ spatially. In general, economic and population growth are lower in rural areas

146

than in urban ones. This is one reason why the spending of
authorities in urban areas needs to be greater. More people in an
area means also more varied and extensive services. However, in
that case also more grants may be obtained. According to Bennett
(1980), rural costs are a combination of high distribution costs,
low labour costs, low costs of living and low levels of externali-
ties from other services. High city costs combine agglomeration
economies with high labour and floor space costs, and low costs of
public services. A fourth point to note is that differences in
government spending are partly the result of differences in
revenues and local taxes. In the Netherlands these differences are
mainly the result of differences in the level of specific grants.
Local taxes are not very important in the Netherlands because only
a small percentage of total revenue is the result of local taxes.
Also decisions by authorities may take effect over different periods
of time. That means that when certain expenditures have been made,
their effects may be detected for several years. Furthermore,
spending in one area can influence other areas by means of spill-
over and back-wash effects, although these are very difficult to
measure. A final point concerns the budgets of municipalities and
provinces in the Netherlands. Even within a standard framework for
budgets there are differences which are the result of different
methods of bookkeeping.

It is not possible to indicate the scale of all these problems
because little investigation has been done yet. Nevertheless,
anyone who is confronted with absolute figures about government
spending should be very careful. For that reason this analysis
will be at a relatively high level of abstraction.

GOVERNMENT STRUCTURE AND PUBLIC EXPENDITURES

The government structure of the Netherlands consists of three tiers
namely state, province and municipality. It is important to
remember that a municipality can be either very small, e.g. 1000
inhabitants, or very big like Amsterdam with 800 000 inhabitants.
The system of public finance is quite complicated (Hoogerwerf,
1981). State, province and municipality all have their own budgets
and responsibilities. The lion's share of the budget of a province
or municipality comes from the national government by means of a
distribution system. For a municipality, for example, the criteria
for the distribution are intensity of building, population size and
expenditure on social welfare. Provinces and municipalities also
get money from the national government by means of specific grants.
State, province and municipality can give grants directly to an
individual citizen or to a group of citizens. On the other hand a
citizen receives money more indirectly from the municipality
because he or she benefits from the construction of roads, the
maintenance of the fire brigade and so on. In the next section an
analysis will be made of the expenditures of state, province and
municipality combined which can be allotted directly to individual
households. The budgeted expenditures of the municipality are
examined in a later section.

EXPENDITURES OF AUTHORITIES IN RURAL AND URBAN AREAS

The first analysis of the regional distribution of governmental
expenditure in the Netherlands was published in 1981 (Sociaal en
Cultureel Planbureau). In this section we shall discuss that
analysis. Table 15.1 shows for regions and sectors such expenditure by central government, provinces and municipalities as can be
allotted to individual households.

147

Table 15.1   Indices of regional distribution of expenditure per household by central government, province and municipalities combined in the Netherlands in 1977

	Rural areas	Urbanised countryside	Commuter areas	Cities	Total	The provinces of Friesland and Groningen[1]
Housing	102	108	117	92	100 (Dfl. 659)	101
Education	96	101	114	96	100 (Dfl.3085)	95
Public transport	98	75	107	108	100 (Dfl. 248)	72
Social welfare	105	82	51	120	100 (Dfl. 198)	144
Culture and recreation	94	86	112	104	100 (Dfl. 308)	97
Public health	36	93	34	133	100 (Dfl. 320)	95
Other[2]	55	99	108	108	100 (Dfl. 71)	88
TOTAL	92	99	106	100	100 (Dfl.4887)	97

Notes: 1 Zuidwest-Friesland included
        2 expenditures under certain social security laws

Source: Sociaal en Cultureel Planbureau, 1981

The differences in expenditure between the regions are considerable in absolute terms, hence the profits per household from governmental grants differ from region to region. In the provinces of Friesland and Groningen, in the rural areas and the urbanised countryside respectively, Dfl. 151, Dfl. 373 and Dfl. 65 less per year per household was spent than nationally in 1977. On the other hand, in the commuter areas Dfl. 3O7 more than nationally was spent while the cities match the national level. These absolute differences can not be neglected.

The greatest spatial differences are for education and public health. One explanation may be the presence or absence of higher education and specialised health services which are in general concentrated in the cities. The provinces of Friesland and Groningen and the rural areas show the greatest differences compared with the national figures for education, public health, social welfare and public transport.

In general we can conclude that the distribution of the seven blocks of expenditure does not differ much between the five regions or between rural and urban areas. These differences seem less than in other countries (e.g. the United Kingdom (Dunn *et al.*,1981; Association of County Councils, 1979).

BUDGETED EXPENDITURES OF MUNICIPALITIES

IN ZUIDWEST-FRIESLAND AND RIJNMOND

Zuidwest-Friesland (a peripheral rural area) and Rijnmond (an urban area) have been chosen as examples of two contrasting regions. Zuidwest-Friesland covers 90 900 ha. and has 110 000 inhabitants, while Rijnmond covers 68 200 ha. and has a population of 1 046 000.

In this section the *budgeted* expenditure of the municipalities in these regions is the primary object of investigation because it

expresses more clearly than *actual* expenditure the political wishes
and visions of the policy makers. In general, budgeted expenditure
does not actually differ much from the actual expenditure.

A municipality has considerable freedom in fixing that part of
its expenditure which is not controlled by national government.
The close relationship between budgeted expenditure and population
size is indicated by the high correlation between them. The average Spearman correlations are 0.97 for Zuidwest-Friesland and 0.91
for Rijnmond. In this section expenditure on public health services,
public security, public works, education, social welfare, culture
and recreation is considered (see Table 15.2).

Table 15.2   Indices of average budgeted expenditure per head
              for municipalities in Zuidwest-Friesland, Rijnmond
              and nationally, 1970-1980[1]

	Zuidwest-Friesland	Rijnmond	The Netherlands
1970-71	100 (Dfl. 553)	100 (Dfl. 1053)	100 (Dfl. 696)
1975-76	221	166	226
1978	286	250	275
1979-80	320	272	298

Notes: 1 Zuidwest-Friesland: 15 municipalities; Rijnmond: 16 municipalities;
          The Netherlands: 774 municipalities
Source: Van Bemmel, 1981a

In Zuidwest-Friesland it is clear that in general considerably
less has been budgeted in total than nationally, but these differences are decreasing. However, it is not only total budgeted
expenditure that is important. The next question concerns the
distribution of budgeted expenditures for the different sections of
public finance. We see that the differences in the patterns of the
budgeted expenditure of municipalities between Zuidwest-Friesland
and Rijnmond are greater than between Zuidwest-Friesland and
nationally. It is remarkable that in recent years these differences
have again become greater; the gap has widened slightly. This may
be an indication of a change in regional policy.

We have already noticed that there are no major differences
between Zuidwest-Friesland and Rijnmond with respect to the relationship between budgeted expenditure and population size. The
conclusion must be that the present differences in budgeted expenditure between both regions are more or less the result of differences
in policy between the municipalities in the two regions.

This conclusion becomes even clearer when we study the results
of a discriminant analysis for the period 1975-6 to 1979-80.
Discriminant analysis is a technique by which it is possible to
distinguish between groups of cases (e.g. municipalities) by testing
an *a priori* classification. Nine variables were used (average Wilks-
Lambda 0.91). These variables are related to budgeted expenditure
and budgeted revenue (grants and taxes). The results of this
analysis indicate that 90 per cent of the 31 municipalities have
been correctly classified with respect to their patterns of budgeted
expenditure and revenue. There are significant differences between
municipalities in Zuidwest-Friesland and Rijnmond with respect to
their pattern of budgeted expenditure and revenue.

CONCLUSION

It has been shown that there are differences in the regional distribution of public expenditure per household between the provinces of Friesland and Groningen, the rural areas, the urbanised countryside, the commuter areas and the cities. In the two provinces and especially in the rural areas less has been spent than in other areas. This is particularly true for public health, culture and recreation, and education.

It has also been shown that in Zuidwest-Friesland, a peripheral rural area, consistently less was budgeted per head by the municipalities between 1970 and 1980 than in Rijnmond (an urban area). It is concluded that there are significant differences between municipalities in Zuidwest-Friesland and Rijnmond with respect to their patterns of budgeted expenditure and revenue.

This chapter started with a summary of the difficulties in interpreting financial data. For these reasons it is not justifiable to say that the differences in expenditure per head between rural areas and urban areas are a definite proof of spatial inequality. However, there are strong indications that rural areas systematically get relatively less money than urban ones. For this reason I would recommend more research in this field by geographers in close cooperation with their colleagues in related disciplines.

REFERENCES

Association of County Councils, 1979. *Rural deprivation*. (London).

Bemmel, van A.A.B. 1981a. *Inkomsten en uitgaven van gemeenten in landelijke en stedelijke gebieden*. (Geografisch Instituut, Utrecht).

Bemmel, van A.A.B. 1981b. *A welfare geographical approach of living in rural areas - some remarks*. (Paper XI European Congress for Rural Sociology, 9-15 August 1981, Helsinki).

Bennett, R.J. 1980. *The geography of public finance*. (Methuen, London).

Boaden, N. 1971. *Urban policy-making influences on County Boroughs in England and Wales*. (University Press, Cambridge).

Bradshaw, T.K. and Blakely, E.J. 1979. *Rural communities in advanced industrial society*. (Praeger, New York).

Cloke, P.J. 1980. New emphases for applied rural geography. *Progress in Human Geography*, 2, 181-218.

Dunn, M. *et al*. 1981. *Rural housing: competition and choice*. (George Allen & Unwin, London).

Gilder, I.M. 1979. Rural planning policies: an economic appraisal. *Progress in Planning*, 11, 213-271.

Hauer, J. and Veldman, J. 1980. *Nota no.2: Kenmerken van landelijke gebieden op COROP-niveau*. (Geografisch Instituut, Utrecht).

Hoogerwerf, A. 1981. Relations between central and local governments in the Netherlands. *Planning and development in the Netherlands*, 2, 215-236.

Johnston, R.J. 1977. Environment, elections and expenditures: an analysis of where governments spend. *Regional Studies*, 11, 383-394.

Johnston, R.J. 1979a. Governmental influences in the human geography of developed countries. *Geography,* 1, 1-11.

Johnston, R.J. 1979b. *Political, electoral and spatial systems.* (Clarendon Press, Oxford).

Jones, G. 1973. *Rural life.* (Longman, London).

Ministerie van Economische Zaken, 1981. *Nota regionaal sociaaleconomisch beleid 1981-1985.* (Staatsuitgeverij, Den Haag).

Smith, D.M. 1977. *Human geography.* (Arnold, London).

Sociaal en Cultureel Planbureau, 1981. *Profijt van de overheid in 1977.* (Staatsuitgeverij, Den Haag).

Warner, W.M. 1974. Rural society in a post-industrial age. *Rural Sociology,* 3, 306-318.

Chapter 16

Population trends and community viability on the Danish small islands

Tony Martin

INTRODUCTION

The land area of metropolitan Denmark (43 075 sq km) consists of the peninsula of Jutland (28 548 sq km) and 406 islands, 87 of which are currently inhabited. The inhabited islands range in size from Zealand (7026 sq km) and Funen (2984 sq km) to smaller islands distributed across the national area from the Wadden Sea to the Baltic outpost of Bornholm (588 sq km). Many of the smaller islands are microcosms of the large ones and are a framework for a mosaic of small, isolated, and often highly individualistic communities with seemingly limited possibilities in a modern economy for any kind of economic activity other than agriculture. The remainder are utilised for agriculture, summer grazing and tempory residence wherever local conditions and landing facilities permit.

The problems currently facing the small island communities are a relatively recent phenomenon. Formerly the islands not only participated in the cultural and economic developments of the rest of the country, but often occupied a position of advantage. Their economic diversity enabled them to adjust to changing economic circumstances. Although agriculture was the keystone of the island economies, shipping and trade were important assets and fostered cultural connections that enriched island life. The isolation which presently characterises them was not so important in an era when sea communication was easier than overland transport. The strict geographical limits of the islands have tended to preserve tradition and reinforce the special character of the island environment.

The island communities reached their peak at the beginning of the present century. Former advantages have since been lost due to greater mobility, to technological changes emphasising large-scale developments and centralisation, and to the widening gap in socio-economic conditions and levels of opportunity between peripheral areas and centres of modern development. Consequently, the small islands have witnessed constant and accelerated economic decline and population loss since World War 1. Similar problems are in evidence in peripheral mainland locations, but the problems are distinctive and more acute on the islands. Without exception, they suffer from declining and ageing populations, limited service facilities and employment opportunities, and difficult communication with the mainland.

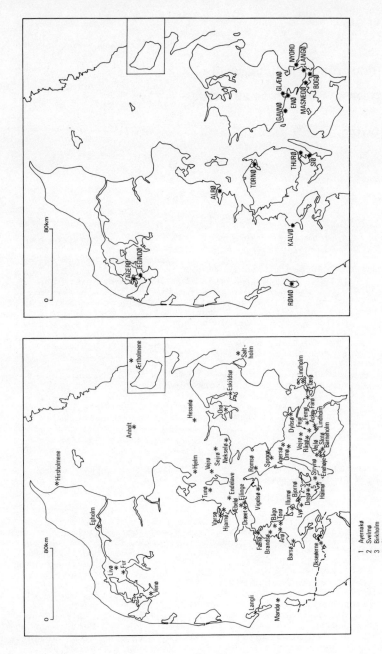

Figure 16.1   Distribution of   a) small islands and   b) linked islands

The islands retain one important advantage in terms of possible planning strategies for the future. They preserve relics of former life-styles, offer small-scale living and unspoilt, idyllic environments, and are perceived as peaceful havens by urban dwellers with increasing demands for areas for recreation and leisure activities. Problems relating to conservation and environmental protection are therefore also significant. The impact of these complex problems is experienced with varying intensity in different parts of the Danish archipelago. Unless remedies to these problems are found and the island economies re-invigorated, many insular communities will cease to be viable and depopulation, already in evidence, will quicken. The aim of the present paper is to discuss the problems facing the small island communities in Denmark by reference to examples of the main categories of island.

In order to provide an operational framework for the present study, the term 'small island' must be defined. The larger islands, Zealand, Funen, Lolland, Falster and Bornholm, and medium sized islands such as Langeland, Møn, Mors and Samsø are excluded. In 1970, the Danish Ministry for Cultural Affairs employed a definition of 'small island' that encompassed a total of 65 inhabited islands under 5000 ha and uninhabited islands over 50 ha. Islands joined to the mainland by bridge or causeway were excluded, although most possess economic structures and problems similar to those of the small islands. Saltholm and Egholm (North Jutland) were also omitted due to their special problems related to transport and urban planning. To facilitate comparisons, this investigation utilises a broader definition than that used by the Ministry. Using the same areal parameters, 15 islands with a bridge or causeway link to the mainland are included as are Saltholm and Egholm. The 82 small islands examined here are shown in Figure 16.1.

THE ISLAND POPULATIONS

The small islands represent only a small fraction of the population and national area of Denmark. In 1976, 11 152 inhabitants were distributed among 82 islands with a combined area of 313 sq km. Table 16.1 indicates that most islands have very small populations, with only 31 islands exceeding 50 persons and those with populations in excess of 500 accounting for 64 per cent of the small island total.

Table 16.1   Population and area of the small islands and linked islands, 1976

Population category	Islands			Linked islands		
	number	population	area(ha)	number	population	area(ha)
Uninhabited	12	0	939	1	0	70
1-19	27	168	4098	2	13	148
20-49	5	145	970	4	138	1109
50-149	10	995	4192	3	209	1398
150-249	7	1390	6242	2	341	919
250-499	2	596	1822	0	0	0
over 500	4	3046	6568	3	4111	2852
Totals	67	6340	24 831	15	4812	6496

The demographic situation is central to the problems faced by the small islands. Population decline was already in evidence by 1920. During the 1940s and 1950s, migration continued at a rate comparable to that experienced in rural areas on the mainland. A marked change took place after 1960 with the pace of outward migration quickening on the small islands while the situation in rural Denmark became more stable. The larger islands and those linked to the mainland by bridge or causeway have retained their population better than the smaller communities. The population loss in rural districts and various island groups is illustrated in Table 16.2.

Table 16.2   Population loss - small islands and rural districts, 1921-70

Island group by 1976 population	Population loss, 1921-70	% loss
Over 500 inhabitants	- 1019	21.6
250-499 inhabitants	- 1072	40.5
150-249 inhabitants	- 623	34.9
60 - 149 inhabitants	- 534	40.1
20 - 59 inhabitants	- 207	40.6
1 - 19 inhabitants	- 502	75.0
Rural districts	- 318 620	21.9

During the 1960s annual net migration represented 1.5 per cent of the population of the small islands, a rate 50 per cent higher than that in other rural areas. A problem related to migration is that of community stability. An annual turnover of over 5 per cent of the population was recorded on islands with populations in excess of 150 and one of almost 25 per cent on those with smaller populations. The change-over of personnel at public undertakings greatly influences migration rates and the stability of the smaller insular communities. Outward migration is dominated by the 15-39 age group and is particularly marked among females of this group. An examination of the age-sex structure of the island populations indicates that the proportion of children on the islands almost reaches the national average. The 15-39 age group constitutes only 23 per cent of the population compared with 35 per cent in the rest of Denmark. The female component in this group is only 9.5 per cent compared with a national figure of 17 per cent. The 40-64 age group is slightly larger than the national average, but the disparity is more marked among the over 65 age group, with several islands having two or three times the national average. Consequently, there is a natural decrease of 0.6 per cent per annum on the small islands compared with an increase of 0.7 per cent nationally. The rate of population decline on the small islands is thus 2.1 per cent per annum.

The demographic experience of the 1960s was used to produce two official population projections for 1980. The first, based on a nil migration hypothesis (M=0), predicted a decline of 3.5 per cent in the small island communities in the decade 1970-80. The second, based on the assumption that migration would continue at the rate experienced during the 1960s (M 1960-69) produced an overall reduction of 31.2 per cent. The situation in 1976 compared with these two projections is shown in Table 16.3. The largest islands have already reached a level below the M 1960-69 figure, while migration appears to have slowed in the other groups, although there is evidence to suggest that the problem has been arrested.

Table 16.3   Island populations in 1976 compared with projections for 1980

Island group	Population 1976	Population projection for 1980	
		M 1960-69	M=0
Over 500 inhabitants	3046	3203	3645
250-499 inhabitants	1289	922	1482
150-249 inhabitants	945	729	1101
60-149 inhabitants	638	278	675
20-59 inhabitants	208	160	277
1-19 inhabitants	108	0	170

The demographic situation dominates every aspect of island life. Geographical isolation, poor employment opportunities and a reduction in services have fostered migration from the islands. This in turn has given added impetus to further contraction of the narrow economic base and increased outward migration. This situation is symptomatic of imbalance in communities generally lacking the essential preconditions for survival and development and where expectations are not fulfilled. Depopulation must also be viewed against a background of structural changes resulting from the modernisation and centralisation of many social, cultural and economic facets of island life. Many of these, formerly taken for granted, now require larger population thresholds for their survival. In Denmark, such changes occurred simultaneously with local government reform and the resultant transfer of decision-making from island to mainland municipalities in 1970 (see the paper by Groot in this book). Remoteness has therefore been reinforced by fears that the special problems experienced by the small islands would be neglected. Despite the fact that more resources are now available to assist island communities, such fears contributed to the formation of the Confederation of Danish Small Islands in March 1974.

THE ISLAND ECONOMIES

Traditionally the island economies were dominated by agriculture and this, with a few exceptions, still holds true. Average farm size on the islands differs little from that on the mainland, but the area under cultivation is generally less due to the higher proportion of meadow, pasture and coastal marsh. The most important problems are related to recent rationalisation and centralisation in the agricultural sector resulting in the closure of island dairies, agricultural contractors and suppliers. Agricultural materials and products are therefore subject to increased transport costs. The net result has been a change from traditional dairying to corn production, often by absentee farmers, and declining investment due to the uncertainty surrounding the future of the ferry and agricultural services. Femø, Fejø and Askø, with an emphasis on fruit, vegetables and seed crops, have been more successful in weathering these changes than more traditional enterprises. Isolation has assisted agricultural use on three small islands. Lindholm (Møn) is owned by the Ministry of Agriculture and used as a research station by the Danish Institute for Viral Research, Livø is an agricultural college for the mentally-retarded, and Strynø has a poultry-breeding research centre. Elsewhere agriculture is in retreat and, in the case of Anholt, abandoned altogether.

Fishing, formerly an important subsidiary occupation on most islands, is now most important on the Kattegat islands, Anholt and Sejrø, the Great Belt islands Omø and Agersø, and on Arø and Christiansø. However, vessels are generally smaller than those operating from mainland ports and rationalisation has resulted in the closure of shore-based facilities and ancillary industries, so that island catches are landed on the mainland. Few of the preconditions for a modern fishing industry exist on the small islands and unless they are adjacent to rich fishing grounds, future development must be viewed with considerable uncertainty.

The industrial structure of the small islands is dominated by service and handicraft enterprises and by public utilities. Most of the latter are small scale, for example electricity generating stations, and will eventually close as rationalisation progresses and mainland links are developed. The sole major industry is located on Fur, where the extraction of diatomaceous material from Lower Eocene Mo Clay deposits and the manufacture of refractory bricks employ about 80 people. An onion processing plant on Fejø and a gas cylinder plant on Fænø, both small in scale, are the only other instances of manufacturing.

Service provision is limited and many facilities operate below generally accepted population thresholds. In most cases, services have contracted in line with population decline to a point where only basic requirements are met and they are dependent on income generated from visitors for their existence. Table 16.4 illustrates this situation. It should be stressed that the island population

Table 16.4   Population base of services, 1970 - a comparison of
             the small islands and rural areas of central Jutland

	central Jutland	small islands
Grocer	292	129
Butcher	1268	507
Motor mechanic	1311	383
Doctor	4778	537
Hairdresser	1191	1206

is not constant in that only the population with access to the relevant service is used in the calculations. On Fur, for example, the number of service establishments declined from 30 in 1950 to 19 in 1975. The number of grocers and butchers was halved, seven traditional trade services were lost and were replaced by five new trades reflecting more general use of motor vehicles and electrical appliances, and an increase in summer visitors (see the paper by De Haard in this book).

The situation in the public service sector is similar. Only Anholt and Fur have schools offering education to the end of the eighth school year and nine islands cater for a seventh year. The rest have either no school or a small, single-teacher primary school with secondary education provided on the mainland. Only 23 islands have a church, but fewer than half are served by an island-based priest. Nine islands have resident doctors but all lack a dispensing chemist and any provision for old people. The poorer educational and health facilities on the islands are important factors in migration. The level of elementary services is generally high considering the population base, a situation unlikely to endure for long if present migration trends continue.

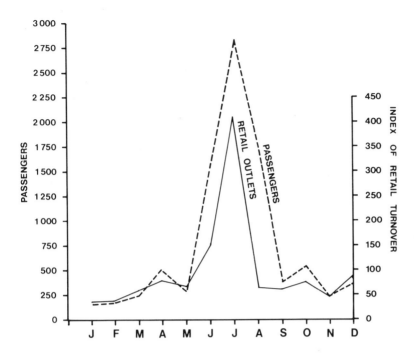

Figure 16.2   Ferry traffic and retail turnover on Anholt, 1970-75.
The retail turnover figures relate to the co-operative
retail outlet only

THE ISLAND FERRY SERVICES

Communication with the outside world is a problem shared by all
island communities. Only 33 islands are served by a daily ferry,
so without a regular ferry service or all-weather link with the
mainland, many islands are destined to remain backwater areas unless
isolation is a prerequisite for development. The smallest inhabited
islands rely on irregular sailings of locally-owned fishing or
supply vessels.

Transport costs represent an additional burden for the island
communities, the main exception being the freight pool scheme
operated by co-operative undertakings. Generally costs are
increased by 10-20 per cent for many agricultural raw materials
compared with the mainland (Ø-Posten, 1974 no.5). Retail prices are
not influenced to the same degree by freight charges due to the co-
operative policy of maintaining standard prices nationally. In
addition to the economic burden of transport costs, several island
communities live under a constant threat that the ferry might be
withdrawn. Those to Fur, Orø, Anholt, Askø and Livø receive considerable public subsidy, but there is evidence that an increase in
the extent and geographical limits of this form of support will be
required in future.

Summer timetables normally provide a more frequent ferry service, especially in holiday periods. The impact of summer visitors
on both the ferry and retail outlets is clearly seen in Figure 16.2.

It is apparent that the economics of island ferry services and
retail co-operatives would be seriously undermined were it not for
seasonal traffic. The continuation of ferry services is also
dependent on the nature of the harbour facilities provided on the
islands. Ten islands have harbours of a high technical standard,
with at least 200 m of quay. Among these are fishing harbours such
as Anholt and purely traffic harbours on Femø and Tunø. The five
largest islands, half those in the 200-400 population range, and
only Omø, Tunø and Christiansø of the smaller islands have harbours
of an acceptable standard. Harbour ownership varies and most
receive subsidy, but many of the smaller harbours are in need of
modernisation and repair.

THE FUTURE OF THE ISLANDS

The future prospects of the small islands vary considerably with
island type, population levels and the resources available for
development. The institutionally owned islands, those with nature
reserves and islands of strategic value for shipping will
undoubtedly continue to support small populations. Generally, only
the larger islands possess any hope of continued viability.
Increasingly, the islands are looking to tourism to sustain the
community by increasing employment opportunities, albeit seasonally,
to maintain existing service levels and to safeguard ferry routes.
In most cases, existing conditions will not permit the development
of tourism unless it takes the form of second homes or is based on
camping, sailing or day visitors.

The islands of the Kattegat, Limfjord, Belts and Wadden Sea have
the best tourist potential in that they possess good quality
beaches and major land-use elements with a recreational value. However, other factors affecting tourist developments require consideration. The islands most suited to recreational use contain sites
of scientific value, nature reserves and conservation areas.
Planning regulations place further restrictions on potential
development and the area affected by them often accounts for a
large proportion of the available land on an island. A further
problem relates to water supply. A report prepared by Danmarks
Geologiske Undersøgelse has shown that possibilities for increased
water extraction are very limited on most small islands. Many are
situated in low rainfall areas and underground reservoirs are
small, but the danger of saltwater penetration with increased
utilisation constitutes the most serious problem due to the close
proximity of the sea and the low level of the water-table. Since
most islands rely on well water, there is also a danger of pollution
from fertilisers, septic tanks and waste water.

A more concentrated development of tourism, with its pronounced
seasonal peaks of load and demand, will first require improvements
in water supply and sewage disposal. Apart from the investment
involved, there are limited possibilities for accomplishing this on
many of the islands. Good quality water is present on Orø, Arø,
Venø, Fur, Femø and Hjarnø, but the best potential for increasing
the supply occurs on Orø, Avernakø, Lyø, Arø, Anholt, Drejø and
Venø. Orø, Arø and Venø are therefore the only islands where both
qualities coincide.

Anholt, Fur and Orø have the main concentrations of second
homes at present, but Anholt, Sejrø, Askø, Fejø, Omø and Femø have
sites planned for a total of 2800 second homes, twice the total on
all islands in 1970. The regulations pertaining to the transfer of
land from agricultural to other uses mean that it is difficult to
acquire a steading for use as a second home. Most second homes

are therefore additional to the housing stock and, since they are regulated by modern building regulations, can be used to justify expenditure on improvements to infrastructure, services and public utilities by local authorities. Contrary to British experience, there is little resentment between incomers and island inhabitants over the second-home issue (see the paper by Clark in this book).

The number of pleasure craft is increasing sharply, although the level of boat ownership in Denmark is much lower than in Sweden. The yachting season is short and during July most island harbours are full. An increase in German and Swedish visitors will lengthen the season, as their holiday period follows the Danish season. Anholt and Tunø are especially well placed to provide facilities for the sailing fraternity. Eight islands have camping sites, and seven possess a hotel, restaurant or cafeteria, and others are planned on the larger islands.

It has been demonstrated that tourism can provide the means for stabilising the small island economies, whereas it is doubtful that the situation relating to agriculture and fishing could be improved even with considerable public subsidy. A systematic development of tourist facilities will also demand considerable public investment to provide the necessary infrastructure - access roads, sewage disposal, water supply and other utilities. Despite its seasonality, it is believed that tourism will result in improved living standards on the small islands and thereby remove the threat of depopulation. Only the largest islands have sufficient population resources and service potential to provide a basis for the more concentrated forms of tourism such as second-home developments. It is also claimed that only second-home concentrations will generate sufficient resources to assist the resident population by underwriting services.

One major problem remains. The ideal of maintaining viable island communities must not be allowed to change the essential character of the island environment. Similarly, the expectations of islanders and visitors must be fulfilled. Most islanders perceive the islands as 'their own little world'. They see the scenery, small dimensions, and bracing environment as the main advantages of island life. Very few are 'islanders at any price' and the majority cite the problems relating to the ferry, employment and lack of medical facilities as disadvantages which could cause them to consider leaving. To the visitor, the islands present picturesque, idyllic and relaxing settings, far removed from the strains and stresses of everyday life. Island environments must therefore be safeguarded since it is the basic attraction for resident and visitor alike.

A questionnaire survey, carried out on islands south of Funen and on Anholt in 1976, indicated that the islanders had a marked preference for second homes and longer-stay visitors. These are considered to be less intrusive socially and less disruptive in environmental terms. No doubt this feeling has developed since many islands offer little scope for large-scale developments, so the impact of tourism on their closed communities would be minimal. A common fear is that the islands might be viewed as living museums, an outcome islanders believe would be encouraged by tourist developments directed more towards day visitors.

It is clear that tourism, if developed sympathetically, could help stem outward migration and provide a means of maintaining viable island communities. Specialist forms of agriculture and fish farming present other possibilities, but their impact would be more limited. It is possible also that land reclamation schemes might, in future, improve links with the mainland, a fact reinforced by the nearness of many islands to mainland coasts. Positive results have been achieved by projects on Rømø and Alrø, while the road link between Funen and Langeland has improved the position of the intervening islands. In the case of such large-scale projects, much will depend on planning strategies adopted at antional level. In the meantime, tourism appears to hold the key to success or failure for many small islands and their inhabitants.

REFERENCES

Det sydfynske ø-udvalg 1972. *ø-problemmer i det sydfynske område.* (Odense).

Engell, B. 1977. Befolknings- og erhvervsudviklingen på Rømø. *Geografisk Tidsskrift,* 76, 63-67.

Friis, A. 1926-28. *De danskes øer.* (Gads Forlag, Copenhagen).

Grenå kommune 1968. *Dispositionplan for Anholt.*

Jacobsen, N.K. 1953. Mandø: en Klit-markskø i Vadehavet. *Geografisk Tidsskrift,* 52, 134-146.

Jacobsen, N.K. 1977a. Rekreation i Vadehavet. *Geografisk Tidsskrift,* 76, 52-58.

Jacobsen, N.K. 1977b. Rømø: naturvurdering på geografisk-økologisk basis. *Geografisk Tidsskrift,* 76, 69-62.

Landskabsplanudvalgets sekretariat 1966. *Strandkvalitet og fritids-bebyggelse.* (Copenhagen).

Ministeriet for kulturelle anliggender 1970. *Danmarks små øer.* (Copenhagen).

Mørch, H. 1975. Population and resources on the minor Danish islands 1860, 1900 and 1960. *Geografisk Tidsskrift,* 74, 21-35.

Ravnsborg kommune 1973. *Dispositionplan 1973-85.*

Sammenslutningen af danske småøer, Ø-Posten, no.1 (1974) proceeding.

Schrøder, N. and Øbro, H. 1976. A ground water model for the island of Anholt, Denmark. *Nordic Hydrology,* 7, 281-92.

Struer kommune 1973. *Dispositionsplan for Venø.*

Wessing, H.C. 1974. *Planlaegning for Mandø.* (Ribe).

Chapter 17

A comparison between objective and subjective measures of quality of life

Jaap Gall

INTRODUCTION

In 1980 the Sociaal en Cultureel Planbureau of the Netherlands published a study of the ranking of Dutch neighbourhoods and communities with respect to 'social disadvantage'. Social disadvantage was defined as a below average share in the distribution of scarce social goods within a given neighbourhood or community. As such, social disadvantage is not only an indicator of standard of living, but also of quality of life or well-being (cf. van de Lustgraaf and Huigsloot, 1979, 11).

This study is based on the Dutch census of 1971. Out of the data from this census, eleven variables indicating scarce social goods were selected. The original units of analysis were households or individuals, but the data were aggregated to the level of the neighbourhood, the unit of analysis in the study under discussion. Parenthetically, this implies that the relationships between the indicators are stronger than those at household or individual level (Robinson, 1951).

Three of these eleven indicators are about education, namely the proportion of working people having only primary education, the proportion of working people having been to grammar school at least, and the proportion of people over 14 at school who are at or above grammar school level. Two indicators concern occupations, one also measuring income, namely the proportion of breadwinners who belong to the group of 'higher employees', and the proportion of bread-winners who are blue-collar workers with an income below £12 000 a year. Income is covered by two indicators, namely proportion of working breadwinners having an income below £8000 a year, and proportion of breadwinners living only on a state pension. Consumption is represented by two indicators, namely proportion of working people having a car, and proportion of breadwinners having a telephone. The last two indicators are about quality of housing, namely proportion of dwellings having a rent of less than £480 a year and proportion of dwellings without bath or shower. Although somewhat materialistic in nature, these aspects are regarded as indicating the more social aspects such as the possibility of development in employment, variation in leisure activities, mental and physical health, and political participation (Sociaal en Cultureel Planbureau, 1980, 9-12).

The eleven indicators were subjected to factor analysis. All indicators load highly on the first unrotated dimension (factor), with coefficients ranging from 0.59 to 0.93. In factor analysis the first unrotated dimension is the most explanatory (van de Geer, 1971, 135) and therefore the most general dimension. This dimension is considered as a measure of social disadvangage. All Dutch neighbourhoods may be graded according to their position on this dimension. The result is a gradation on the basis of an objective measure of social disadvantage.

The Sociaal en Cultureel Planbureau (1980, 6) has thought about the possibility of constructing a ranking by means of a subjective measure of social disadvantage. Because of three objections, this possibility was rejected. One objection was trivial: subjective data were not available. The other two objections were more fundamental.

According to the Bureau, social and cultural policy cannot restrict itself to those people who subjectively experience social disadvantage. In one respect this is correct. Sometimes policy has to ignore subjective differences: some people do not like washing themselves, but all houses should have sufficient equipment to allow people to wash if they wish. In another respect it is wrong. One can just as well argue that a good social and cultural policy is not possible without taking into account the subjective experiences of social disadvantage.

The third objection is that subjective evaluations of living conditions are not stable. I do not agree with this objection. Although subjective evaluations do change with time, they are rarely excessively variable and may even be quite stable (Bradburn, 1969, 43 and 79; Robinson and Shaver, 1978, chapter 2). Secondly, the change can often be related to objective changes in living conditions (Campbell, Converse and Rodgers, 1976, 207). Thirdly, even when these evaluations are unstable, this instability will be much less after aggregation to the level of the neighbourhood.

I think these arguments support the necessity and possibility of constructing a ranking based on subjective measures of quality of life, not as a replacement for the objective one, but as an addition to it. The subjective ranking can be considered as an important external criterion of the objective ranking, i.e. the validity of the objective ranking can be assessed by comparing it with the subjective ranking (cf. van de Lustgraaf and Huigsloot, 1979, part III). Such a comparison is made in this article.

The subjective ranking can be constructed by means of the subjective measures of quality of life that were used in two studies of Dutch villages (Gall *et al.*, 1977; Gall, 1978). In both studies we tried to assess the 'liveability' of villages as experienced by the inhabitants. The first study covered six small villages in the eastern part of the province of Zuid-Holland, namely Amerstol, Hoornaar, Kedichem, Leerbroek, Noordeloos and Rijnsaterwoude (see Figure 1.1). In each village, a representative sample of fifty inhabitants was interviewed. Two subjective measures of quality of life were used, a direct assessment of the liveability of the village and a measure of psychosomatic feelings of stress like headache and stomach ache (taken from Dirken, 1967). Both measurements are related to specific subjective indicators, mainly of a social nature. One third of the variance of the measure of liveability is explained by these indicators, especially by the quality and quantity of contacts with local friends and neighbours. About one quarter of the variance of the measurement of stress (26 per cent)

is explained by the involvement in local politics (the more inducement, the less stress), by the social help one could receive and
give (the more help, the less stress) and by the quality of the
physical environment (the better the quality, the less stress).

The second study was directed at twelve villages in the southern
part of Zuid-Holland, namely Achthuizen, Dirksland, Goedereede,
Middelharnis, Ouddorp and Stellendam on the former island of
Goeree-Overflakkee, and Goudswaard, Klaaswaal, Oud-Beijerland,
Piershol, Strijen en Zuid-Beijerland, all located in the Hoeksche
Waard (see Figure 1.1). In each village a representative sample of
45 inhabitants was interviewed. Again, two subjective measures of
quality of life were applied, a direct assessment of the liveability
of the village, similar but not identical to the one used in the
first study, and a measurement of feelings of well-being, both
positive and negative, e.g. feelings of happiness, of satisfaction
with life, of depression and worries (Hermans and Tak-van der Ven,
1973). About forty percent of the variance of both measurements was
explained specific subjective indicators, in particular by the quantity and quality of contacts with local friends and neighbours and by
the quality of the physical environment.

Before the various measurements can be compared, some
difficulties should be mentioned. Firstly, the ranking is possible
only with respect to the six and twelve villages presented above.
This means a restriction of the objective ranking, which covers all
Dutch neighbourhoods in all municipalities.

Secondly, the measurements differ in the time of the survey.
The census took place in 1971, our studies in 1977 and 1978. In the
comparison in which our second study is involved this time gap can
be partly corrected. Only the data concerning those people who
were born in or near the village were taken into account. Most
lived in the same village in 1971 and 1978. The remainder were newcomers. Moreover this problem is not very serious, because we are
not comparing absolute differences between 1971 and 1977-78.

Thirdly, the Sociaal en Cultureel Planbureau did not analyse
neighbourhoods with less than 1000 inhabitants situated within the
same municipality. These neighbourhoods were taken together by
means of hierarchical cluster analysis, which may be correct
statistically, but certainly is not correct in practice. A lot of
rural communities in the Netherlands consist of several villages
with frequently very different characteristics. In the aggregation
no attention was paid to this fact. This resulted in forced
combinations between parts of one village and parts of another. For
example, the new part of Dirksland, in its region an important
village where the local hospital is situated, was sampled along with
Herkingen, a small village on a minor road. When a village was
divided into several parts, each part with or without part of
another village, I have weighted the score of each part on the basis
of its number of inhabitants and averaged the score.

Fourthly, the present studies used restricted samples, 45-50
inhabitants per village, while almost the complete population
participated in the census. The results are therefore less reliable
than those for the census.

Spearman rank correlation coefficients were computed between social disadvantage, liveability and stress/well-being with the village as the unit of analysis (Tables 17.1 and 17.2). Although all correlations are in the right direction, i.e. more social disadvantage is accompanied by more stress, less liveability and less well-being, only the correlations between liveability and stress/well-being are significant on the 10 per cent level.

Table 17.1   Spearman correlation coefficients between social disadvantage (based on Sociaal en Cultureel Planbureau, 1980) and stress and liveability (based on Gall *et al.* 1977).

	Measure of stress	Measure of liveability
Social disadvantage	0.49	-0.60
	(0.164)	(0.104)
Measure of stress	coefficient	0.77
	(level of one-tailed significance)	(0.036)

Table 17.2   Spearman correlation coefficients between social disadvantage (based on Sociaal en Cultureel Planbureau, 1980) and well-being and liveability (based on Gall, 1978).

	Measure of well-being	Measure of liveability
Social disadvantage	-0.25	-0.26
	(0.215)	(0.208)
Measure of well-being	coefficient	0.48
	(level of one-tailed significance)	(0.059)

How can the weak relation between social disadvantage and liveability/stress/well-being be explained? I suppose only part of it is due to the previously mentioned difficulties. The rest could be explained by a real difference between social disadvantage and the subjective assessments. Although I cannot assume that the subjective measures are perfect, I direct my criticism at the measurement of social disadvantage.

Two somewhat related objections against this measurement can be formulated. The content of social disadvantage is too restricted and the gradation of neighbourhoods and communities does not pay attention to ecological factors. These objections need some amplification.

As said before, social disadvantage is related to (lack of) well-being. The stress theory of Howard and Scott (1965; Scott and Howard, 1970; see also Sociaal en Cultureel Planbureau, 1978, appendix II) can therefore be applied to social disadvantage. They consider human behaviour as the processing and solving of problems. To solve problems a wide variety of resources is more or less available. These resources range from the general (intelligence or important social contacts) to the specific (tools or particular skills). The operationalisation of social disadvantage does not cover this variety sufficiently. In particular the social resources of quality of life, are represented only partly and indirectly by this measure, while economic resources get much attention. From this viewpoint the assessment of social disadvantage is incomplete.

Furthermore, the comparison of social disadvantage between neighbourhoods does not take into account important ecological factors. If the measurement of social disadvantage were more complete, this would be no problem. Now, however, the comparison is distorted as may be shown.

Gall (1978) concluded that inhabitants of small villages and villages which have the physical shape of a circle, have more and better social contacts than inhabitants of larger villages and villages which take the form of ribbon development. These social contacts affect the subjective evaluations.

In peripheral areas housing costs less than in central areas (see Ostendorf's paper in this book). In our second study, the Hoeksche Waard is a less peripheral area than Goeree-Overflakkee. One indicator of the quality of housing in the social disadvantage study is the cost of housing. The result is a systematic distortion of the quality of housing in favour of the villages in the Hoeksche Waard.

Low socio-economic status of the population is chosen as a component of social disadvantage partly because it is an indicator of low political participation (Sociaal en Cultureel Planbureau, 1975, 195). Although the original inhabitants of rural areas have a relatively low status, their voting percentage, another important indicator of political participation, is higher than that of the urban population (Heunks, 1973, 310).

In all cases rural areas and smaller communities are discriminated against in comparison with urban areas and larger communities. The best way of resolving this problem of discrimination is by taking a wider range of indicators of social disadvantage, for instance adding indicators on social contacts and political participation (voting or not voting) and by taking a different indicator of quality of housing, e.g. the ratio of rooms to persons. With such extensions and changes, the influence of the ecological factors can be detected.

My conclusion is that the assessment of social disadvantage in its present form is not suited to measuring social disadvantage as defined above. It gives too fragmentary and rough an indication of social disadvantage to reach solid conclusions on which policy can be made. A further theoretical and practical elaboration of the assessment is necessary, especially with regard to the social aspects of life and to the possible influences of ecological factors. This elaboration can only be successful when the measurement of social disadvantage is compared continuously with subjective measures of the quality of life.

REFERENCES

Bradburn, N.M. 1969. *The structure of psychological well-being.* (Aldine, Chicago).

Campbell, A., Converse, P.E. and Rodgers, W.L. 1976. *The quality of American life.* (Russell Sage, New York).

Dirken, J.M. 1967. *Het meten van 'stress' in industriele situaties.* (J.B. Wolters, Groningen).

Gall, J.C., Midden, C.J.H., Nederhof, A.J., Ritsema, B.S.M. and Swelm-Duijndam, J.Th. van, 1977. *Leefbaarheid in kleine kernen.* (Rijksuniversiteit, Leiden).

Gall, J.C. 1978. *Dorpsleven.* (Rijksuniversiteit, Leiden).

Geer, J.P. van de, 1971. *Introduction to multivariate analysis for the social sciences.* (W.H. Freeman, San Francisco).

Hermans, H.J.M. and Tak-van der Ven, J.C.M. 1973. Bestaat er een dimensie positief innerlijk welbevinden? *Nederlands Tijdschrift voor de Psychologie,* 28, 731-754.

Heunks, F.J. 1973. *Alienatie en stemgedrag.* (Katholieke Hogeschool, Tilburg).

Howard, A. and Scott, R.A. 1965. A proposed framework for the analysis of stress in the human organism. *Behavioral Science,* 10(2), 141-160.

Lustgraaf, R.E. van de and Huigsloot, P.C.M. 1979. *Sociale indicatoren, een bewste keuze?* (Sociaal en Cultureel Planbureau, Rijswijk).

Robinson, J.P. and Shaver, P.R. 1978. *Measures of social psychological attitudes.* (Institute for Social Research, Ann Arbor).

Robinson, W.S. 1951. Ecological correlations and the behavior of individuals. *American Sociological Review,* 16, 812-818.

Scott, R.A. and Howard, A. 1970. Models of stress. in *Social Stress,* Levine, S. and Scotch, N.A. (eds), (Aldine Publishing Company, Chicago).

Sociaal en Cultureel Planbureau, 1975. *Sociaal en cultureel rapport 1974.* (Staatsuitgeverij, 's Gravenhage).

Sociaal en Cultureel Planbureau, 1978. *Sociaal en cultureel rapport 1978.* (Staatsuitgeverij, 's Gravenhage).

Sociaal en Cultureel Planbureau, 1980. *Sociale achterstand in wijken en gemeenten.* (Sociaal en Cultureel Planbureau, Rijswijk).

SECTION THREE:

RURAL PLANNING

Chapter 18

Planning small villages in the Netherlands:

a comparative approach

Herman Kiestra

INTRODUCTION

At the end of the Second World War the Netherlands was an impover-
ished and destitute country. After the liberation, the restoration
of employment and a destroyed infrastructure, and the rebuilding of
houses in the stricken areas received full attention from the
authorities. Initially priority was given to economic planning but
in the course of time physical planning gained in importance.

In the fifties, many Industrialisation reports were produced.
The Government Department for the National Plan (now called the
Government Physical Planning Department) directed its main efforts
at that period to the three provinces in the west of the Netherlands
(Noord-Holland, Zuid-Holland and Utrecht) with its large 'Randstad'
conurbation, while in the sixties more attention was paid to the
rest of the Netherlands.

For those parts of the Netherlands where developments were less
dynamic or where the growth of employment even stagnated (e.g.
farming areas), the provincial planners in the sixties were mainly
engaged in drawing up village and agrarian welfare plans. This is
exemplified by the interim report *Spatial development of the Frisian
countryside* (P.P.D. in Friesland, 1962), and the second version with
the same title which appeared in 1966. The Frisian village plan
was complemented by a structural outline sketch for the province,
entitled *Friesland in 2000* (P.P.D. in Friesland, 1966).

In Gelderland structural outline sketches (called 'concept outlines') were drawn up for Rivierenland, Veluwe and Eastern Gelderland. The structural outline sketch *Friesland in 2000* and the three
Guelders concept outlines were published in 1966 and served as the
basis for a dialogue with the national government about the *Second
Report on Physical Planning in the Netherlands* (M.V.R.O., 1966).

Key publications, from which applied research in the field of
rural physical planning derived much benefit, were the theses of
Constandse (1960) and Groot (1972). Constandse carried out comprehensive field work in the Noordoostpolder. His study is important
for its theoretical content and its use and elaboration of the
'liveability' concept. Groot commenced his field work in 1963. The
investigation was carried out in part of the Frisian clay/grassland
area, the northern part of the Achterhoek, the agrarian central
area of the Alblasserwaard and the north-eastern part of Noord-

Brabant. This extensive field work was commissioned in 1961 by the Government Department for the National Plan. The results became available in 1968 with the publication of the preliminary report *Scale enlargement and village ties* and with his thesis *Small rural centres in Dutch society* (Groot, 1972). Groot also elaborated the 'liveability' concept and made it operational. Later, the same concept was frequently used in many comprehensive (notably official) literature studies. In this category falls also the report *If you don't value small things*... (V.N.G., 1979), which reflected the views of the municipalities.

However, Constandse and Groot could not foresee the changes that were to have such a great impact on Dutch society in the sixties. By the mid-sixties car ownership had become widespread in the Netherlands. The car was a major factor in the design of the settlement pattern envisaged for the Oostelijk Flevoland polder by Constandse's employer - the Government Department for the IJsselmeer polders. They decided to build three villages instead of the five foreseen by Constandse and which were still to be found in the blue-prints for Oostelijk Flevoland as late as 1959.

The great prosperity enjoyed in the Netherlands since the mid-sixties also made itself felt in the small villages. Much money was spent on the building of community centres and the construction of sports facilities, so that village life became very attractive for many people throughout the Netherlands (suburbanisation). There-fore, commuting has become much more common, both absolutely and relatively. Hence the rather gloomy picture given by Groot in his thesis is not in agreement with the present situation, although for some years now we have been planning for a decline in village facilities because of the general economic downturn.

Concentration of population in larger settlements with a greater chance of attracting industry is a planning tactic that is still used frequently. However, in the Netherlands, which is so densely populated, there still are very large areas with a typically agrarian character and a settlement pattern of numerous small villages.

For a more thorough analysis, two areas have been selected; a part of Friesland and the regional planning area of Rivierenland in Gelderland. Both have many small rural centres and the provincial governments' views concerning physical planning are somewhat divergent. Three questions form the starting point for this investigation:
1) How has the population developed in the two study areas since 1945?
2) Which planning goals have been set for the small villages in the regional plans?
3) What are the provincial governments' policies regarding facili-ties (including governmental and municipal policies)?

VILLAGES AND SERVICES

On February 28th 1971, the Frisian study area (including the small towns of Bolsward, Franeker and Harlingen) had a population of 85 335. Of this number 10 350 (12 per cent) were dispersed in the smallest settlements over the area. On the same day 151 425 people were living in the Guelders area and the number of people dispersed over the area was 31 350 (21 per cent), including a small number living in caravans, house boats and barges. So the dispersed popu-lation of Guelders Rivierenland was considerably greater than that in the Frisian-clay study area (Figure 18.1).

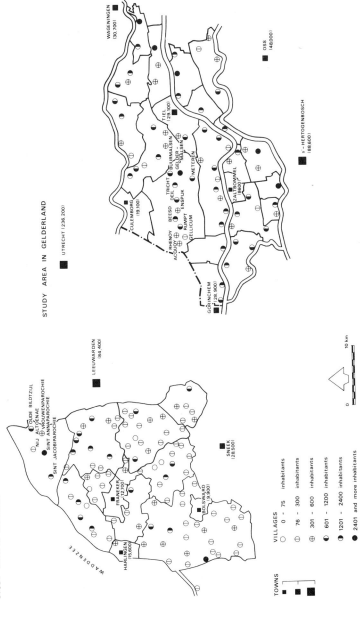

Figure 18.1 Study areas

All eleven Dutch provinces have their own definitions of a village and attempts by the Government Physical Planning Department to arrive at a uniform definition have been unsuccessful. A few years ago the province of Gelderland revised its list of villages, and the following criteria were used (Kiestra, 1978):
i) the presence of more than 25 dwellings that form a unit;
ii) the unit should have a name;
iii) the presence of two of these facilities:
    a school (primary or nursery school)
    a community centre or public building with hall accommodation
    a shop for daily needs
    a sportsfield.

The province of Friesland considers as villages settlements which are very small and do not meet the Gelderland criteria. According to the Guelders criteria, a village should have at least 25 dwellings (that is, 75 inhabitants). The range of services for a village, the so-called local service level, is considered reasonable if the following facilities are available:
a) schools - nursery school, normal primary school, secondary modern school;
b) sports - two sports fields, hall, tennis court;
c) medical and social facilities - full-time general practitioner, full-time dentist, health centre, and old people's home;
d) cultural facilities - community centre of at least one public building with at least one large hall and two or more smaller rooms, public library;
e) shops - shops for primary needs with a score of 41 or more as defined in *Service structure in central Gelderland,* (P.P.D. van Gelderland, 1974).

Table 18.1   Number of villages in size classes, 1971

| Population | number of villages | | | |
| | Friesland | | Gelderland | |
category	abs.	%	abs.	%
Under 76	5	5	none	none
76-300	54	55	7	10
301-600	18	18	20	28
601-1200	11	11	21	30
1201-2400	9	9	17	24
2401 and over	2	2	6	8
	99	100	71	100

Source: C.B.S., Census 1971

A survey of villages classified according to size in 1971 (Table 18.1) shows that those with a population up to 300 formed the majority in the Frisian study area, whereas this category is a small minority in the Guelders area. Large differences between the areas occur in the categories with a population of over 601. In Rivierenland villages in 1974 services were strongly concentrated in villages falling in the 1201-2400 category. This leads to the assumption that the local service level will be poorer in the Frisian area than in the Guelders area, although the areas are of

174

more or less the same size. A clear relationship between the population of a village and its service level has already been demonstrated in Gelderland (Kiestra, 1978).

POSTWAR POPULATION DEVELOPMENT

Another striking feature is the great difference in population development between the two areas. The Frisian area comprises farmland, two small towns (Bolsward and Franeker) and the small seaport of Harlingen. Farming villages in Friesland showed out-migration which changed after 1971 into a slight growth for most of the agrarian municipalities. This development closely resembled the growth pattern for Friesland; population loss before 1971, then population gain (Table 18.2).

Table 18.2   Development of population in the two study areas by degree of urbanisation

Frisian study area

| | inhabitants | | | |
| | abs. | index | | |
	1950	1960	1971	1981
rural municipalities	61256	93	88	93
towns	27738	107	115	138
Frisian study area - total	88994	97	96	107
Guelders study area				
Middle Betuwe	17433	111	135	155
West Betuwe: rural part	32218	105	124	139
West Betuwe: towns	25214	120	149	191
West Betuwe: total	57432	111	135	162
Bommelerwaard: rural part	20320	122	137	155
Bommelerwaard: Zaltbommel	4933	121	145	173
Bommelerwaard: total	25253	121	138	158
Land van Maas en Waal	12710	109	121	132
Guelders Rivierenland - total	112828	114	134	158

The development of the population in the Guelders Rivierenland differs from the Frisian study area in that the postwar migration losses change into migration gains in 1962. Moreover, the 1960 census shows that Rivierenland is an area with much long-distance commuting, mainly to the Rijnmond area (Kiestra, 1967). Cars have been replacing public transport for home/work travel since the mid-sixties. As a result commuting became more attractive and the number of immigrant commuters rose substantially in rural municipalities as well as in the towns.

OBJECTIVES FOR THE SMALL VILLAGES IN THE REGIONAL PLANS

When the *Second report on physical planning in the Netherlands* was prepared, the provinces of Groningen, Friesland and Drente voiced their preference for enhanced population growth and this preference was taken into account in the report. However, population forecasts were soon being lowered because of reduced birth rates. Friesland was favoured in this respect because the shift in birth rates made it the province with the highest birth rate. Combined with net in-migration, this made the provincial government of Friesland adopt a more sympathetic attitude towards population growth in the Frisian villages.

In drawing up the Frisian regional plan various alternatives were found.
- Alternative 1 is 'directed outwards': 'Friesland should offer space for an overspill of population and activities from the rest of the country'.
- Alternative 2 is 'directed inwards': 'Friesland should remain an open, relatively thinly populated, green province'.

From the report *Starting points and basic decisions* (Provinciale Staten van Friesland, 1979) it is evident that the Provincial States favoured alternative 2. The report observed, 'In any case the natural growth, estimated at 42 000 for 1980-1995, must be absorbed. However, for the time being, a positive migration balance will continue to exist. There are no objections to this provided that migration is limited and does not exert a negative influence on the composition of the population nor on the labour market'. Since the publication of the *Second report on physical planning in the Nether-lands* (M.V.R.O., 1966), Frisian planning policy has changed drasti-cally. Much lower population growth is now considered desirable, while it is recognised that the earlier population losses in the villages must be converted into slight growth. On the basis of consultation and citizen involvement, the following population figures were established for the 'A' category municipalities in the study area (Table 18.3). In addition, population was often allocated to villages. The criterion was that population growth would mainly be concentrated in villages with schools.

Table 18.3   Population of the municipalities in the Frisian study area in 1980 and planning goals for 1990

municipality	population 1980	planning goal, 1990 (index, 1980=100)
Baarderadeel	5044	108
Barradeel	6657	110
Het Bildt	8348	104
Franekeradeel	3883	107
Hennaarderadeel	4835	109
Menaldumadeel	12934	107
Rauwerderhem	2859	113
Wonseradeel	11870	107

For Rivierenland one main objective was chosen. 'Starting from the retention of open space, special attention should be paid to improvement of the socio-economic structure and to welfare'. The report *Alternative Preliminary Drafts,* (P.P.D. van Gelderland, 1974) was published with two alternatives.

Alternative 1

Here emphasis was placed on retaining local community relations and reinforcing local services. The interests of the local community were best served by fulfilling the housing needs arising from natural population growth and with a slight growth in the whole Rivierenland area because of net in-migration. This also has a favourable influence on the social climate in that this growth will contribute to a better balanced age structure and a stronger social life, while some in-migration will reduce social tensions generated by an unbalanced age structure. A balanced population also means that the basis for service facilities in the small villages is enlarged. A number of centres have been selected for additional growth to widen their facilities.

Alternative 2

Here the central focus is improvement of the employment situation and of regional services. Attempts were made to create such planning conditions that a stimulus would be given to employment and the accessibility of places of employment would be improved. It has been found that there is a relationship between the development of modern industry and urban environments. Hence, a concentration of population growth will create the best conditions for a quantitative and qualitative growth of the service sector.

Furthermore, population growth in larger settlements may be expected to have a positive influence on the development of the industrial sector. Therefore, it is proposed to let the population grow substantially in excess of natural growth. A considerable increase of population in the larger centres will reinforce regional service provision.

The next step in the planning process was the *Options Report* (Provinciale Staten van Gelderland, 1977) which attached most weight to creating conditions for the proper functioning of the local community. Furthermore, it observed that the functioning of the local community is largely determined by village ties. By providing housing to persons with local ties, the most important planning condition is fulfilled. The municipalities can indicate one or, if necessary, two villages where a wider interpretation can be given to the 'village ties' concept. In doing so due allowance should be made for such factors as nature and landscape, service provision and the agricultural situation. Thus the net out-migration in some villages can be offset. On these grounds a zero migration balance is taken into account in determining the number of houses to be built in each municipality. Therefore, the options report was based on a lower positive migration balance than the Preliminary Draft 1.

It was also considered desirable to improve the employment situation in Rivierenland for the local population. One of the conditions for achieving this is a concentration of population in the larger centres. Too great a dispersal over the area was to be avoided in view of the lower net in-migration. The centres for growth were Tiel and, to a lesser extent, Zaltbommel and Culemborg. The report also stated that the Draft Regional Plan should further improve the accessibility of minimally required facilities so as to enhance community relations.

The reduced population targets in the Draft Regional Plan for Rivierenland in 1990, including the municipalities of Culemborg, Tiel and Zaltbommel, are 185 000 and 190 835, respectively (Table 18.4).

Table 18.4   Populations of the municipalities in the Guelders
              study area in 1980 and planning goals for 1990

municipality	population 1980	planning goal, 1990 (index, 1980=100)
Echteld	6193	106
Kesteren	8777	106
Lienden	6202	104
Maurik	5773	103
Buren	8905	107
Geldermalsen	20373	107
Neerijnen	10116	102
Herwijnen	2412	105
Vuren	2858	110
Ammerzoden	3963	104
Brakel	6137	106
Hedel	3443	108
Kerkwijk	5761	103
Maasdriel	7908	106
Rossum	2879	99
Heerewaarden	1120	101
Appeltern	4089	105
Dreumel	3156	105
Wamel	9422	109

FACILITIES IN SMALL VILLAGES

The provincial government of Friesland believed that in the coming
years population growth should be absorbed in those villages where
there were schools, whereas in Rivierenland much importance was
attached in the alternative preliminary drafts to reinforcing
either local or regional service provision. In the *Rural Areas
Report* (M.V.R.O., 1977), it was stated that the central government
was prepared to appoint a limited number of 'support centres' in
areas with net out-migration. In the Frisian study area this
involved nine villages. The central government would give such
villages priority in maintaining and providing facilities. For
this purpose measures may be taken such as:
- flexible application of standard pupil numbers for schools;
- providing proper transport schemes for pupils from smaller
  villages to the support centre;
- giving subsidies for social and cultural amenities;
- maintaining the public transport system.

   The *Description of the Preliminary Draft Regional Plan Friesland*
(Gedeputeerde Staten van Friesland, 1981) states that 'efforts will
be made to formulate and execute in the next five years a 'basic
development plan' for every support centre. If the financial
resources can be found, one or more co-ordinators will be

appointed to render assistance with the preparation, formulation and execution of the basic development plan'.

The Guelders Rivierenland did not qualify as a support area, because there has been net in-migration since 1962 in this area, so it does not meet the government's criteria. Nevertheless, service provision has been affected by structural scale enlargement. Therefore, the question is how far this trend can continue without exerting a detrimental effect on living conditions in the villages. To determine what minimum facilities should be provided in small rural centres, a Welfare Working Party was formed within the regional planning office. The following criteria were used in establishing the minimum facilities:

i) the facility should have a high participation rate and a high frequency of use;
ii) the facility should be one of those that one prefers to have in one's vicinity;
iii) the population should place high priority on the accessibility of the facility.

The Welfare Working Party considered the following nine facilities to be essential (Werkgroep Welzijn, 1978):

1. Shop for daily needs
2. Nursery school
3. Primary school
4. Hall for meetings
5. Sports field
6. General practitioner
7. Health centre
8. Post Office
9. Bus stop (link with transport system).

Investigations have been made into whether these facilities are actually available in the villages or in villages within a distance of three kilometres. The findings were summarised in the regional plan for Rivierenland (Provinciale Staten van Gelderland, 1980, 88-92). Certain facilities are lacking in twelve villages especially those with 75-300 and 301-600 inhabitants.

In one field government policy has led to tangible results, namely, the introduction of a 'district bus'. In 1977 five trial projects were undertaken and by the end of 1981 there were 51 district buses in the Netherlands. In the Frisian study area they were found in Ferwoude and Idsegahuizem, while in the Guelders area we find them in Est and Buurmalsen. A study of them by Kropman and Peters (1982) led to the following interesting findings:

i) almost half of the district bus users stated that the bus had enhanced their mobility. This especially applied to elderly people, women and children;
ii) there are two major uses of the bus - shopping and visiting relatives and friends in another region.

In the near future the provision of adequate educational facilities will form a bottleneck. Nursery and primary schools are to be integrated into a single type of school. The lower limit at which small primary schools will be closed has been reduced to 30 pupils and the integration should be completed by 1985 (see the paper by Tricker in this book). For many villages, the continued existence of their school will depend on birth rates in the coming years. In Table 18.5 the provision of schools is shown for the two study areas.

Table 18.5   Schools in the two study areas, June 30th 1982

Population category	Frisian area						Guelders area					
	Under 76	76 to 300	301 to 600	601 to 1200	1201 to 2400	2401 and over	Under 76	76 to 300	301 to 600	601 to 1200	1201 to 2400	2401 and over
	%	%	%	%	%	%	%	%	%	%	%	%
lacking nursery school	100	87	33	9	0	0	-	86	20	0	0	0
lacking primary school	100	52	6	18	0	0	-	29	10	0	0	0
primary school, with 1 or 2 teachers	0	33	39	0	0	0	-	29	20	0	0	0
primary school with 3 or more teachers	0	15	55	82	100	100	-	42	70	100	100	100

Since the Second World War there has been a huge scale enlargement of the retail trade throughout Western Europe. In the Frisian area there are 28 villages without shops against 4 in Rivierenland. Because of ribbon development and good roads, both areas can easily be served by mobile shops and there is a clear relationship between low population and the absence of shops. Since 1976 distribution planning research has been obligatory for regional and land-use plans. In both study areas research has been carried out. No conclusions will be given here, except for the observation that in many villages shopkeepers have additional sources of income which will undoubtedly prolong the existence of the shops in question.

In his thesis Groot (1972) also made a study of scale enlargement processes. This mainly affects shops and employment. Two typical examples of concentration of employment are the brick industry in Rivierenland and the dairy industry in the Frisian area. Another interesting case is that of the distribution of sports facilities. Just after the Second World War indoor pools and tennis courts were only found in the towns. The indoor swimming pool in Leeuwarden has long been the only one of this type in the province of Friesland. A breakdown of sports facilities in the two areas is given in Table 18.6.

Some interesting conclusions can be drawn from this table. The more favourable settlement pattern in Rivierenland is apparent from the average number of sports facilities per village - 3.35 in the Guelders area and 1.95 in the Frisian area. In addition, the table shows the importance of skating and 'kaatssport' in Friesland. In the Frisian area 25 of the 54 villages with a population between 75 and 300 in 1971 had a 'Kaats' club (Pietersen, 1956). This illustrates that villages can have a flourishing social life even if certain facilities are lacking.

Table 18.6   Provision of sports facilities in the two study areas, 1980

Sports facilities	number		average per village	
	F area	G area	F area	G area
Outdoors				
football fields	34	94	0.34	1.36
tennis courts	23	43	0.23	0.62
'kaats' fields	46	-	0.47	-
korf ball fields	7	4	0.07	0.06
ice rinks	35	13	0.36	0.19
horse riding	4	13	0.04	0.19
swimming pools	5	8	0.05	0.12
subtotal	154	175	1.57	2.54
Indoors				
gymnasium	27	46	0.28	0.67
gymnasium halls	-	2	-	0.03
sports halls	5	1	0.05	0.01
swimming pools	1	3	0.01	0.04
horse riding	4	4	0.04	0.06
subtotal	37	56	0.38	0.81
Total	191	231	1.95	3.35

Sources: CBS (1981)

MUNICIPAL PHYSICAL PLANNING

Het Bildt

On the basis of a discussion report (Gemeente Het Bildt, 1976) the physical planning problems of this area will be dealt with briefly. Within the municipality there is a distinct tendency to a concentration of population.

The figures in Table 18.7 do not show the strong decline of the population living on the Oude and Nieuwe Bildtdijk in houses built along the dike (ribbon development). The number of one- and two-room dwellings, the so-called dike dwellings, declined substantially. In 1966, there were more than 600 of such dwellings in use as permanent family dwellings; now there are only about 50. Many are second homes; in 1978 there were 143 second homes in the municipality of Het Bildt. Houses of this type are often unoccupied and many of them are offered for sale. On the Oude Bildtdijk a concentration of houses has occurred; around the Reformed Church the new village of Nij Altoenae has arisen. About 300 inhabitants are now living there and they are included in the population figure of Sint Annaparochie. The 'Plaatselijk Belang' (local interest

181

Table 18.7   Population distribution between villages in the
             municipality of Het Bildt in 1950 and 1982

villages	populations	
	January 1 1950	January 1 1982 1950 = 100
Sint Annaparochie	3649	125
Sint Jacobiparochie	3085	69
Vrouwenparochie	939	76
Oude Bildtzijl	1773	61
total	9446	90

group) has gone to much trouble to have new dwellings built there.
The regional facilities are also concentrated in the main centre
Sint Annaparochie. The village has a sports hall, an outdoor/
indoor swimming pool, a secondary modern school and a wide variety
of shops.

The *Village Plan* report (Gemeente Het Bildt, 1976) states that
'a gradual growth in population is desirable in line with the scale
enlargement which is still taking place'. It has already been
mentioned that according to governmental and provincial planning
Oude Bildtzijl is considered as a 'support centre'.

The policy recommendations of the *Village Plan* (Gemeente Het
Bildt, 1976) are as follows:
1. Sint Annaparochie is the central village for Het Bildt. During
   the past ten years this central function has been consolidated,
   and consequently further stimulation of population growth is not
   necessary.
2. In the next few years Sint Jacobiparochie should maintain its
   level of facilities. Population growth will be a prerequisite
   to keep up these facilities, which means that population growth
   should be equal to natural growth.
3. Vrouwenparochie has a number of basic facilities and a flourishing social life, both of which should be retained. Therefore,
   Vrouwenparochie should keep its natural growth.
4. Oude Bildtzijl should keep a level of facilities that befits a
   central village. This is only possible if the population shows
   reasonable growth in the future. This growth would comprise
   population growth in the village itself and the concentration of
   the village-oriented 'dike' population in Oude Bildtzijl.
5. Oude and Nieuwe Bildtdijk and Nij Altoenae; the population of
   Oude and Nieuwe Bildtdijk has shown the greatest decline owing
   to the demolition of dike dwellings and their sale as second
   dwellings. One of the greatest problems here is that the young
   people are leaving. They prefer to live in Sint Jacobiparochie,
   Nij Altoenae and Oude Bildtzijl and to a much lesser extent in
   Sint Annaparochie. This concentration will have to be taken into
   account by the municipality.

In view of the developments outlined above, house-building
targets have been set for the period up to and including 1985
(Table 18.8). The population estimate for the year 1990 is 8700;
this has been agreed upon by the municipal and provincial
authorities within the scope of the regional plan.

Table 18.8   House-building goals in Het Bildt villages for 1976-85

village	new houses	replacement houses	total
Sint Annaparochie	50	30	80
Sint Jacobiparochie	60	60	120
Vrouwenparochie	25	5	30
Oude Bildtzijl	20	20	40
Nij Altoenae	10	20	30
total	165	135	300

Geldermalsen

The new municipality of Geldermalsen was formed on January 1st 1978 from four old municipalities. It now comprises eleven villages which differ considerably in size; the municipality now distinguishes three subareas (Table 18.9).

Table 18.9   Population in the villages of the municipality of Geldermalsen on January 1st 1978

village	inhabitants		village	inhabitants	
Beesd	2815		Geldermalsen	8497	
Acquoy	615		Buurmalsen	1051	
Rhenoy	726		Meteren	1182	
Subtotal		4156	Tricht	2194	
Deil	1551		Subtotal		12924
Enspijk	590				
Gellicum	373				
Rumpt	849				
Subtotal		3363			
total					20443

Source: Gemeente Geldermalsen, 1982

Regarding facilities, the following observation is made in the report: 'The starting point for future development is to provide for the specific situation of a clearly distributed population. Efforts will be made to obtain an optimal distribution of village-level facilities, while higher-level facilities should be placed in the main centres of Geldermalsen or Beesd'. The minimum level of facilities that has already been discussed in the context of the regional plan for Rivierenland, has been translated by the municipality of Geldermalsen into its own specific situation (Table 18.10).

Table 18.10　　Minimum level of facilities for villages in
Geldermalsen according to the regional plan
for Rivierenland

village	facility									
	1	2	3	4	5	6	7	8	9	
Acquoy	M	3	3	x	3	3	3	M	x	1-shop
Buurmalsen	x	x	x	x	3	3	3	3	3	2-nursery school
										3-primary school
Enspijk	3	3	x	x	3	3	3	x	x	4-hall
Gellicum	M	3	x	3	3	3	3	M	3	5-sports field
										6-general practitioner
Meteren	x	x	x	x	x	3	3	x	x	7-health centre
Rhenoy	x	x	x	x	3	3	3	M	x	8-post office
										9-public transport
Rumpt	3	x	x	x	x	3	3	M	x	
Tricht	x	x	x	x	x	x	x	x	x	x-facility available
Beesd	x	x	x	x	x	x	x	x	x	3-facility available
Deil	x	x	x	x	3	x	x	x	x	within 3 km distance
										M-mobile facility
Geldermalsen	x	x	x	x	x	x	x	x	x	

It should be noted that the mobile facilities (library bus, post
office and shop) fulfil clearly felt needs (see the paper by Moseley
in this book). The report states that 'the regional bus between
Tricht, Buurmalsen, Geldermalsen and Meteren forms a welcome
addition to the public transport system'. For the main centre 'the
necessity of a sports hall is evident'. Furthermore, it is observed
that in the long run a socio-cultural centre at Geldermalsen should
replace older accommodation and also cater for juvenile welfare
work, rehearsal rooms for music, dance, drama and perhaps a cinema.
'A sports hall in Beesd can only be realized in the long term. A
second library in the municipality should be placed in Beesd. The
policy continues to be aimed at maintaining community facilities in
the central villages'.

As regards house building in the smaller villages, the report
states: 'The number of new houses is less than that calculated on
the basis of natural growth. Not everybody continues to live in
his own village and wants a house there'. On Gellicum the report
says: 'In order to provide houses for those who have ties with
Gellicum, additional dwellings would be required. The building
possibilities within the existing structure have been exhausted,
and so use should be made of the villages of Deil and Rhenoy'.
In contrast, more new houses than are needed for natural growth
will be built in the three largest villages (Beesd, Deil and
Geldermalsen) (Table 18.11).

Unlike the situation in Het Bildt, house-building in Geldermalsen
will be concentrated in the main centres for the next few years.

Table 18.11   House-building targets for Geldermalsen villages
in the period 1981-88

villages	number of houses	villages	number of houses
Beesd	137	subtotal	362
Acquoy	18	Geldermalsen	424
Rhenoy	32	Buurmalsen	43
Deil	105	Meteren	23
Enspijk	16	Tricht	66
Gellicum	15		
Rumpt	39		
subtotal	362	total	918

CONCLUSIONS

The problems of villages are receiving ever greater attention from physical planners. In the sixties the Government Physical Planning Department placed much emphasis on research (Groot, 1968; 1972), while in the seventies the publication of the *Rural Areas Report* also placed physical planning policy in the limelight.

At a provincial level, too, more attention was paid in the seventies to the problems of villages. In some provinces, including Friesland and Gelderland, regional plans were preceded by village and agrarian welfare plans. An important element in the village plans was the functions performed by the villages. The plans mentioned central and dependent villages; the service function of the village for its own population and for the population in the vicinity received much emphasis.

In the studies undertaken in the seventies attention was also paid to the liveability aspects of a village but other aspects (in-migrant commuters, for example) received perhaps even more attention. As far as citizen involvement and consultation regarding governmental, provincial and municipal reports are concerned, it should be observed that villages have always been a lively topic. From the public's reactions it has become clear that sufficient houses must be built in the villages to keep up the level of services.

In those regions in the Netherlands where village population declined in the past, major efforts will be made to ensure that the present growth of villages will persist. In those regions where villages grew much too quickly because of the large intake of commuters, a much more controlled growth will have to be taken into account. The study areas are typical of these two types of area.

REFERENCES

C.B.S. 1981. *Inventarisatie sportaccommodaties 1980.* (Den Haag)

Constandse, A.K. 1960. *Het Dorp in de IJsselmeerpolders.* (Zwolle).

Gedeputeerde Staten van Friesland, 1981. *Ontwerp-streekplan Friesland.* (Leeuwarden).

Gemeente Het Bildt, 1976. *Dorpenplan, discussienota*. (Het Bildt).

Gemeente Geldermalsen, 1982. *Structuurnota kernen-voorontwerp*. (Geldermalsen).

Groot, J.P. 1968. *Schaalvergroting en dorpsbinding; een onderzoek naar de houding van bewoners van agrarische dorpen tegenover de veranderende spreiding van bevolking en voorzieningen*. (Wageningen, Amsterdam).

Groot, J.P. 1972. *Kleine plattelandskernen in de Nederlandse samenleving; schaalvergroting en dorpsbinding*. (Wageningen).

Kiestra, H. 1967. De ontwikkeling van het werkforensisme inzonderheid naar afstand op de grote stedelijke centra in het westen des lands gedurende de periode 1947-1960. *Tijdschrift voor Economische en Sociale Geografie*, 57, 57-67.

Kiestra, H. 1978. Het verzorgingsniveau van de Gelderse dorpen. *Geografisch Tijdschrift*, 12, 317-325.

Kropman, J.A. and Peters, H.A.J. 1982. *De buurtbus:bijdrage aan de leefbaarheid in kleine kernen. Onderzoek in opdracht van het Provinciaal Bestuur van Gelderland*. (Instituut voor toegepaste sociologie, Nijmegen).

M.V.R.O. 1966. *Tweede Nota over de ruimtelijke ordening in Nederland*. (Den Haag).

M.V.R.O. 1977. *Nota Landelijke Gebieden*. Derde nota over de Ruimtelijke ordening, deel 3a, (Den Haag).

Pietersen, L. 1956. *Friesland in en rond het perk. Enige sociale en sportieve aspecten van het Friese kaatsspel*. (Leeuwarden).

PPD in Friesland, 1962. *De ruimtelijke ontwikkeling van het Friese platteland*. (Leeuwarden).

PPD in Friesland, 1966. *De ruimtelijke ontwikkeling van het Friese platteland*. (Leeuwarden).

PPD in Friesland, 1966. *Friesland in 2000*. (Leeuwarden).

PPD van Gelderland, 1974. *Nota Alternatieve Voorontwerpen, Verzorgingsstructuur in Midden-Gelderland*. (Arnhem).

Provinciale Staten van Friesland, 1979. *Streekplan Friesland grondslagen en basisbeslissingen*. (Leeuwarden).

Provinciale Staten van Gelderland, 1977. *Keuzenota*. (Arnhem).

Provinciale Staten van Gelderland, 1980. *Streekplan Rivierenland*. (Arnhem).

V.N.G. 1979. *Wie het kleine niet eert.... Discussienota opgesteld door de Studiecommissie Kleine Kernen van de Vereniging van Nederlandse Gemeenten*. (Den Haag).

Werkgroep Welzijn, 1978. *Nota Minimumvoorzieningen in de kleine kernen streekplangebied Rivierenland; een toetsing van uitgangspunten*. (Arnhem).

Chapter 19

Key-village and related settlement
policies in Scotland

Douglas Lockhart

Key settlement policies have been the principal agent of planned
change in post-war rural Britain (see the paper by De Bakker in this
book). Such policies have figured prominently in the literature of
rural geography in recent years (Gilg, 1978; Cloke, 1979; Blacksell
and Gilg, 1981), however most authors discuss settlements in England
and Wales, and only Woodruffe (1976) comments on Scottish examples.
The lack of research on Scottish rural settlement planning is also
evident in journals and textbooks and, while settlement studies are
numerous, almost all have concentrated on the historical evolution
of farms and villages. Exceptions such as Turnock's discussion
(1979) of regional planning in Scotland's outer regions are very
rare. The neglect of this topic can in part be explained by the
strong historical bias in teaching and research in Scottish universi-
ties during the 1950s and 60s and also by practical difficulties in
dealing with the literature. Few planning documents were deposited
in local libraries, a problem which has eased considerably since the
reorganisation of local governments in 1975. Perhap, however the
greatest discouragement to researchers was the nature of the source
material, namely county development plans and development plan
amendments, which were largely factual references to locations of
proposed developments which were shown on the county development map.
Such documents, which were the basis of planning until local government reorganisation, rarely discussed issues or suggested alternative
strategies.

THE EVOLUTION OF DEVELOPMENT PLANS

The evolution of settlement planning in Scotland in the post-war
period was broadly similar to that in England and Wales. The
Development Plans prepared by the county planning departments identified where future development could take place and were also a
means of safeguarding resources such as high-amenity areas, minerals
and good agricultural land. However, such plans gave very little
guidance on the future shape of rural settlements. A review of
Selkirk County Development Plan (1952) noted that 'although certain
planning matters relating to the Landward area are covered in general
fashion... there was no detailed analysis of the settlements in the
Yarrow and Ettrick valleys'. Nevertheless, development plans may be
used to investigate the degree of selection for growth. Two main
categories of plan can be distinguished. There are those which
direct new development into larger villages such as that for Angus
(1962) which advocated that building should take place in
development centres rather than on uncoordinated sites throughout

the countryside. Secondly, there are those plans which contained a
more clearly defined category of growth villages. Policies adopted
in East Lothian between 1947 and 1974 were based on a three-tier
system, the first tier, known as growth villages were defined as
locations 'where active development plans for both local authority
and speculative houses were under way or would be encouraged and
where a full range of commercial and community facilities would be
provided'. One such village was Gifford in the foothill zone of
the Lammermuir Hills, an area suffering from depopulation in most
villages in spite of the growth of tourism. Gifford was seen as a
'gathering point' for those living in the area as well as for summer
visitors.

The second stage in the evolution of post-war planning consisted
of reviews of the county development plans during the late 1960s and
early 1970s. At about the same time, a number of counties prepared
reports on major issues, including settlement planning, that had not
been discussed in detail in earlier plans. The influence of the
regional policies of central government should also be noted. In
the Borders, a White Paper (Scottish Office, 1966) proposed an
increase of 25 000 in the population by 1980. How this was to be
achieved was described in *The Central Borders* plan (Scottish Development Department, 1968) which contained proposals for the expansion
of existing towns and the creation of new communities at St Boswells
and Tweedbank. However, apart from the preparation of several non-
statutory village plans (St Boswells and Bowden, for example) the
rural settlement pattern was not examined until 1974 when Roxburgh
County Council published *The Landward Community Development Strategy*
which was intended to supplement *The Central Borders* plan. Rapidly
changing economic circumstances in other counties too were responsible for the production of settlement plans. For instance, in north-
east Aberdeenshire oil exploration was expected to have a wide-
ranging impact upon employment, housing and the overall distribution
of the population. As a result, three sets of proposals were put
forward between 1972 and 1974 for the settlement pattern of the
Buchan district. Such reports provide the first reasoned proposals
for rural settlement and include statements on overall policy and
individual settlements. Again, it is possible to distinguish two
kinds of strategy.
  i) Some policies directed new development towards the larger
     settlements. Thus Roxburgh County Council (1974) proposed a
     four-fold classification in which category A settlements were
     considered suitable for substantial expansion which would
     'probably ... alter the existing character of settlement'.
 ii) Other policies provided more detailed explanation, such as that
     for Buchan which had been by-passed in favour of the Aberdeen-
     Elgin corridor in the Gaskin Report (1969). In the context of
     the late 1960s, it is not difficult to justify this decision
     since Buchan occupied a terminal location in contrast to the
     major road and rail routes which link Aberdeen with Elgin and
     Inverness. The rural areas were experiencing persistent de-
     population and only the towns on the periphery were increasing
     their populations. The study of Deer District (Aberdeen County
     Council, 1972) was the most detailed though two further studies
     containing several amendments were also published. The plan
     dealt with a number of themes at district level before proceeding
     to a classification of individual villages. Physiography,
     district influences, population, communications, industry and
     employment were discussed. The historical legacy was also
     reviewed and the large number of eighteenth- and early
     nineteenth-century estate villages were regarded as a severe
     handicap to contemporary planners. The district was sub-
     divided into settlement groups, two based on large towns
     (Peterhead and Fraserburgh), the third being the central Buchan

area comprising farms and villages. Three alternatives were proposed:
a) concentration of growth in one settlement in each group, with only minimal growth in others;
b) moderate expansion of one settlement in each group with the development of nearby 'satellite settlements' and only minimal growth elsewhere;
c) distribution of growth amongst all the settlements in each group.

Strategy C simply meant persisting with the status quo and was therefore discounted on grounds of cost, its likely failure to halt depopulation and the denial to central Buchan of a potentially dynamic growth centre or key village. In fact, strategy A was chosen and Mintlaw, which had a good record in terms of recent growth, prosperity and accessibility, was selected from a short-list of six places. Subsequent policy statements diluted the primacy of Mintlaw and instead it was envisaged that some growth would be permitted in villages in the area between Mintlaw and Peterhead (Aberdeen County Council, 1973a; 1973b). These studies suggested that communities should be expanded up to the limits of their present service capacity though in the longer term, expansion would be limited to Mintlaw. Table 19.1 demonstrates the scale of growth at Mintlaw which is in contrast to the pattern in other settlements and to the stagnation of the remaining villages. These were all located in the interior of Buchan at considerable distances from Mintlaw and from the buoyant economy of the coastal area which by 1973 was beginning to experience the impact of offshore oil exploration.

Table 19.1   Population trends in key and non-key settlements in Buchan

	1961	1971	1979 (estimate)	1981
MINTLAW	608	657	2250	2283
Fetterangus	280	251	252	-
Longside	446	424	572	-
Old Deer	199	140	187	-
Stuartfield	391	306	432	-
Maud	656	684	630	644
New Deer	619	601	630	619
New Pitsligo	1217	1125	1070	1122
Strichen	967	962	930	873

Sources: Banff and Buchan District Council (1980b);
Census of Scotland 1981; Preliminary Report,
(H.M.S.O., Edinburgh) 1981

The third stage in the evolution of post-war planning was indicated by the reorganisation of local government in 1975. County councils were replaced by a two-tier system comprising regional and district councils. However, three levels of planning may in fact be said to exist.
i) The Regional Report sets out the broad policies which the Regional Council has adopted.
ii) The structure Plan details the land-use and development impli-cations of these policies. This can be regarded as the Statutory Development Plan for the region. It is concerned with the selection of settlements in which new housing or

industry will be located. Structure Plans are prepared in
three stages. First, a survey of the region is prepared and a
report of the survey is published. This is a factual document
which does not propose planning policies. The second stage
involves preparation of an 'issues report' which will be the
subject of public consultation. Third, the Structure Plan is
prepared, drawing on the results of public consultations.

iii) Local Plans provide the detailed basis for development control.
Such plans are prepared by the planning departments of district
councils, except in Dumfries and Galloway Region and Highland
Region where plans are prepared at regional level. Local Plans
embrace a number of different categories: district plans (Banff
and Buchan District, 1980b); action-area plans for individual
villages (Renfrew District Council, 1979) and subject plans
which deal with specific topics (Angus District Council, 1976).
The chronology of Local Plan production is broadly similar to
that of Structure Plans.

In short, the framework of the development plan system is the
policies and proposals set out in the Structure Plan. More detailed
recommendations for individual localities are worked out in the
Local Plans.

CASE STUDIES

The areas chosen, Angus District in Tayside Region and the Western
Rural Area of Central Region share certain characteristics. Both
are located short distances from major cities and have experienced
the pressure of commuter-led housing development on small communities such as Killearn and Strathblane (Central Region) and on the
coast between Monifieth and Arbroath (Angus). However, the north-
west of both areas comprises remote uplands containing few
nucleated settlements and suffering from persistent depopulation.
The choice of two apparently similar areas for study may seem
surprising, however there are important differences in the influence of Dundee and Glasgow in respect of housing demand in the
adjacent rural areas, and the size and distribution of villages.
Nevertheless, the major interest lies in the contrasting settlement
policies that have been adopted; a rigorous classification of
villages and hamlets in Angus and a simpler, three-fold division of
villages in Central Region based on the rate of increase of housing
stock which will be permitted.

Angus District

Four factors led Angus District Council's Planning Department to
prepare a settlement policy in 1976; depopulation in the more
remote areas, pressure for commuter housing in the south-east, the
absence of a comprehensive approach to rural settlement planning
and the likelihood that the Tayside Structure Plan (which was the
responsibility of the Regional Planning Authority) would not be
completed for a number of years. The settlement policy was intended
to establish a settlement hierarchy and to provide an input into
future regional plans. The methodology used to prepare the settlement policy involved a survey of each nucleated settlement with
reference to population, range of facilities, physical character,
distribution, appearance and degree of remoteness. The survey
enabled the identification of five grades of settlement ranging from
large rural centres offering a fairly comprehensive range of
facilities, through to hamlets with virtually no services. The
second stage in the analysis assessed the potential for growth
among the settlements in each grade on the basis of service constraints such as water supply, drainage capacity, and primary and
secondary school capacities. A matrix showing the type and number

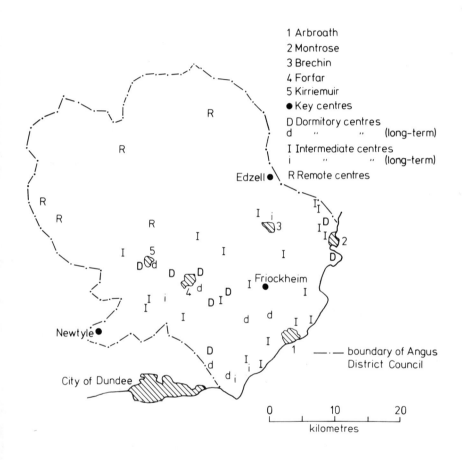

1 Arbroath
2 Montrose
3 Brechin
4 Forfar
5 Kirriemuir
● Key centres
D Dormitory centres
d " " (long-term)
I Intermediate centres
i " " (long-term)
R Remote centres

Edzell ●

Friockheim ●

Newtyle ●

City of Dundee

—.— boundary of Angus
District Council

0   10   20
kilometres

Figure 19.1   Angus District: proposed rural settlement policy

191

of constraints for all settlements was compiled. Several of the larger settlements were only restricted by one constraint, such as Edzell where the sewage works was overloaded. The number of constraints were greater among the smaller villages and hamlets and many were particularly badly placed to absorb even moderate growth. The survey of servicing constraints was complemented by a survey of bus services, outstanding planning permissions (an indicator of demand for housing and commercial premises), and the quality of agricultural land in the immediate vicinity of each settlement.

The rural policy identified five categories of settlement (Figure 19.1).
   i) Key centres were located outwith the primary catchment area of a town, had a good range of services, spare capacity for growth and industry would be encouraged to locate in them.
  ii) Dormitory centres were villages located within the primary catchment area of a burgh and had a basic range of services. These settlements had some growth potential and their dormitory role would be maintained to allow town workers to live outside urban areas. There would also be some encouragement to industry wishing to locate there.
 iii) Intermediate centres tended to be located further from towns and provided a limited range of basic services. Such settlements would only receive infill housing development.
  iv) Remote centres often have a recreational and tourist role. They are unsuited to large-scale growth, and only very limited development related to recreation, agriculture or forestry would be permitted.
   v) Other rural settlements were small hamlets inhabited by agricultural or forestry workers. Infill would not be permitted as sites were to be held in reserve for the housing needs of farm workers.

The performance of these policies was reviewed in 1980 and it was found that several major problems had arisen. The concentration of development in and around many villages had put pressure on good-quality agricultural land, and those villages which had expanded were now suffering from service constraints. Because of servicing constraints and remoteness, the key centres had failed to show any appreciable growth (Table 19.2) and instead growth tended to take place in dormitory centres. This had put pressure on agricultural land and existing service provision.

Table 19.2   Population trends in key and non-key villages in Angus

	1961	1971	plan date	1981
Friokheim (key)	809	807	810	774
Newtyle (key)	655	664	670	638
Edzell (long-term key)	644	658	710	689
Letham (dormitory)	690	804	800	985
Monikie (dormitory)	120	110	320	N/A
Hillside (dormitory)	691	692	N/A	791

Sources: Angus District Council Village plans, 1977-80; *Place names and population, Scotland: an alphabetical list of populated places derived from the Census of Scotland, 1961,* (H.M.S.O., Edinburgh), 1967; *Index of Scottish place names from 1971 Census,* (H.M.S.O., Edinburgh), 1975; *Census of Scotland 1981 - Preliminary Report,* (H.M.S.O., Edinburgh), 1981.

Western Rural Area, Central Region

The northern part of the area has experienced a long period of rural depopulation though since 1971 there has been some recovery due mainly to the expansion of Callander. The south-western part has steadily expanded its population, largely owing to the influx of commuters from Strathclyde Region. The central section is characterised by agricultural employment and depopulation which has only partly been offset by the growth of several villages which have attracted commuters to Stirling. The population trends in the major villages are shown in Table 19.3 which indicates that much of the recent housing development has been concentrated in only a few settlements.

Table 19.3   Population trends in selected villages in the Western Rural Area

	1961	1971	1978 (estimate)	1981
South-west area				
Strathblane/Blanefield	1060	1590	1810	1910
Killearn	689	1086	1435	1757
Balfron	1100	1149	1149	1129
Drymen	474	659	659	737
Gartmore	249	253	253	N/A
Central area				
Doune	786	741	870	1017
Gargunnock	309	457	480	N/A
Thornhill	493	443	460	N/A
Callander area				
Callander	1655	1786	2200	2199
Northern area				
Aberfoyle	853	793	865	N/A
Killin	583	600	665	541
Lochearnhead	202	175	210	N/A
Strathyre	166	155	190	N/A

Source: Central Regional Council (1980b) Census of Scotland (see Table 19.2)

Since 1975 three-quarters of new owner-occupied housing has been built in Strathblane/Blanefield, Drymen, Killearn and Callander. Allied to housing concentration has been the very high level of planning permissions sought from Stirling District Council. Approximately 1000 permissions were submitted between 1976 and 1978. Although not directly comparable, the annual average number of planning applications in five Angus villages (Friokheim, Newtyle, Edzell, Letham and Monikie) during the mid-1970s was only 31. However, there was relatively little local-authority house construction and between 1975 and 1979 only fifty units were built, about half of them in Callander. The Regional Council was faced with a decision whether to continue to cater for population growth of a similar magnitude to that experienced during the 1960s and 1970s or to limit

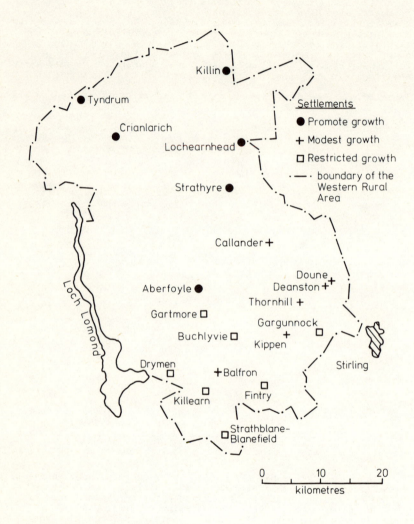

Figure 19.2   Western Rural Area, Central Region: proposed rural settlement policy

this growth and ensure that local needs were satisfied.

A survey of 27 villages and hamlets was conducted which examined population, education, shopping and community services. In contrast to Angus, the problem of village decline was not present to any significant extent and although some settlements had shown fluctuations in their populations, the general trend in the last thirty years was upward. However, given the very different population trends in the three sub-areas, the planning authority made no attempt to produce a rigid classification of villages. Instead the level of development and distribution of services and facilities was 'based as much on local circumstances as on the need for an overall, cost-effective policy' (Central Regional Council, 1979). The policy aimed to bring the rate of house building in the Western Rural Area more into line with the regional average, to introduce stricter controls over the release of land in commuter villages, and to encourage employment and housing development in the northern part of the sub-region. To achieve these aims a settlement strategy comprising three categories of settlement was introduced (Figure 19.2).

First were settlements where future growth would be restricted. These places had experienced considerable recent growth and, because of infrastructure constraints, it was proposed to limit growth to 5 per cent between 1981 and 1991. Apart from Gartmore, the settlements in this category are associated with commuter housing in large estates. While the advantages of an influx of young families was recognised, high growth rates had put pressure on educational and recreational facilities and it was also the view of Community Councils in public consultations that rapid population growth had adversely affected the quality of 'village life'. In future, greater emphasis would be placed on a more gradual approach to housing development.

The second category of villages was scheduled for modest growth. These include several large settlements which had absorbed growth in the past and, because of their size and level of services, were suitable for further expansion (e.g. Callander, Doune). Also included were villages where infrastructure constraints had previously limited expansion (e.g. Balfron), but where capital would be available to remove any bottlenecks to development. Two villages in the central area which had had static populations (Thornhill, Kippen) were also included in this category which had an envisaged growth rate of between 5 and 10 per cent.

The third category comprised settlements where housing growth would be encouraged. These occur in the most remote areas where the emphasis is on population retention.

The strategy will effectively reduce the amount of new housing required in the sub-region and alter the distribution of future housing growth among the sub-sections (Table 19.4). The actual shape of development within each community will become apparent once Local Plans prepared by Stirling District Council are adopted. At present only the Callander Local Plan has been approved and it indicates that new housing will be accommodated in several small areas, a contrast to the practise during the 1970s when relatively large parcels were allocated to builders of 'suburban' estates that could not be assimilated easily into either the physical or social fabric of small communities.

Table 19.4   House completions 1975-81 compared to estimated
              completions 1981-91 in the Western Rural Area

Local Plan Area	Yearly average house completions, mid-1975-81	Estimated yearly average house completions, mid-1981-91	Percentage change
South-west area	61	34	-44%
Central area	21	24	+14%
Callander area	33	20	-39%
Northern area	9	10	+10%

Source: Central Regional Council (1982)

CONCLUSIONS

Although Scottish rural settlement plans have received relatively
little attention from geographers, the present paper has demonstrated that a considerable amount of county and district planning
publications are available. Unlike England where the conceptual
basis for key settlements can be traced back to the early Development Plans, key settlements in Scotland only become prominent at
Review stage during the early 1970s. Analysis of a selection of
plans however, revealed many characteristics which were in common
with previous studies in England. For example, the classification
of settlements in Angus bears a close resemblance to that of West
Sussex (1968) while the proposals of Warwickshire C.C. are similar
to those in the Western Rural Area of Central Region (Woodruffe,
1976, 32; Cloke, 1979, 82). Problems encountered in the implementation of settlement policies such as sewage and water-supply
constraints are similar to those experienced in several English
counties.

Finally, a more detailed analysis of rural settlement policies
will be possible once the current round of Structure Plans is
completed. Urban planning problems have had priority in several
regions. However, the first Structure Plan documents for the rural
area of Grampian Region were published in February 1983 and the
Written Statement is expected to be published in spring 1984, while
detailed rural settlement proposals in Tayside Region should be
available later that year.

ACKNOWLEDGEMENTS

I would like to thank the Director of Planning and his staff in the
undermentioned local authorities for their advice and assistance:
Grampian Regional Council, Tayside Regional Council, Central
Regional Council, Borders Regional Council, Angus District Council
and Stirling District Council.

REFERENCES

Aberdeen County Council, 1972. *Deer District Strategic Plan*.

Aberdeen County Council, 1973a. *Buchan: the next decade*.

Aberdeen County Council, 1973b. *Buchan: the next decade, modified
    recommendations*.

Angus County Council, 1962. *County Development Plan.*

Angus District Council, 1976. *Settlement and development in the countryside.*

Angus District Council, 1980. *A policy for housing development in the countryside.*

Berwickshire County Council, 1972. *A rural policy for Berwickshire.*

Banff and Buchan District Council, 1980a. *District Plan vol.1: survey summary and key issues.*

Banff and Buchan District Council, 1980b. *District Plan vol.6: district local plan.*

Blacksell, M. and Gilg, A. 1981. *The countryside: planning and change.* (Allen and Unwin, London).

Central Regional Council, 1979. *Western Rural Area Structure Plan - consultative survey report.*

Central Regional Council, 1980a. *WRASP - report of initial consultations.*

Central Regional Council, 1980b. *WRASP - draft written statement.*

Central Regional Council, 1981. *WRASP - publicity and consultations report.*

Central Regional Council, 1982. *WRASP - written statement.*

Cloke, P. 1979. *Key settlements in rural areas.* (Methuen, London).

East Lothian County Council, 1974. *County planning policy.*

Gaskin, M. (ed) 1969. *North-east Scotland: a survey of its development potential.* (H.M.S.O., Edinburgh).

Gilg, A.W. 1978. *Countryside planning: the first three decades 1945-76.* (David and Charles, Newton Abbot).

Lanark District Council, 1977. *Local plans in Lanark District.*

Renfrew District Council, 1979. *Neilston Centre Local Plan - written statement.*

Roxburgh County Council, 1969. *Bowden village study.*

Roxburgh County Council, 1969. *St Boswells planning report.*

Roxburgh County Council, 1974. *Landward community development strategy.*

Scottish Development Department, 1968. *The Central Borders.* (H.M.S.O., Edinburgh).

Scottish Office, 1966. *The Scottish economy 1965-1970: a plan for expansion.* Cmnd.2864, (H.M.S.O., Edinburgh).

Stirling County Council, 1956. *Development Plan - written statement.*

Stirling County Council, 1974. *County Development Plan - supplementary written statement.*

Stirling District Council, 1981. *Callander Local Plan.*

Tayside Regional Council, 1980. *Tayside Structure Plan - written statement.*

Turnock, D. 1979. *The New Scotland.* (David and Charles, Newton Abbot).

Woodruffe, B.J. 1976. *Rural settlement policies and plans.* (Oxford University Press, London).

Chapter 20

Optimising the settlement pattern:
review and research design

Dinny de Bakker

INTRODUCTION

Part of the Zuidwest-Friesland research project was a study of
optimising the settlement pattern. By evaluating spatial policy
for Zuidwest-Friesland (a remote rural area in the Dutch context),
it is hoped that some general conclusions can be drawn about the
functioning of the settlement pattern in Zuidwest-Friesland and the
roles of municipal, provincial and central government. After 1966
the policy in Zuidwest-Friesland was one of concentrating housing,
employment and services in selected centres. An examination of the
effect of this 'centrumdorpenbeleid' is the focus of this study
(see the paper by Lockhart in this book). In the first part of
this paper 1 will describe the Dutch spatial context in which
processes of change within the settlement pattern have taken place.
Thereafter I will present a broad outline of literature in this
field. A research design for this study will emerge from this
outline.

THE SPATIAL CONTEXT

The transformation of the rural areas in the Netherlands after 1950
led to a growing interest by spatial planners in the functioning of
four major processes. First, there was the rapid decline in employment in agriculture due to farm enlargement caused by mechanisation
and rationalisation. Second, processes of scale enlargement raised
the thresholds for the establishment of commercial services. Third,
public authorities were confronted with extremely high costs of
service provision in the relatively sparsely populated rural areas.
Fourth, there was a major increase in the mobility of the people in
rural areas which was made possible by the motor car.

Within this context the attitudes of people in rural areas
changed. The growing awareness of life and opportunities in the
cities led to higher expectations with regard to the provision of
services. Thanks to their increased mobility, people in pressured
rural areas could get a job in the city without leaving their home
in the countryside. Many people also left the city to live in
rural areas while keeping their jobs in the city.

This study was undertaken in a remote rural area, outside the
reach of the labour and housing markets of the major cities.

People seeking work or education had to move to the city. The net
migration loss which resulted was selective. As a consequence,
remote rural areas were not only marked by depopulation but also
by an ageing of the people who stayed behind. Depopulation and
ageing, in turn, influenced the provision of services in a negative
way; a lack of employment and services stimulated out-migration,
the resulting depopulation causing a further reduction in services.
At the national level the Dutch government reacted by using regional
economic policy to disperse economic activities to the more peripheral parts of the country. The aim of this policy was twofold;
first, to diminish the problems in the overcrowded western part of
the Netherlands and, second, to relieve the loss of farm employment
in the peripheral regions.

At the regional level, the policy with regard to the settlement
pattern was one of concentration of activities in selected centres.
In remote rural areas the aim of this policy was to bring about
growth. It was hoped that concentration of population and services
in selected centres would attract industry, maintain the provision
of commercial services and diminish the costs of providing public
services.

In the seventies opposition grew against the policy of concentration. Reactions came in particular from the small villages which
felt themselves harmed by the concentration policy (see the paper
by van der Meulen in this book). They argued that the social
advantages of living in small settlements should be taken into
account. Shortage of services could be compensated by a stronger
sense of community. Government action should be directed towards
maintaining or raising the living conditions of the small villages
(Vereniging van Nederlandse Gemeenten, 1979).

In central and regional government reports we find the effects
of the complaints from the small villages. Interest in the
functioning of the settlement pattern moved into the background.
Attention shifted from the regional to the village level. Villages
were supported by such measures as grants for local shopkeepers,
flexible application of closing standards in the case of services
with a low level of use, small-scale housing projects, experiments
in public transport and mobile services. This meant that the policy
of concentration was replaced by one of dispersal.

In the eighties a renewed interest in the settlement pattern is
probable. The economic recession has caused economic ways of
thinking to predominate over social concerns. The financial side
of policy is of increasing importance. Small villages are likely
to be the victims of government cut-backs because measures taken
for them are mostly experimental or *ad hoc* in character. The
fragmentation of government measures over different departments will
also act in a negative way for small villages. Different government
departments are responsible for different aspects of policy.
Decisions are not taken in the spatial context of one village but
in the context of the budgetary policy of one or more government
departments.

Planners themselves may realize that a policy of dispersal
directed towards the individual village will be too costly. A
return to a policy of concentration is one possibility for the
future.

The tension between the economic, regional and sociological points of view is the central theme of this study. The spatial policy in Zuidwest-Friesland in the period from 1966 until the late seventies was a clear example of economic priorities. At that time a Village Plan was formulated by the Provincial Planning Agency. The thought behind this plan was that the level of service provision on a regional scale could only be maintained by selective concentration of growth in selected villages. Therefore a three-tier settlement hierarchy was constructed. Central villages ('centrumdorpen') and independent villages ('zelfstandige dorpen') were selected by such criteria as the number of inhabitants in the village and its service area, the number of service elements in the village, the distance to villages of the same or higher order and the village's potential for growth (P.P.D., 1966, 127). Central villages were to be centres for their area in the provision of daily services and also some important non-daily services. Independent villages were to be independent from other villages with respect to the provision of daily services. The rest of the villages could be more or less dependent on higher-order places for the provision of services.

The central theme of this study is a twofold evaluation of spatial policy in Zuidwest-Friesland at the regional and local levels. There are three topics for research at the regional level. How much concentration has taken place in the selected villages during the planning period? Are the selected villages acting as centres for their environs in the provision of services and employment? To what extent are the actual developments the result of the Village Plan?

At the local level the main aim of the research was to discover the effects of the policy for different types of settlement, grouped on the basis of the actual developments which have taken place during the planning period. In the following pages I will discuss the theoretical background to these research questions.

The first three questions need a deeper insight into the theoretical background to central-village policy. Cloke (1979) argues that key settlement policy in England was strongly influenced by theoretical considerations for two reasons. First, there were close links between planning practitioners and geographers and, second, there was a need for theoretical justification of key-settlement policy. The two most important theories in this respect were growth-centre theory and central-place theory.

The fact that many geographers were employed by the Provincial Planning Agencies in the Netherlands, also led to a strong integration of geography and planning practice. The Frisian 'centrumdorpenbeleid' has clearly been influenced by both central-place and growth-centre theories. The use of concepts like threshold, range of a good and hierarchy prove the impact of central place theory. The impact of growth-centre theory was much less marked, though the use of potential for growth as a selection criterion points to the influence of this theory.

The empirical validity of central-place theory and growth-centre theory is rather questionable in the context of small central villages; in Zuidwest-Friesland the population size of the central villages varies between 1500 and 4000. Moseley (1974) saw four possible growth-centre mechanisms for rural service centres; economies of infrastructure and service provision, economies of agglomeration, the spread of development to peripheral areas and the introduction of intervening opportunities in areas of depopulation. Cloke concluded from an overview of literature that key settlements are not of sufficient scale to provide a full range of growth-centre attributes. In his opinion, however, a full range is not necessary to cope with rural-scale problems. Villages with a population between 3000 and 5000 would function as centres for service and infrastructure provision, if necessary with some additional small-scale employment. All but one of the central villages in Zuidwest-Friesland have less than 3000 inhabitants. So, they are too small even to act as growth centres in the very restricted sense that Cloke proposed.

There are also serious doubts about settlements of the size of central villages functioning as central places. Functional hierarchies designed for the Netherlands (Keuning, 1971; Buursink, 1971; R.P.D., 1974) or for the northern part of the Netherlands (I.S.P., 1976) stop at the level of the regional centres. At an intra-regional level the only research is that by Buursink who studied central-place structures within the city. He concluded that, within the city, central-place theory works only weakly because distances from consumers to shopping centres were all so small. For rural areas there is some evidence from Canada, however. Dahms (1980) concluded that the spatial organisation of rural areas had become more similar to that of urban areas. Functional relations between settlements and their specialisation became more important, creating a 'dispersed city'.

Lewis (1979), too, saw a more complicated relationship between population size and the number of functions in a settlement. He argued that consumer behaviour deviates from that postulated by theory and he cited studies which showed differences in consumer behaviour between social groups. Using evidence from studies which show the functional interdependence between settlements, he concluded that there is more differentiation between lower-order settlements than theory suggests.

Data for Zuidwest-Friesland suggest that the settlement hierarchy, which was clearly present at the time that the Village Plan was formulated, is now slowly disappearing. Table 20.1 shows

Table 20.1   The changing distribution of population, 1960-80

	1960	1965	1970	1975	1980
Regional centres	35.7%	38.5%	41.8%	43.1%	42.7%
Central villages	16.1%	16.0%	16.4%	17.4%	17.9%
Independent villages	13.4%	13.2%	13.3%	13.6%	14.2%
Dependent villages	34.7%	32.3%	28.4%	25.9%	25.1%
Zuidwest-Friesland	100.0	100.0	100.0	100.0	100.0
(absolute)	94 960	96 263	99 783	104 619	108 689

Source: Sudergoa/P.P.D.

the changes in the spread of population over the settlement categories. Here we see that some concentration has taken place. The proportion of the population in dependent villages has diminished while that is regional centres has grown. The relative growth of the central villages is not impressive. Two explanations are possible here. The first is that the central villages functioned as intervening opportunities for people who would have moved to the city. In this case policy has had a positive effect in stopping the depopulation of the rural areas in Zuidwest-Friesland. The second and more probable explanation is that the population growth in the area was in the central villages. In this case, the effect of policy has been a faster decline in the dependent villages than would have been the case if there had been no central-village policy.

Table 20.2   The number of retail service elements for each settlement category, 1964-80

	daily retail services			Non-daily retail services		
	1964	1972	1980	1964	1972	1980
Regional centres	242	171	120	317	319	375
Central villages	131	105	71	43	73	143
Independent villages	130	101	73	42	64	102
Dependent villages	243	178	99	0	53	126

Source: Schouten/Witmer-Oor, 1982

That the settlement hierarchy is fading is clear when we combine the conclusions from Table 20.1 with those from Table 20.2. We see in Table 20.2 that many daily retail services have disappeared in each settlement category. The number of non-daily retail services, however, has increased rapidly in every category. Particularly striking is the increase in the dependent villages from zero in 1964 to 126 in 1980. This points to more functional interdependence between villages. Formerly there were a limited number of centres which offered a complete range of non-daily services. Now, there are a lot of villages which offer only one or two non-daily retail services. The policy has not suceeded in concentrating the non-daily services in the central villages. Despite a relative loss of population from the dependent villages, their number of services increased.

The second part of the study looks at the central-village policy from a local perspective. The effects of policy will be evaluated for different types of settlement. In essence, the effects of the whole policy can be seen in the outcome of a decision to concentrate or disperse one activity in one settlement. Three types of effects can be distinguished:
1) effects on the distance between people and activities;
2) qualitative effects;
3) financial effects.

The distance between people and activities will generally grow under a concentration policy. What is important is how far increasing distance affects the access of villagers to services (see the paper by Huigen in this book). Physical access can be defined as the ability of an individual to move himself in order to reach the activities he desires. It depends not only on the means of transportation, but also on characteristics of the individual, such as age, position in the household and participation in the labour market. A second important effect of increasing distances is that participation in an activity decreases as average distance to the consumers increases. This friction of distance differs between types of activity.

Three types of qualitive effects can also be distinguished.
There are, firstly, effects on the activities themselves. Concentration leads to bigger units. In general, it is supposed that
bigger is better because this enhances professionalism. Small units
can, however, offer specific advantages. Much research has been
done into the effects of school size on the quality of education by,
for example, Tricker (in this book), and Gilder (1978). No firm
conclusions can be drawn from this research. Secondly there are
the effects on the differentiation of activities. If one activity
disappears from a village, this can have an extra negative effect
on the frequency with which other activities are used in the village
because it diminishes the possible number of multi-purpose trips
and, thereby, the attractiveness of the settlement for consumers
(see the paper by de Haard in this book). Finally there are the
social effects on village society. The disappearance of vital services (for example, the primary school or the last shop) can have an
extra negative effect on village society. By way of contrast, a
rapid increase in population can also lead to social problems.

The financial effects can be subdivided in two groups. There
are the effects on the costs of the activities themselves. The
general supposition is that bigger is cheaper because of internal
economies of scale. That internal economies of scale do exist in
the service sector can be proved from the literature. Gilder
(1978) found this in West Suffolk for primary education and sewage
works. He found insufficient proof, however, for medical services.
Two comments can be made about the economies of scale. First, as
concentration continues, costs will eventually rise again. In the
long term, the relationship between average costs (Y) and output (X)
is a U-shaped curve of the form: $Y/X = a - b.X + c.X^2$
Moseley (1974) found agreement among authors about the existence of
internal economies of scale. Considerably less agreement was found
about the question of when diseconomies start to operate. The
second comment is that there is the problem of fixed capital assets
in existing settlements. The relative costs of investment in the
existing settlements pattern should be weighed against investment
in some new pattern. Gilder (1978) used a Present Discounted Value
to charge for fixed costs. · This means that costs will rise yearly
in the future because a service must be replaced by an equivalent
service. Other possible approaches are yearly depreciation of
capital on a nominal basis or taking into account rent and
redemption only.

One must also consider the effects on the users of activities.
Lower costs for the activities themselves can mean higher costs for
the users. Gruer (1971) undertook a cost-benefit analysis to prove
that out-patient care was cheaper through small, dispersed clinics
than in big, central hospitals.

CONCLUSION

In this paper I have proposed a theoretical basis for an evaluative
study of the policy of service concentration. From a description
of processes which took place in rural areas of the Netherlands, it
can be concluded that there was a conflict of interest between the
regional and the local levels. Therefore, a twofold approach was
introduced for this study. At the regional level, the central issue
is the functioning of the settlement hierarchy. Some empirical data
suggest that the hierarchy in the settlement pattern is gradually
becoming more diffuse. At the local level, an analysis of three
types of effects (distance effects, qualitative effects and

financial effects) should give a clear picture of the results of
the settlement policy on individual villages.

REFERENCES

Buursink, J. 1971. De Nederlandse hiërarchie der regionale centra
    een institutionele wijze van hiërarchisering. *Tijdschrift voor*
    economische en sociale geografie, 76-81.

Cloke, P.J. 1979. *Key settlements in rural areas.*
    (Methuen, London).

Dahms, F.A. 1980. The evolving spatial organisation of small
    settlements in the countryside: an Ontario example.
    Tijdschrift voor economische en sociale geografie, 5, 295-306.

Gilder, I.M. 1978. Rural planning policies: an economic appraisal.
    Progress in planning, 11(3).

Gruer, R. 1971. Economics of outpatient care. *The Lancet,* 20,
    (February, 1971).

Integraal Struktuurplan Noorden des Lands, 1976. *Het sociaal-*
    ekonomisch beleid voor het Noorden des Lands.
    (Staatsuitgeverij, 's-Gravenhage).

Keuning, H.J., 1971. Spreiding en hiërarchie van de Nederlandse
    verzorgingscentra op de grondslag van hun winkel apparaat.
    Tijdschrift voor economische en sociale geografie, 3-17.

Lewis, G.J. 1979. *Rural communities: a social geography.*
    (David and Charles, London).

Moseley, M.J. 1974. *Growth centres in spatial planning.*
    (Pergamon Press, Oxford).

Moseley, M.J., Harman, R.G., Coles, O.B. and Spencer, M.P. 1977.
    Rural transport and accessibility. vol.1, main report,
    (Centre of East Anglian Studies, Norwich).

Provinciale Planologische Dienst Friesland, 1966. *De ruimtelijke*
    ontwikkeling van het Friese platieland. (Leeuwarden).

Rijksplanologische Dienst, 1974. *Hiërarchie van kernen.*
    ('s-Gravenhage).

Vereniging van Nederlandse gemeenten, Studiecommissie kleine kernen,
    1979. *Wie het kleine niet eert ... Diskussienota over kleine*
    kernen in Nederland. (Uitgeverij, 's-Gravenhage).

Chapter 21

Planning policy and socio-economic changes in post-war Montgomeryshire

David Grafton

The aim of this paper is to monitor the intra-regional effects of development planning policy in a remote rural area. Whilst a considerable amount of literature exists at the regional level concerning such areas, there is a relative dearth of research at a finer spatial scale. Mid-Wales was chosen as a case study area following an analysis of 1971 census data at the rural district level (Grafton, 1981) which showed that mid-Wales and the Welsh Borders exhibited remote rural characteristics to a very high degree (Figure 21.1).

This index was derived from a range of socio-economic variables that a literature survey indicated as important characteristics of remote rural areas. The technique used was a summed-rank analysis whereby each rural district was assigned a rank according to its relative position on each of the variables, and a total rank calculated by summation of the individual rank scores. The rank correlation coefficients between the individual variables and the final index are shown in Table 21.1.

Mid-Wales has long been recognised as a classic remote rural area with a history of depopulation (Ashton and Long, 1972; Bowen, 1962). Although generally regarded as a problem region, it is important to make a distinction within the area between the scattered settlements of the uplands and the larger, more nucleated settlements of the valley floors. In the latter, particularly the small towns and their commuting hinterlands, a relatively diverse and growing range of employment is available. Depopulation in such areas has not been a problem since the 1950s and indeed was less apparent than in the rural areas prior to that.

The largest population losses have occurred in the more remote uplands, associated with the continuous decline of a traditional, agriculturally based economy, in places exacerbated by a concomitant reduction in employment in mining. The relatively harsh physical conditions of the uplands, small farms and lack of accessibility to the small towns have combined here to produce standards of living well below twentieth-century expectations (Thomas, 1972).

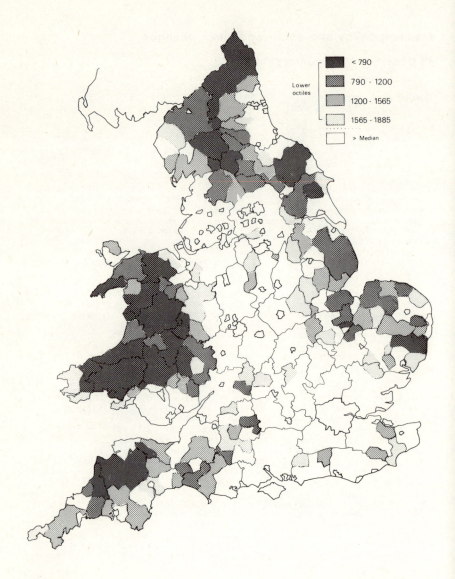

Figure 21.1   Rural Districts with remote rural characteristics in 1971

Table 21.1   Rank correlation coefficients between the indicator
              variables and the final index

Variable (1971)	Rank correlation with final index
Net outmigration, 1966-71	0.58
Net outmigration of persons aged 15-44 (1966-71)	0.65
Population change, 1921-71	-0.87
Population change, 1951-71	-0.86
Persons retired	0.67
Population density	-0.84
Proportion employed in agriculture	0.83
Economic activity rate	-0.18
Domestic rateable value per head	-0.37
Commercial rateable value per head	-0.65
Industrial rateable value per head	-0.59
Total rateable value per head	-0.70

Note: In compiling the index, all variables with a negative coefficient
      were transformed to be positive, to enable sensible summation of ranks.
      All coefficients significant at the 95% level.

TOWARDS A SOCIO-ECONOMIC INDEX

In attempting to portray post-war changes in socio-economic conditions within the region, an approach is needed that can encompass
the important intra-regional variations discussed above and will
allow comparisons to be made through time. The traditional
indicator of population change may be employed, but clearly a wider
range of variables would be of value. In terms of spatial scale,
the most detailed level for which comparable information is available during the post-war period is that of the civil parish, and
the most useful data source that of the census. Whilst published
statistics are not available at the parish level for socio-economic
data, this information is held at the Office of Population
Censuses and Surveys (O.P.C.S.). Unfortunately, no data are held
for 1951, and the 1981 results are not yet available. The need for
comparable information at a fine spatial scale dictates therefore
that this analysis will be concerned with changes in the decade
1961-71.

    A range of indices was chosen on the same basis as for the rural
district analysis. These are shown in Table 21.2 for 1961. By
combining the variables in the manner described for the rural
district analysis, an aggregate index score was calculated for
each parish in mid- Wales, although the results presented here will
in the interests of clarity only relate to Montgomeryshire
(Figure 21.2). This county is typical of mid-Wales in its clear
division between 'urban-influenced' and 'rural' areas, and the
general comments pertaining to Montgomeryshire are also valid for
the other parts of the region.

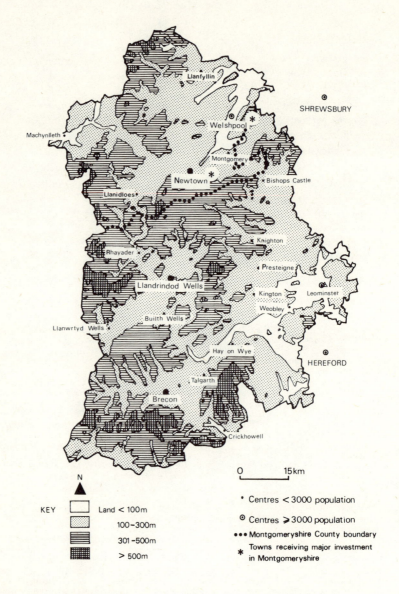

KEY

Land < 100 m
100–300 m
301–500 m
> 500 m

N

O   15km

• Centres < 3000 population

⊚ Centres ⩾ 3000 population

••• Montgomeryshire County boundary

* Towns receiving major investment
   in Montgomeryshire

Figure 21.2   Mid-Wales and the Borders: topography and major
settlements

Table 21.2   Socio-economic variables 1961

Variable	Rank correlation with final index
Unemployment	0.3
Primary sector employment	0.4
Secondary sector employment	-0.4
Tertiary sector employment	-0.4
Commuting	-0.5
Elderly	0.2
Male activity rate	0.0
Female activity rate	-0.3
Population change, 1951-61	-0.4
Population density	-0.7

Note: In compiling the index, all variables with a negative coefficient
were transformed to be positive, to enable sensible summation of
ranks.
All coefficients significant at the 95% level except male activity
rate.

Figure 21.3 shows the distribution in 1961 of socio-economic
scores in Montgomeryshire. The areas with high scores are in
general those with relatively low unemployment, a balanced employment structure and an increase in population during the decade 1951-
61. To a large extent such areas comprise the small towns and
their commuting hinterlands located on flatter, low-lying land and
with generally good accessibility. In contrast, those areas with
low socio-economic scores are predominantly upland (compare with
Figure 21.2), with the worst conditions being found in areas remote
from the small towns, and with a dispersed settlement pattern.
Intra-regionally, then, the index confirms that the major problems
are associated with the more remote rural areas. Against this background, the development-planning policies in the post-war period
can now be discussed.

POST-WAR DEVELOPMENT-PLANNING POLICY IN MID-WALES

The initial period of development planning immediately after the
Second World War was intended primarily to attack the problem of
rural depopulation. It was characterised by small-scale and piecemeal attempts at development by individual counties as described in
the county development plans for Breconshire, Radnorshire and
Montgomershire which were produced in the late 1940s and early
1950s. The first moves towards a regionally co-ordinated policy
were made by the Mid-Wales Industrial Development Association. This
was formed in 1957 and comprised representatives from all the
county councils within mid-Wales. It sought to identify areas
suitable for small-scale industry and to allocate sites as and when
an enterprise was attracted to the region. These early planning
policies were poorly financed and led to spatially dispersed
development, having little impact on the key problems of rural
depopulation and weak employment structure (Grafton, 1981). These
policies were superseded by a range of new initiatives during the
1960s.

211

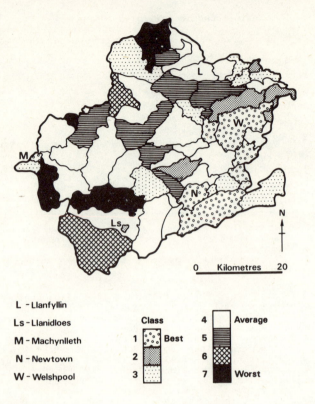

L - Llanfyllin
Ls - Llanidloes
M - Machynlleth
N - Newtown
W - Welshpool

Class
1 · Best
2
3
4 Average
5
6
7 Worst

Figure 21.3   Socio-economic scores in Montgomeryshire parishes, 1961

It is important to stress that the principal development-planning strategies of this later phase were based on the growth-centre concept.   These policies originated from an extremely influential government report published in 1964 by the Beacham Committee.   This report was commissioned to suggest ways of halting depopulation in mid-Wales.   It made two main recommendations: 'To effect any worthwhile improvement in the economy of mid-Wales two things are essential.   The first is that a policy of reducing the existing scatter of population by nucleation into fewer and larger settlements should be implemented.   The second is that more varied employment opportunities should be provided.'   In practice this meant an emphasis on attracting manufacturing industry to the twelve largest towns of the region, with Welshpool being selected as the centre most likely to be able to generate 'spread' effects within Montgomeryshire and thereby improve conditions in the county as a whole.

Three main reasons for the emphasis on manufacturing industry were advanced.   Firstly, it was argued that manufacturing incomes were needed to supplement the low level of regional income in mid-Wales.   In particular it was hoped that, following export-base theory, the development of indigenous exporting industries would have beneficial multiplier effects to the local economy.   Secondly,

the rapid decline of all sectors of the economy with the exception
of tourism (which was not seen as capable of providing a secure
year-round basis for regional development) meant that economic
diversification was needed, and the manufacturing sector offered
the best option in this respect. The third justification for a
policy based on the development of manufacturing industry was the
existence of untapped 'slack' in the regional economy, through
underemployment of male and female labour and the very low female
activity rate.

From the Beacham Committee's premise that, 'the basic cause of
mid-Wales' problems is the out-dated settlement pattern, combined
with a generally low level of population', it was suggested that to
perpetuate the scattered settlement pattern by financing widely
dispersed development would serve only to reinforce the problem that
intervention was designed to overcome. Much hinges on the validity
of the premise that mid-Wales's problems do indeed stem from the
nature of the settlement pattern. It would seem reasonable to
suggest that this is a simplistic view, but one which it is
necessary to take if, as the Beacham Committee further suggested,
the orientation of policy was to become more concerned with the
promotion of regional growth rather than the prevention of rural
decline.

Financial support to further these aims was granted after 1966
when mid-Wales was designated a Development Area. This meant that
a wide range of grant and tax concessions became available to
industrialists wishing to locate in mid-Wales (for a review of U.K.
regional policy incentives see Manners *et al.*, 1980).

The growth-centre approach to regional development in mid-Wales
was further manifest, in an extreme form, in the proposals of a
planning consultancy (Economic Associates Ltd., 1966) which
suggested creating a major new town of 70 000 people stretching
20 km along the upper Severn valley and centred on Caersws, near
Newtown in Montgomeryshire. This report was clearly in favour of a
radical growth-centre strategy: 'The alternative to scattered,
small-scale action under existing or even modified powers is a much
more massive form of development which would positively reverse the
downward trend.'

The principle of concentrating growth on Newtown was confirmed
when it was designated as a New Town in 1968; the first use of the
New Towns Act as an instrument of rural development. The size of
expansion was reduced however, with Newtown scheduled for growth
from a population of 5500 in 1966 to 11 000 by 1980. The growth-
centre approach was again justified on the basis that wide-reaching
spread effects would emanate from Newtown to improve the economy of
the region as a whole. This strategy was also endorsed by two
government reports which again identified Newtown as the centre to
receive by far the greater proportion of public investment (Welsh
Office, 1967; 1969).

The overall effect of these policies has been to increase
industrial employment within Montgomeryshire in general and Newtown
in particular. The major phase of industrial expansion came after
1968 with 55 government-financed factories in Newtown providing
approximately 50 000 m^2 of floorspace by 1977. Taking the growth in
private and public factory floorspace together, Newtown accounted
for 75 per cent of the total Montgomeryshire increase.

Despite the success in generating growth at Newtown, the extent
to which these large-scale developments were able to generate

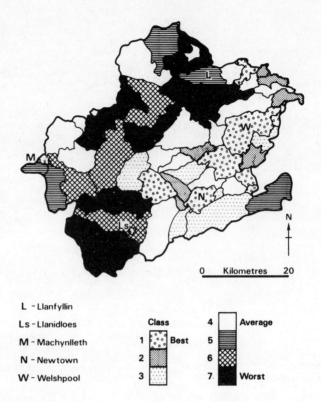

Figure 21.4   Socio-economic scores in Montgomeryshire parishes, 1971

spin-off effects proved limited, since the population losses from
many of the rural districts in mid-Wales continued throughout the
1960s. Most of the country towns grew during this period, again in
part due to their success in attracting new industries. Nevertheless, the pattern of overall decline in the rural districts does not
provide evidence of the hoped for 'spread effects' bringing a
reversal in population trend, particularly in the most remote
districts.

SOCIO-ECONOMIC INDEX IN 1971 AND 1961-71 CHANGE

A quantitative assessment of the effects of the growth-centre
policy can be made by comparing the 1961 and 1971 socio-economic
indices. Figure 21.4 shows socio-economic scores in Montgomeryshire
in 1971 (using the same range of variables as for 1961) and comparison with the 1961 distribution (Figure 21.3) shows little overall change. The same general pattern of relative prosperity in the
small towns and lowlands is revealed. Similarly the pattern of low
index scores in 1971 conforms broadly to that of 1961 with the
scattered settlements of the uplands still recording the lowest
values.

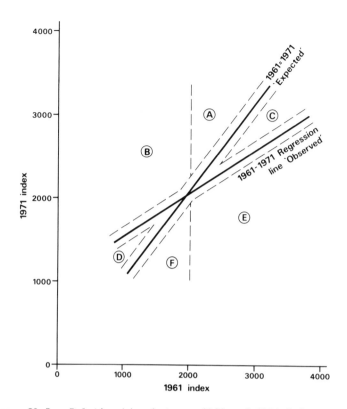

Figure 21.5   Relationship between 1961 and 1971 index scores for Montgomeryshire parishes

A more rigorous comparison of these two scores may be made by plotting the 1961 score against the 1971 score as shown in Figure 21.5. As this regression line is less steep than 45°, it can be concluded that the changes between 1961 and 1971 were in the direction of equalisation. The overall pattern shows parishes with small scores in 1961 to be somewhat higher in 1971, and parishes with large scores in 1961 to have become less dominant in 1971. Thus it is tempting to suggest, on the basis of this evidence, that planning policy helped towards equalising living conditions between 1961 and 1971. However, it is essential to note that the regression line is not a perfect representation of change between 1961 and 1971. Considerable variation exists around the line (r^2 = 0.45) and the importance of the residuals must not be overlooked.

The discussion thus far has neglected an important spatial element which may be incorporated if the residuals are plotted. Figure 21.5 also shows a classification of residuals based on their position on the graph. The residuals of greatest interest are groups A, B and F. Group F represents the problem parishes in 1961 for which planning policy made a negative impact during 1961–71. The presence of group A residuals implies a trend towards a greater concentration of benefits rather than equalisation during the ten

215

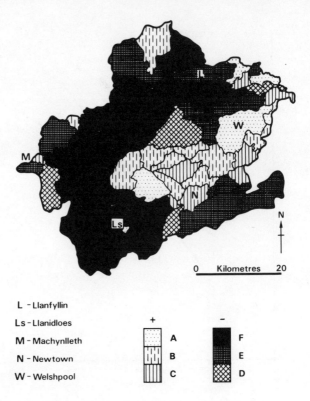

L - Llanfyllin
Ls - Llanidloes
M - Machynlleth
N - Newtown
W - Welshpool

+

A
B
C

-

F
E
D

Figure 21.6   Residuals from regression

year period. Group B residuals are important because they represent
the low scoring areas in 1961 that had considerably improved their
position by 1971. If substantial spread effects had been generated
during this period, then the presence of this group in the remoter
areas would be anticipated.

The spatial pattern of residuals is shown in Figure 21.6. This
map shows clearly that, during the currency of the development-
planning policies discussed earlier, the position of the growth towns
and their immediate hinterlands had been enhanced (group A residuals)
and the position of the more remote, upland areas worsened (group F).
The group B residuals are almost entirely confined to parishes close
to Newtown and Welshpool, indicating the presence of some spread
effects, although these are clearly limited in spatial extent.
There is little evidence therefore to support the spread effect of a
growth centre beyond a very small radius from Newtown and Welshpool.
Whether these changes in socio-economic position can be ascribed
simply to planning policy is a contentious issue. What may be concluded from this analysis is that, in a negative sense, the development-planning policies did not achieve success in their stated aim
of benefiting the least advantaged rural areas during the decade.

The crucial point in explaining the lack of spread effects lies in recognising the presence of two distinct types of economic activity within the region. The first is the traditional, declining, agriculturally-based economy, and the second is typified by the newer economic activities generated by development policy in the post-war period. For a growth-centre policy to be successful, as Moseley (1974) has shown, growth centres must not only develop linkages with closely allied economic activities in 'economic space' (Perroux, 1955) but must also develop linkages in a spatial sense. In the case of Montgomeryshire, the economic links of the new industries have been with other urban centres, particularly Birmingham, and not with the surrounding rural economy. Further, evidence from Jones (1965) would suggest that the rate of depopulation from the rural hinterlands of Newtown and Welshpool accelerated partly as a result of the new development at those towns. Thus, while Newtown and Welshpool have grown and diversified their economic structure, this development has failed to regenerate the remoter areas largely because of a lack of functional integration between the traditional and new economic activities.

The policy implications of this analysis would suggest that a more dispersed pattern of investment, concentrating on smaller centres, may be of greater effect in improving conditions in the remoter areas. This view is central to a number of other European countries' policies for remote rural areas (Grafton, 1981) and has been given support by Powys County Council, in their *Growth towns study* (1978) which stated that, 'The remoter parts of the County are unlikely to benefit by investment in the growth towns... Large areas of the County are beyond the ambit of growth towns and there is scope for developing policies which benefit other areas as well'.

There may even be a case for moving away from the growth-centre idea completely, especially in the large areas of the county with a highly dispersed settlement pattern. Here, the lack of even low-order centres implies the need for a policy which encourages spatially dispersed rather than concentrated development, and the Powys County Structure Plan (1979) stresses the importance of activities such as small craft workshops and farm tourism as means of achieving this end. Such policies, favouring small-scale dispersed development need not be seen as an alternative to the traditional forms of growth centre policy, but rather, as complementary to them.

Unfortunately, there are four major difficulties associated with a more permissive attitude towards development in areas of scattered settlement. Firstly, there is a deeply ingrained planning doctrine that, in rural areas, there should be a presumption against development in the open countryside. Whilst for most of lowland England this seems a sensible policy, it is more questionable when applied to the dispersed settlement pattern of upland Wales. Nonetheless, the Powys Structure Plan did encounter difficulties in this respect when submitted to the Secretary of State for approval.

Secondly, the existing pattern of residential planning permissions, resulting from growth-centre policy in previous years, is dominated by Welshpool and Newtown, which together account for approximately two-thirds of all outstanding planning permissions in Montgomeryshire. These are likely to be able to accommodate the population growth of Montgomery for at least the next ten years; the number of permissions in the remoter areas is correspondingly very small (Powys County Council, 1979). It may be expected

217

therefore that the future pattern of population change will not differ radically from that of the recent past.

The third problem arises from the policies of the Development Board for Rural Wales (D.B.R.W.). This authority was created in 1977 to take over the activities of all development agencies in rural Wales, including the Newtown Development Corporation. Although its policy documents mention the need for small-scale development in some selected small towns, the emphasis of the Board's efforts remains directed towards large scale development, particularly at Newtown (D.B.R.W., 1979). For the period 1977-78, £3.2 million of the Board's total budget of £3.7 million was allocated to Newtown. No mention was made of specific aid for the areas of scattered population. Since the D.B.R.W. controls to a large extent the finance available for creating new employment in rural Wales, there would seem at least the potential for conflict between the Board and the County Planning Department over the scale and location of development to be promoted.

A fourth and more general difficulty likely to affect the future pattern of economic development in mid-Wales as a whole is the recent (1982) removal of development area status from much of the region. This will probably have a greater impact on Newtown, although the prospects for the remoter parts of the region are clearly not enhanced by such a policy change.

The conclusions of this analysis are not restricted to Montgomery. According to the aggregate 1981 census returns available to date, many of Britain's remoter rural areas have increased in population and employment, and Montgomery has shared in this trend (Champion, 1981; O.P.C.S., 1981). Whilst some authors have interpreted this reversal of traditional decline as evidence for 'counter-urbanisation', it must be stressed that this is a highly misleading term for two reasons. Firstly, the spatial scale used is that of the county or district, and this is likely to blur important differences between small-town growth and rural decline. Secondly, 'counter-urbanisation' implies processes at work that are independent of urbanisation. As can be seen in Montgomery and other areas the idea that the rural economy is reviving is false (Hodge and Whitby, 1981). Extra-urbanisation rather than counter-urbanisation is being imposed on, rather than generated in, the remoter regions, with urban-based employment and population growth largely restricted to the major rural settlements. The traditional pattern of stagnation or decline is still apparent in the more remote, agriculturally-based areas outside the commuting hinterlands of the growing centres.

REFERENCES

Ashton, J. and Long, W.H. (eds) 1972. *The remoter rural areas of Britain*. (Oliver and Boyd, Edinburgh).

Beacham Committee, 1964. *Depopulation in mid-Wales*. (H.M.S.O., London).

Bowen, E.G. 1962. Rural Wales. in *Great Britain: geographical essays*, Mitchell, J. (ed), (Cambridge University Press, Cambridge).

Champion, A.G. 1981. Counterurbanisation and rural rejuvenation in rural Britain - an evaluation of population trends since 1971. *Department of geography seminar paper 38*, (University of Newcastle upon Tyne).

Development Board for Rural Wales, 1979. *Board policy statement.* (D.B.R.W., Newtown, Powys).

Economic Associates Ltd. 1966. *A new town for mid-Wales.* (H.M.S.O., London).

Grafton, D.J. 1981. *Geography and planning in remote rural areas with special reference to mid-Wales and the Borders and S.E. Switzerland.* (unpubl. Ph.D thesis, University of Southampton).

Hodge, I. and Whitby, M. 1981. *Rural employment: trends, options and choices.* (Methuen, London).

Jones, H.R. 1965. Rural migration in central Wales. *Transactions of the Institute of British Geographers,* 37, 31-46.

Manners, G., Keeble, D., Rodgers, B. and Warren, K. 1980. *Regional development in Britain,* 2nd ed., (Wiley, London).

Moseley, M.J. 1974. *Growth centres in spatial planning.* (Pergamon, Oxford).

Office of Population Censuses and Surveys, 1981. First results from the 1981 census. *O.P.C.S. Monitor,* July 21, (O.P.C.S., Titchfield, Hampshire).

Perroux, F. 1955. Note sur la Nation de 'pôle de croissance'. *Economie Appliquée,* D(8).

Powys County Council, 1978. *Growth towns study.* (Powys C.C., Llandrindod Wells).

Powys County Council, 1979. *Structure plan written statement.* (Powys C.C., Llandrindod Wells).

Thomas, J.G. 1972. Population change and the provision of services. in *The remoter rural areas of Britain,* Ashton, J. and Long, W.H. (eds), (Oliver and Boyd, Edinburgh).

Welsh Office, 1967. *Wales: the way ahead.* (H.M.S.O., London).

Welsh Office, 1969. *Industrial growth in mid-Wales.* (H.M.S.O., London).

Chapter 22

Land speculation and the under-use

of urban-fringe farmland in the

Metropolitan Green Belt

Richard Munton

INTRODUCTION

Idle or under-used[1] farmland occurs in the fringes of many Western
cities (O.E.C.D., 1979). Three main reasons are usually given to
account for its presence. First, land close to the edge of urban
areas is subject to urban intrusion of various kinds, such as
trespass and pollution, which militates against efficient farming
(A.C.A.H., 1978). Second, the urban fringe contains a disproportion-
ate number of hobby farmers who are not especially interested in the
full agricultural use of their land (C.R.C., 1978). Third, farmers
located in the fringes of expanding cities often experience a combination of high municipal taxes and an uncertain future in farming.
These act as disincentives to investment in fixed improvements and
lead to farming of low intensity and the premature sale of land to
urban developers (Sinclair, 1967; Brown et al., 1981; Bryant, 1974;
1982).

The individual farmer's response to the insecurity of his
farming situation is affected by two separate but related sets of
issues. At a general level, his response is influenced by the
proximity of his farm to the urban edge and the level of demand for
urban land, as reflected in the difference in price between urban
and rural land and the rate of transfer of rural land to urban uses.
More specifically, the occupier's decisions on when to sell his
land and how to manage it prior to its sale are crucially affected
by, among other matters, whether he owns the freehold to the land,
whether he has a secure lease and, if he is a tenant, by the aims
of his landlord.

In the absence of either land-use or land-market regulations (or
both), the under-use of farmland may be a rational economic response
to the price signals emanating from the urban-fringe land market.
Indeed, substantial areas of under-used farmland in the urban fringe
are often regarded as indicating inefficiency in the land conversion
process and a wide range of land policies are employed in many
Western countries to minimise the extent of under-use (Lichfield and
Darin-Drabkin, 1980; Pearce, 1980). In spite of this public concern,
there are relatively few empirical analyses of the relations between

---

[1]Under-use is a vague term and one that is employed inconsistently in the
literature. Here it is applied to farmland of low productivity in which the
low productivity is not the result of any particular physical constraint or
lack of economic potential.

land-market forces, landowner behaviour and the under-use of farmland. This is especially so in Britain where the planning system
is usually viewed as keeping under-used land to a minimum, although
this position has been increasingly questioned in recent years
(Coleman, 1976; Moss, 1981). A further practical reason for the
lack of research is the absence of a land register for England and
Wales which is open for public inspection and in which beneficial
interests in land are recorded. This means that such data can only
be obtained from the details of public auction sales, which are
unrepresentative, or through time-consuming field investigation, the
reliability and completeness of which are dependent on the goodwill
of individual landowners. As a result, there are many studies of
land use and land-use change in the urban fringe in Britain (for
example, Low, 1973; Coleman, 1976; Thomson, 1981), but few attempts
to relate the nature and structure of the urban-fringe property
market to the use and management of land.

This paper seeks to throw some light on these issues through a
case study from the Metropolitan Green Belt (M.G.B.) around London.
The analysis focuses on the standards of maintenance of farmland
adjacent to urban areas and how, in particular, these standards are
influenced by land tenure differences.

THE USE OF FARMLAND IN THE URBAN FRINGE

Much of the recent discussion in the geographical literature stems
from a paper by Sinclair (1967). He argues that although the total
value of land declines with distance away from the urban edge, the
proportion of its value that is attributed to its agricultural use
rises both absolutely and proportionately with distance away from
the city boundary to a point where its urban speculative value is
nil and its value is determined by agricultural considerations alone.
The reason for this is that as the urban development potential of
the land rises close to the city edge so uncertainty over the future
of farming increases. Those farming enterprises demanding greatest
fixed capital investment, which Sinclair assumes make the most
intensive use of the land, are the most sensitive to this uncertainty
and therefore he postulates a number of land-use rings of increasing
farming intensity *away* from the urban edge (Sinclair, 1967, 79-81).
Land right on the urban edge will, it is argued, have little if any
agricultural value.

The explanatory value of this argument depends on the assumption
that the uncertainty arising from urban development is the exclusive,
or at least the dominant factor determining farmers' decisions on
investment and choice of enterprise, and that all farmers respond in
a similar manner. Unfortunately, Sinclair does not provide any
empirical evidence in support of his position and this, Bryant has
maintained, is because he failed to explain the mechanism by which
farmers' anticipations of urban growth could be linked to changes
in their farming practices (Bryant, 1974). This led Bryant to try
and define farmers' behaviour under conditions of uncertainty, and
as a first approximation he used classic profit-maximising
principles. Subsequently, he relaxed his assumptions about farmers'
behaviour accepting that not all of them would respond in the same
way (Bryant, 1981). Some farmers perceive urban development as a
threat to their farming livelihoods and withdraw resources from
their farms, as Sinclair assumed, but others make additional
investments, in farm-gate sales for example, in order to capitalise
on new market opportunities. Bryant also argued that the
consistency of farmers' responses would depend very largely on their
proximity to the urban edge and thus to the perceived immediacy of

urban development. His empirical analysis of farming change around Paris provided some support for his ideas, but wide variations in response at the individual farm level meant that his results lacked statistical significance. It is quite possible that the lack of statistical confirmation is attributable to the small size of his sample (only 63 farms) and to local variations in farming conditions within his study areas.

Bryant's inability to produce more than limited empirical support for Sinclair's conceptual framework cannot be taken to mean that the anticipation of urban development does not affect farm patterns, merely that it is not the only factor at work (Munton, 1974) and that it does not lead to the kind of agricultural land-use zonation around cities postulated by Sinclair. This assertion is supported by Thomson's analysis of farming patterns around the main conurbations of England (Thomson, 1981). Although his findings reveal that farming productivity is low close to the urban edge, he is unable to establish any zonation of farming activities away from the urban boundary and he concludes that the agricultural land-use pattern around each conurbation owes more to regional differences in farming than to urban proximity (see also Bryant and Greaves, 1978). This conclusion may also be a result of the tight planning policies in operation around English conurbations and, in particular, the extensive use today of green-belt restraint as a planning tool. Green-belt restraint should increase the security of farmers and reduce the amount of speculative land market activity (Boal, 1970). Indeed, Thomson (1981) is able to demonstrate that farming within green belts is marginally more intensive at comparable distances from the city edge than farming outside green belts.

To conclude, the existing empirical literature
  i) provides only partial support for Sinclair's postulates on the patterning of farming around urban areas, but does indicate the existence of areas of extensive farming close to the urban edge;
 ii) suggests that the anticipation of urban development only exerts an important effect on farming practice right on the immediate urban edge, especially in areas of urban restraint;
iii) indicates wide variations in response to urban proximity at the individual farm level, and
 iv) does not examine at all thoroughly the effects of tenurial differences on the management of urban fringe land (the work of Sharp, see this volume, is an exception).

A CASE STUDY FROM THE METROPOLITAN GREEN BELT: BACKGROUND

The case study concerns the maintenance of urban-fringe farmland within the statutorily approved Metropolitan Green Belt, the 1982 boundaries of which are shown on Figure 22.1. Central government regards the green belt as a strategic planning tool designed to restrain urban development, in this case the outward expansion of London, and to help re-direct urban growth to other planned over-spill schemes (S.E.J.P.T., 1970; D.O.E., 1978a). Most forms of urban land use are prohibited inside green belts (M.H.L.G., 1955, para. 4), a restriction that is primarily implemented by local authorities through the use of their development control powers.

Following the approval of the county development (structure) plans for the area around London between 1978 and 1982, the area of the statutory M.G.B. increased from approximately 3000 km^2 to 4300 km^2. Inevitably, given a green belt of this size, the effectiveness of restraint has not been total. Nevertheless, restraint has been

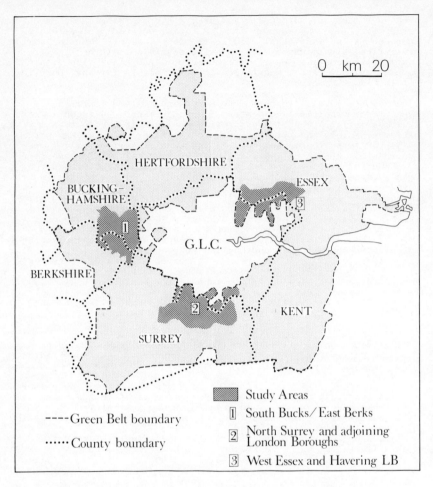

Figure 22.1   Area of the approved Metropolitan Green Belt in 1982
              and location of the study areas

implemented tightly. In 1969 about 17.5 per cent of the M.G.B. (as
approved at that time) was covered by urban land uses[2] (D.O.E.,
1978b), but many of these urban uses were already in existence when
the Green Belt was formally established in the early 1950s. Thus
the increase in these uses within the M.G.B. between 1947 and 1969
amounted to only about 2.5 per cent of the Green Belt area or

---

[2] These urban uses, termed 'developed area' by D.O.E. (D.O.E., 1978b), cannot be
equated exactly with prohibited green-belt land uses as they include urban open
space, reservoirs and hospitals which are acceptable land uses in green belts.
Nevertheless, together these uses impart a good sense of the urbanisation of
the Green Belt.

approximately 350 ha per annum. Except for the construction of the M25 Motorway which encircles London wholly within the M.G.B. and will require about 1900 ha of land, the rate of land conversion to urban uses within the M.G.B. has not increased greatly during the 1970s. There is also general agreement among local planning officers that green belt restraint has prevented scattered development (Munton, 1983).

Partly because the rate of growth of urban uses has been fairly well restricted and partly because of increasing concern over environmental issues, the attention of local planners and politicians has become increasingly focused on the appearance of open land and the way that it is being managed. To local people it is no longer sufficient for green-belt land to be kept open and free of buildings. The land must be positively maintained so that local amenities are enhanced and farmland kept fully productive. It is for these reasons that a number of reports have noted with growing concern low levels of agricultural productivity in the M.G.B. (Standing Conference, 1976; 1977). What these studies do not do, however, is to analyse in any depth the causes of low agricultural productivity (Munton, 1983).

URBAN ANTICIPATION - THE LAND MARKET

Even in areas of tight planning control, urban anticipation affects the willingness of some owners to invest in the long-term maintenance of their farmland. By maintaining an area of open land on the edge of the city, green belts are an increasingly attractive location for certain essential city functions that cannot be accommodated within the urban area; and, by reducing the supply of housing land in areas of desirable residential development, green belts have the effect of raising the value of that small amount of land which is released for development. In this situation, to be largely successful in preventing development is not sufficient to eliminate speculative activity. The possibility of acquiring planning permission is kept alive, even if restraint increases the risk of planning refusal and, in spatial terms, restricts the possibility of development to serviced land adjacent to existing settlements.

In these circumstances there are, effectively, three markets in urban-fringe land if special markets, such as that for mineral land, are excluded. The three markets are those for agricultural land, land with planning permission for residential development, and land with 'hope' value or the land upon which speculative urban value may be said to fall. In the M.G.B. the area of 'hope' value land is limited, although its extent and price have varied according to the state of the development land market generally and in response to short-term adjustments in central government policy toward the release of green belt land (for example, D.O.E., 1973). Generally, however, the price of this 'hope' land is relatively low. For example, land that is expected to acquire planning permission over the next two or three years, perhaps following a planning application and appeal, is usually valued at about one-quarter of that with planning permission, reflecting the substantial risk of failure to obtain permission. Even so the average value of 'hope' land was always 15-20 times that for farmland during the 1970s.

As a result, there is an immense difference between land with planning permission for residential development and land valued simply for agricultural purposes. In 1980, for example, land sold with planning permission for residential development in the

225

Outer Metropolitan Area[3] averaged nearly £180 000 per ha whilst
farmland in south-east England only fetched £3390 per ha. Inside the
M.G.B., development land prices are even higher than in the Outer
Metropolitan Area as a whole. There is no source of data for the
Green Belt itself but a survey of district valuers suggests that
land sold for residential development in the M.G.B. during the
1970s had an average price 40-50 per cent above that for the Outer
Metropolitan Area as a whole (Munton, 1983).

The incentive to speculate is considerable. The above data
consist only of gross values and in practice owners have had to pay
capital and development gains taxes of various kinds and at varying
marginal rates since 1965. Indeed, the tax situation has changed
rapidly in recent years but at no time has the marginal rate of tax
on the capital gain approached 100 per cent. At present, for
example, there is a standard rate of 60 per cent on all development
land gains. This is an apparently high rate of tax until the
absolute size of the gain, net of tax, is taken into account.
Speculation therefore continues even if the large amounts of money
needed to buy the freehold of 'hope' value land in recent years,
and the uncertainty over obtaining planning permission, have
encouraged developers not to buy the freehold but to acquire other
interests in land such as an option or conditional contract. These
allow the developer to seek planning permission without a large
initial capital outlay. They also allow the owner to carry on
farming without loss of income and to retain a non-returnable
deposit if the developer fails to obtain planning permission within
a specified period. Developers' interests in land thus extend well
beyond the land they own. Although they are reluctant to discuss
the extent of their land interests, interviews with developers leave
no doubt that they either have, or have had during the last decade,
an interest in M.G.B. land running into hundreds and probably
several thousand hectares. Furthermore, where they do own the freehold they almost always let the land to farmers on short-term,
insecure leases.

URBAN ANTICIPATION AND FARMLAND MAINTENANCE

Other studies reveal a wide range of individual responses to urban
proximity in terms of enterprise selection. It has proved
necessary, therefore, to define and measure another indicator of
continuing commitment to farming that is transferable between farms.
Standards of land maintenance have been selected to represent this.

Any attempt to assess standards of land maintenance is bound to
include an element of subjectivity. It is impossible to define
precise and absolute standards against which to relate field data.
What is required of land by one farming enterprise is not
necessarily required of it by another. As a result, the survey only
determined whether the occupier was maintaining his land in such a
manner as to avoid any long-term deterioration in its condition.
It was decided that a sufficiently reliable assessment could be
achieved based on a visual evaluation of certain estate management
and farm husbandry features, cross-checking these against the data
collected at interview with the occupier. To be certain of a high
degree of accuracy a detailed analysis of farm investment would be
necessary, but many farmers do not keep records of this kind.

---

3 The Outer Metropolitan Area covers a larger area than the M.G.B. and consists
  of south Bedfordshire, east Berkshire, south Buckinghamshire, west Essex,
  Hertfordshire, north-east Hampshire, west Kent, Surrey and the north-east of
  West Sussex.

Moreover, this level of detail is only necessary when seeking to resolve a particular farm problem and not, as here, when trying to reach an informed view about general standards of land maintenance on urban-fringe farms.

Simplicity has been the primary guide to procedure both in terms of the number of categories used (there are only three) and in the general criteria employed to allocate land to them (Table 22.1).

Table 22.1   Land maintenance categories and criteria

Category I: Areas exhibiting high standards of crop husbandry and farm maintenance; rough margins to fields, small patches of weeds or thin areas in standing crops ignored. Field boundaries and farm roads well maintained, and hedges fully stock-proof where appropriate. Woodland managed; few signs of poor field drainage.

Category II: Areas indicating management problems - crops thin and weedy, with permanent grass deteriorating to rough grazings for no apparent environmental reason. Signs of poor field drainage: farm 'infra-structure' inadequately maintained; woodland receiving little or no attention.

Category III: Derelict or semi-derelict land with either no, or extremely limited, agricultural value at the present time. Occasional grazing and the odd hay crop taken. Field boundaries and farm roads badly maintained unless for the benefit of a neighbouring field.

Most areas exhibited more than one attribute characteristic of their category.
_____

On the one hand, the general condition of growing crops, grazings and woodlands was examined to gain an impression of husbandry standards. On the other, the state of repair of farm roads, hedges and field boundaries, and evidence of poor field drainage, were assessed. Judgements on these separate elements were then combined qualitatively to produce a single value. Category I land gave the appearance of good or even adequate maintenance, the benefit of the doubt always being given to the farmer. Category III land, or land of extremely limited or no agricultural use in its present state, was only assigned sparingly and it represented a minimum estimate of the area experiencing wholly negligent maintenance. All the land in category II may reasonably be adjudged to be poorly maintained and often under-used.

The study was conducted on 185 farms in three representative areas within the inner half of the Metropolitan Green Belt (Figure 22.1). Some farms, or at least parts of them, were sited right on the urban edge and none were more than 10-15 km beyond the city boundary. In each area the sample consisted of approximately 40 per cent of all holdings in excess of 4 ha as recorded in the 1977 Agricultural Census. Each occupier was also interviewed in order to provide the necessary data from which to test a number of hypotheses. These hypotheses stated that lower than average stand-ards of land maintenance were associated with:
  i) land subject to a high risk of urban development;
and that on land of high risk especially low standards of maintenance were associated with:
 ii) owner-occupation;
iii) ownership by non-farming companies;
 iv) the short-term letting of land.

The area surveyed amounted to 14 113 ha and nearly two-thirds of this exhibited few, if any, maintenance dificiencies (Table 22.2). Major areas of farmland did not lie derelict but substantial tracts

Table 22.2   Standards of land maintenance and the effect of urban proximity

Maintenance category	I		II		III		TOTAL
	ha	%	ha	%	ha	%	ha
Planning permission sought	331	31.7	570	54.7	142	13.6	1043
Planning permission not sought	396	38.9	356	35.0	265	26.0	1017
All land next to residential development	727	35.0	926	45.1	407	19.8	2060
All other land	8404	69.7	3274	27.2	375	3.1	12,053
Total	9131	64.7	4200	29.8	782	5.5	14,113

Percentages rounded to one decimal place

were poorly maintained. There is no reason to assume that a proportion of all farmland in the countryside is not poorly maintained.
In the absence of a control survey, however, it was impossible to
put these findings into a broader context but farms even a small
distance away from the immediate urban edge had higher standards of
maintenance than those right beside it (see below).

Farmland maintenance on the urban edge and attempts to acquire
    planning permission

Occupiers were often reluctant to discuss their attempts to acquire
planning permission even though these are recorded publicly in
planning applications registers in local planning offices. They did
not want to appear to be speculating in land. Nevertheless, virtually all were prepared to indicate whether attempts had been made
to obtain planning permission between 1970 and 1977, even if only
about half would furnish details as to the precise parts of their
farms concerned. This meant that it has proved impossible to con-
duct a field-by-field analysis of land maintenance standards in
relation to attempts to obtain planning permission. But those who
gave details confirmed the view that in the M.G.B. it was very rare
to seek planning permission for land that did not immediately abut
existing residential development. Thus what follows is an analysis
of the standards of maintenance of all land next to existing residential development on those farms where the occupier admitted to
seeking permission. These results are compared with those for all
other land on the urban edge where permission had not been sought.
The findings are given in Table 22.2.

On average, fields next to residential areas were demonstrably
less well maintained than other land, but there is no significant
difference in standards for these fields between occupiers who had
sought planning permission and those who had not. Inadequately
precise data may be concealing the full effects on land maintenance
standards of seeking planning permission but other problems, such
as urban intrusion, are obviously important to the maintenance of
land next to residential areas. Moreover, the increasing use of
options by developers, leaving them rather than the landowner with
the responsibility for acquiring planning permission, means that
the landowner may not feel compelled to run down the use of his
land as a way of strengthening his case with the local planning
authority. Some further insights are obtained from examining the
tenure of land right on the urban edge.

Land maintenance and land tenure on the urban edge

Many occupiers were both owners and tenants, but of the land surveyed here only 36.5 per cent was owner-occupied compared to
approximately 57 per cent for England and Wales as a whole (Northfield Committee Report, 1979). Of the remaining 63.5 per cent that
was let, 57.6 per cent was leased on full, secure agricultural
tenancies and 5.9 per cent was leased short term.

When asked their views on the advantages and disadvantages of
farming in the Green Belt, the primary division of opinion came
between those who were largely or wholly tenants and those who
owned the majority of the land they occupied. Tenants approved of
the increased security from development that green-belt restraint
brought them, while owner-occupiers were prone to complain of their
loss of development prospects - an indicator in itself of their
mixed motives for occupying green-belt farmland. Financially,
tenants had little to gain and potentially a great deal to lose from
the development of their land (ibid., 41-9) while the reverse was
true for landlords and owner-occupiers. Many tenants were aware
that if they could demonstrate at a planning inquiry their farming
efficiency and the serious consequences for them of the loss of
their land then their landlords might fail to get planning
permission. For them, the incentive was to farm their land well
while owner-occupiers might improve their chances of planning
permission by running down their land. Local planning officers
admitted that from time to time this did strengthen the applicant's
case. If this phenomenon were at all widespread then it should be
reflected in lower standards of maintenance on owner-occupied land
than on land with full agricultural tenancies. The results in
Table 22.3 do show a small difference in favour of tenanted land
but this crude tenurial distinction adds little to our understanding
of differences in land maintenance standards on its own. Again,
either the ability to agree options with developers might be mini-
mising the difference, or these findings might be said not to bear
out the opinions expressed at interview.

Table 22.3   Standards of land maintenance, land tenure and urban
proximity

Maintenance category	I		II		III		TOTAL
	ha	%	ha	%	ha	%	ha
Owner-occupied land	211	35.3	292	48.9	94	15.7	597
Land let on full tenancies	505	42.1	532	44.3	163	13.6	1200
Land let to companies	22	19.1	71	61.8	22	19.1	115
Land let short-term	11	4.2	102	38.8	150	57.0	263
All land next to residential development	727	35.0	926	45.1	407	19.8	2060

Percentages rounded to one decimal place

Within this category of land let on full tenancies, land let by
companies with no primary commitment to farming was separately
identified. Much of this land was owned by gravel companies and
developers, and the standard of maintenance achieved on this land
was well below the average even for land on the urban edge (Table
22.3). Owners were disinclined to invest in the agricultural
improvement of this land and the occupiers faced an insecure future

as planning permission for urban development or mineral extraction provides the basis for a valid notice to quit the farm.

Finally, those owners unsure about the future agricultural use of their land would avoid letting it on a full agricultural tenancy if they could possibly do so. The short-term letting of land, usually for 6-12 months, allowed the owner to re-possess his land quickly and many of these agreements were informal and technically outside the legislation on agricultural tenancies; but they suited both parties in a situation of land hunger. Vacant possession land also commanded a premium equivalent to approximately 30 per cent of the vacant possession price. Obviously, neither tenants nor owners would invest in improvements to land let short term and this was borne out by the field survey (Table 22.3). The land was very poorly maintained: over half lay derelict or semi-derelict and virtually none was in full productive use.

CONCLUSIONS

Significant areas of farmland on the edges of Western cities lie under-used or even derelict as a consequence of the conversion of rural land to urban uses. This phenomenon is widely reported and it occurs even where tight planning controls designed to prohibit urban development are in operation. In Great Britain, the links between the operation of the urban-fringe land market and the use of farm-land have not been subject to detailed examination; and, more generally, the importance of land tenure to the explanation of individual farmer's responses to urban growth has not been explored.

Data presented in this paper for farmland close to the edge of London support the hypothesis that farm tenure characteristics are important in accounting for differences in land maintenance standards. They are especially significant when the land is owned by non-farming companies and is let short term. It may be concluded that these characteristics, when combined with the effects of urban intrusion, are particularly potent forces encouraging poor land maintenance, even within the Metropolitan Green Belt where urban restraint has been firmly enforced. The effect of green-belt restraint is to confine the spatial extent of land speculation largely to fields immediately adjacent to the urban edge, but not to eliminate it altogether.

Unlike some parts of North America, the owners of urban-fringe farmland in Great Britain are not liable to high municipal taxes assessed on the market value rather than the agricultural worth of their land. Indeed, farmland in Britain incurs no annual local tax (i.e. rating) at all. As a consequence, most owners are under no pressure to sell their land to developers, even if they may be tempted to do so by the large development gain (net of tax) they would make from the sale. This fact, combined with the current reluctance of developers to buy the freehold to such land, ensures a fuller use of the land than would otherwise be the case. Never-theless, there is still a financial incentive to run-down the use of the land, as and when it suits the owner's interest to do so, because many owners believe that this improves their chances of acquiring planning permission for urban development. Anecdotal, rather than hard, factual evidence supports this belief, and the retention of a free market in development land will guarantee specu-lation and the under-use of land. Whether the public benefits to be derived from the fuller use of agricultural land in the urban fringe, and whether the reduced inequity between the owners of urban-fringe land and other members of society, combine to warrant the complete regulation of the land market and the elimination of private betterment, must be a largely political judgement which lies beyond the scope of this paper.

ACKNOWLEDGEMENTS

I would like to thank Professor Chris Bryant (University of Waterloo) and Elizabeth Sharp (University College London) for their valuable comments on a draft of this paper.

REFERENCES

Advisory Council on Agriculture and Horticulture (A.C.A.H.), 1978. *Agriculture and the countryside.* (Robendene, Chesham).

Boal, F.W. 1970. Urban growth and land value patterns. *Professional Geographer,* 22, 79-82.

Brown, H.J., Phillips, R.S. and Roberts, N.A. 1981. Land markets at the urban fringe. *Journal of the American Planning Association,* 47, 131-44.

Bryant, C.R. 1974. The anticipation of urban development: part I. Some implications for agricultural land use practices and land use zoning. *Geographia Polonica,* 28, 93-102.

Bryant, C.R. 1981. Agriculture in an urbanizing environment: a case study from the Paris region, 1968-1976. *Canadian Geographer,* 25, 27-45.

Bryant, C.R. 1982. *The rural real estate market.* Department of Geography Publication Series no.18, (University of Waterloo, Waterloo).

Bryant, C.R. and Greaves, S.M. 1978. The importance of regional variation in the analysis of urbanization-agriculture inter-actions. *Cahiers de Géographie du Québec,* 22, 329-48.

Coleman, A. 1976. Is planning necessary? *Geographical Journal,* 142, 411-30.

Countryside Review Committee (C.R.C.), 1978. *Food production in the countryside: a discussion paper.* Topic paper no.3, (H.M.S.O., London).

Department of the Environment (D.O.E.), 1973. *Widening the choice: the next steps in housing.* Cmnd. 5280, (H.M.S.O., London).

D.O.E. 1978a. *Strategic plan for the South-East review: government statement.* (H.M.S.O., London).

D.O.E, 1978b. *Developed areas 1969: a survey of England and Wales from air photography.* (D.O.E., London).

Lichfield, N. and Darin-Drabkin, H. 1980. *Land policy in planning.* Urban and Regional Studies no.8, (George Allen and Unwin, London).

Low, N. 1973. Farming and the inner Green Belt. *Town Planning Review,* 44, 103-16.

Ministry of Housing and Local Government (M.H.L.G.), 1955. *Green Belts.* Circular 42/55, (H.M.S.O., London).

Moss, G. 1981. *Britain's wasting acres: land use in a changing society.* (Architectural Press, London).

Munton, R.J.C. 1974. Farming on the urban fringe. in *Suburban growth: geographical processes at the edge of the western city,* Johnson, J.H. (ed), (Wiley, London), 201-23.

Munton, R.J.C. 1983. *London's Green Belt: containment in practice.* (George Allen and Unwin, London).

Northfield Committee Report, 1979. *Committee of inquiry into the acquisition and occupancy of agricultural land.* Cmnd. 7599, (H.M.S.O., London).

Organization for Economic Co-operation and Development (O.E.C.D.), 1979. *Agriculture in the planning and management of peri-urban areas. vol.I: Synthesis.* (O.E.C.D., Paris).

Pearce, B.J. 1980. Instruments for land policy: a classification. *Urban Law and Policy,* 3, 115-56.

South East Joint Planning Team (S.E.J.P.T.), 1970. *Strategic plan for the South-East: report.* (H.M.S.O., London).

Sinclair, R. 1967. Von Thünen and urban sprawl. *Annals of the Association of American Geographers,* 57, 72-87.

Standing Conference, 1976. *The improvement of London's Green Belt.* SC 620, (Standing Conference, London).

Standing Conference, 1977. *The improvement of London's Green Belt - a second report.* SC 860R, (Standing Conference, London).

Thomson, K.J. 1981. *Farming in the fringe.* CCP 141, (Countryside Commission, Cheltenham).

Chapter 23

Rural settlement planning:

too great an expectation?

Patrick Hanrahan and Paul Cloke

Many commentators on statutory planning in Britain agree that the
Town and Country Planning Acts of 1947 and 1968 heralded false dawns
in the sense that hardship, which seemed capable of significant
reduction, continues entrenched. Recent Ministerial advice has
reasserted land-use priorities for planning. In rural areas,
material deprivation continues to cast a shadow over the lives of a
significant proportion of the inhabitants (Walker, 1978; Moseley,
1979; Shaw, 1979). Perhaps the chief criticism that can be levelled
at rural settlement planning practice in Britain since the War is
that it has lacked an appreciation of the heterogeneity of the population, with their variety of needs and varying abilities to cope
with the problems facing them. Rural planning policies generally
have appeared insensitive to the plight of the deprived, concentrating as they have done on maintaining the agricultural and amenity
monopolies of vast areas. They have tended to display a seemingly
intractable stance against development except for overseeing the
concentration of opportunities and resources in certain focal
settlements.

Prima facie the statutory planning system has been misapplied
in rural areas, but what is also questionable is whether or not the
existing apparatus itself provides an efficient adaptive instrument
for dealing with disadvantage. We suggest that it does not. Where
attention has been paid by planners and local politicians to the
plight of the deprived, it amounts only to limited potential
provision or has proven largely ineffectual in practice. The
obstacles to progressive social planning appear insurmountable.
What exactly are the obstacles? Precisely where are the planners
located within the wider sphere of public policy processes in Britain
and what are the limitations inherent in the formulation and
execution of their plans? Answers to these questions will aid our
understanding of current and past deficiencies in planning responses
to rural disadvantage.

THE ROLE OF THE STATE

Planning and the application of policy decisions by local authorities
are aspects of state activity and *ipso facto* circumscribed. This is
not a novel assertion, but its past advocacy has rested largely upon
assessment of the relationship between central and local government
rather than that between the state in its entirety and its host
society. Nonetheless, prompted by Miliband's theory of the

capitalist state with its contentious pivotal theme that the
principal functionaries are drawn from and act in co-ordinated
fashion to maintain the dominance of an élite (Miliband, 1969), a
lively debate has grown up around this topic amongst European
sociologists focussing primarily upon urban issues (overviews are
presented by Dunleavy, 1980; Saunders, 1980). However, much of the
British literature on public policy *per se* has eschewed robust
analysis of the state-society connection, tending to subscribe,
often implicitly, to a liberal political theory of the state as an
independent entity arbitrating conflict between divers interests
within society all of which are deemed to have a representative
voice in the democratic system of government. This pluralist
conception provides a theoretical underpinning to the original
'managerialist' thesis in which any question of the state's
impartiality is bracketed (Pahl, 1970).

Following Miliband, the notion of the capitalist state's
neutrality has also been rendered problematical by an analytical
perspective, derived from Althusserian 'structuralism', which
identifies it instead as an epiphenomenon of a power differential
between social classes. Thereby its activity is seen to maintain
the social formation from which it derives authority, with the state
itself having no autonomy outwith limits cast by the economic
structure it exists to preserve. It is thus perceived as essentially conservative, its *raison d'être* being ultimately to sustain the
disparity characteristic of the society of which it is born
(Poulantzas, 1973; Castells, 1978).

The foregoing is an obvious conflation of a complex theoretical
and meta-theoretical critique of 'uncritical' appraisals of the
state's role, but it indicates an alternative starting point for
examining public policy and the restricted outcomes of state intervention. Although recent urban research has flown in the face of
the 'liberal' convention (Cockburn, 1977; Lambert *et al.*, 1978;
Blowers, 1980; Saunders, 1980), rural local government, including
its statutory planning, has yet to be placed within a framework of
critical political economy (excepting the work of Newby *et al.*,
1978). If the policy at all levels including the local one proceeds
on an acceptance of a delimited range of policy options, then it is
appropriate to establish how democratic and reputedly independent
local government is thus confined. Change effected through redistributive state activity may be seen to be piecemeal and superficial
but, as Saunders (1981) argues, the description of the impact of
state intervention has largely failed to present empirical evidence
to show *how* the outcome is reached. The state is not a monolith
and there are readily apparent conflicts between and within
different echelons and agencies making up its institutional
arrangement. Nor do its actions necessarily serve the immediate
interests of the powerful in society, although ultimately they
might. Whilst responsive in theory to the local electorate, local
government is in reality hindered by the environment in which it
operates - an environment within which it is engaged in three sets
of imbalanced relations all of which compress any potential ability
to engineer progressive social change.

DECENTRALISATION AND LIMITED AUTONOMY

The first of these relations concerns the administrative core and
devolved authority. Rhodes (1980) aptly described the relationship
between central and local government in Britain as being characterised by 'ambiguity, confusion and complexity'. Public administration at local and national scales, as well as regional and area
ones, produces a dense network of linkages between departments and

agencies supervising the allocation of resources. Thus different sectors of sub-national government enjoy some leeway beyond meeting their basic statutory obligations,this leeway being manifest spatially in the extent and quality of service provision and in imperfect levels of expenditure (see the paper by Van Bemmel in this book). Nevertheless, the discretion of local government is narrowed by the overriding influence of central government which subordinates locally identified need to its own definitions of priorities. Despite the apparent delegation of power to local authorities, their policies for education, transport, housing, health and social services have to be endorsed by the respective government ministries and, similarly, development plans have to obtain the approval of the relevant Minister. His ability to alter the emphasis of planning strategies was applied recently, for example, to the structure plans for North Yorkshire, Gloucestershire and the Lake District (see the paper by Clark in this book).

The centre has wide powers of veto, *de jure* and *de facto,* which allow it to regulate activity on the administrative periphery, with its controls being channelled through both legal and financial avenues. The realms within which local and regional authorities function, and their duties therein, are determined by law. In addition, non-statutory directives or advice notes significantly augment the established code. The possible application of the *ultra vires* principle and the invocation of default powers by central government can also have a deterrent effect. Financially, too, the heavy dependence of public authorities upon the Exchequer markedly restricts expenditure and shapes its distributon, reinforcing legislative controls. Auditors acting for central government can recommend redress for what is deemed unreasonable expenditure. Thereby authorities are open to challenge on grounds of fiduciary irresponsibility and elected members of local government to the threat of being personally surcharged for unlawful spending. The budgets of public agencies are subject to central government review, and levels of expenditure by local authorities are bound by the centre's benchmarks upon which its grant contribution is based. Not only are targets for revenue expenditure stipulated annually, with financial penalities for spending above these limits, but also the loans with which local government finances the bulk of its capital programmes are liable to central sanction. This dependence and attendant accountability is further underlined by recent measures introduced to prohibit local authorities from levying supplementary rates (i.e. local property taxes) in order to meet financial deficits. This tightening of the grip has been justified by the Minister concerned in terms of recognising a 'difference between freedom and licence' (The *Guardian,* 28.7.82), a phrase which encapsulates the tenor of the central-local relationship and indicates the restriction under which local authorities are placed.

THE CONFUSION OF SUB-NATIONAL GOVERNMENT

By ordinance and financial contingency, local government is affectively impelled to act in compliance with exhortation from the centre, and this control is compounded by the second dimension to the public policy environment - that of overlapping administrative responsibilities between different bodies (see the paper by Sharp in this book). District and county councils, sometimes of different political colours, are required to liaise with each other and with area and regional authorities over the provision of public services in areas common to their geographical boundaries. A frequent result is antagonism at their interface (Leach and Moore, 1979). Districts and counties have to achieve consistency between their plans for land use and general development, and clashes can arise where,for example, a county plan specifies settlements to be given preference for development and the choice of settlements is disputed by the districts concerned. Thus, in Gloucestershire all six districts objected to the inclusion in the county's draft structure plan of lists of 'main villages' for development priority. They were particularly perturbed by the fact of the selection which they interpreted as an abrogation of their responsibilities. In this case the county capitulated and rewrote the plan. Elsewhere (for example, Glyndwr district in Clwyd) county/district conflict over the content of the established structure plan has led to an open challenge to approved policies by the district. Again, the onus upon districts to fit local plans to county structure plans can be made additionally heavy where proposed road developments are specified by the county without guarantee of completion owing to uncertainty over the availability of funds. Consequently, plans resting upon an extended transport infrastructure have to be formulated on an 'if and when' basis.

Moreover, local policies impinging upon different fields such as health services and public utilities are required to complement those of the area and regional administrations directly concerned. One illustration of this is provided by Glyn-Jones's(1979) account of the shortfall of housing development in Hatherleigh (Devon) in part attributed to a failure by the regional water authority to increase the capacity of the sewage disposal system for the town despite its key settlement status in the Devon county plan. This situation was made even more complex by the intervention of one of the settlement's major employers which wished to use its influence in favour of additional housing development.

Even when public agencies are compelled to collaborate by law or practicality, they reflect the general organisational tendency of jealously guarding their own decision-making domains rather than willingly submitting to co-operative and harnessed effort. To deviate from an established course or to compromise in anything other than loose or temporary liaison might belie future justification for maintaining the separate status of bodies obliged to underline their legitimacy in perennial competition for resources. With regard to rural areas particularly, there is a general paucity of co-operative ventures beyond unavoidable joint action, notwithstanding limited experiments such as those undertaken in Hereford and Worcester (Hereford and Worcester County Council, 1978) and Powys. Therefore, calls for widespread co-ordination of effort display an ingenuousness. As Leach (1980) put it: 'authorities will only co-operate when it suits them, or when they have to, and then very much on their own terms, and in line with their own interests'. Nonetheless, the central government-sponsored Working Party on Rural Settlement Policies (1979) could still advocate concerted action as the main policy tool for rural areas. Sharp's paper in this book would suggest that these observations were as true of the 1930s as they are today.

Conflict of interest is also a feature of intra-organisational relationships. Given the considerable variation in local authorities' internal structures (Greenwood et al., 1980), arguments for a 'corporate' approach in local government planning have been predicated on a recognised need to overcome departmentalism. The emergence of 'executive' officers and 'policy and resources' committees has heralded attempts to integrate the activity of otherwise separate individuals and sections. At both district and county levels, planning officers encounter impediments to achieving strategies compatible with the policies of other departments within the same authority, even though the structure plan era has produced some improvement in organisational co-operation over that demonstrated in the earlier development-plan period. The county planners' assumptions concerning personal mobility in rural areas could run counter to the surveyor's department's concentration upon road building and maintenance to the detriment of public transport provision. For the district planner, changes to the housing department's investment programme might appear similarly disruptive to the key settlement policy. Comprehensive policy statements will generally rebound upon their authors, proving inapplicable because of their 'in-built tendency to invade the empires' or to 'threaten the claims to professional exclusiveness' of others who have had little or no say in their composition (Davies, 1972).

Additional complications arise within local government out of relations between the full-time officials and the part-time elected membership. The former's claim to technical expertise may influence the issues considered and the decisions taken by the latter. The power of departmental heads, senior management, committee chairmen and leading local political party members may therefore be of significance (Gyford, 1976; Newton, 1976; Blowers, 1980; Flynn, 1981). Turnover of personnel and changes in the political complexion of some councils can also undermine the consistency and continuity of policies. The variegated composition and internal relations of non-elected public resource agencies also affect patterns of distribution and expenditure. This is exemplified by the levels of provision made by health authorities which have historically varied across the country.

The institutions involved in policy and action at the local level function from positions of differing power and with a zeal exposing their functionaries' attitudes towards the role of their organisation in relation to specific issues. When actors perceive an external challenge to their authority over particular matters, they tend to close ranks which serves in turn to preclude co-operative action on any scale. Thus both overlapping interests and coterminous policy areas as well as disagreements and contradictions are common to the administrative periphery.

PUBLIC PLANNING AND PRIVATE INTEREST

There is a further crucial dimension to the environment within which planning strategies are drafted and pursued. The amorphousness or ambiguity of much planning policy is founded as much upon an awareness of the uncertainty of appropriate financial backing by private interests as upon any aim to provide flexibility in the face of any other unanticipated complication. Town and country planning can be negative in proscribing change, passive in officially allowing private developments without the involvement of the authorities themselves as underwriters, and positive in committing resources to projects which they are prepared to undertake or actively co-ordinate. With the often stultifying effect of externally determined budgetary limitations, the negative and, to a lesser extent, the passive modes have predominated in rural areas. As a result local-government development policy is largely reliant upon the availability of private investment. Councils' failure to realize what potential the 1975 Community Land Act afforded is evidence of this.

Much of what appears in local authority plans as being positive or potentially constructive is necessarily couched in tame rhetoric, expressing intentions to 'encourage', 'promote' or merely 'permit' private developments within their boundaries. Relatively scant financial resources are available to local government for capital projects and for expenditure on service provision. In the case of the latter, the high unit costs encountered in rural areas with small and dispersed settlement have prompted concentrated deployment based upon calculations involving the presence of existing facilities and population thresholds, together with campaigns for deficiencies to be offset by community 'initiatives' and self-help in lieu of official intervention. Such economic rationalisation, the legitimacy of which has actually been partially questioned (Gilder, 1979), rests within a situation in which it is private and not public finance which supports the development that planners are charged with managing.

In this respect too, then, their policies amount to sketches of possibilities rather than blueprints, often specifying controls to be imposed upon hypothetical development which they are unable to guarantee anyway. Such is exemplified by Cambridgeshire County Council's attempts in the post-war period to steer housing and manufacturing development into some outlying parts of the county which met with a mixed success, even allowing for variations in the availability of supportive facilities. Areas with similar public utility provision and proximity to main communication networks and urban centres fared differently, principally because the owners of investment capital had their own reasons for choosing some places and not others (Cambridgeshire County Council, 1966). Such selectivity indicates the fundamental weakness of any positive part that local authorities might play in this regard. They enjoy little more than a supervisory capacity *vis-à-vis* speculative activity. A comment by Murie (1980) referring to their housing record, sums it up: 'Much housing planning and policy-making gives little attention to the private sector and perpetuates a myth of competence and control. One consequence is often a very wide gap between plans and practice and between the intention and consequence of policy.' One such gap has been revealed by Blacksell and Gilg (1981) whose research in Devon found that, despite intentions to impose strict planning controls, new residential development may not be resisted in pressured rural areas when developers decline to undertake schemes in locations preferred by the local authority but opt instead for sites from which they would otherwise have been debarred.

Where central government does intervene to stimulate growth, for example through the provision of incentives in the form of regional grant aid or under the auspices of the Council for Small Industries in Rural Areas, entrepreneurial interest often remains elusive as is shown by the boom in high-technology industries in counties such as Berkshire rather than in centrally designated growth areas, including 'enterprise zones' in depressed conurbations. Capital remains largely independent of public manipulation and is deployed according to a market logic, with profit as the key criterion and the accumulation imperative absolute. Statutory planning can only be neo-planning in such circumstances. Whether capital is active or inert in the first instance lies beyond the planners' immediate sphere of control. There are essential limits to the current relationship between state and society.

CONCLUSION

Although rural settlement planning in Britain has persistently demonstrated an insensitivity towards disadvantage, it has been argued that any failure to confront the matter is as much a consequence of the essential limitations to the planning system as it is of obvious complacency. Those who are instrumental in formulating policy and charged with overseeing its application do have scope for initiative but its scope is narrow. Planning is a restricted exercise as it is an aspect of state activity, and as such the brief accorded planners is limited by the nature of the relationship between devolved and central government, between planning agencies and public resource bodies, and between all these and private interests.

The haggling, trade-offs and compromises which characterise both the formulation and application of policies are constrained by the statutory duties of the participants. The content and impact of planning policy need to be seen, therefore, as the legacy of conflict, and official plans viewed as catalysts for the further interaction of disparate interests as well as being the product of contradictions. The statutory planning system as we know it in Britain is an inappropriate device for procuring progressive social change. If nothing else, its apparently ineluctable failure to ameliorate significantly the situation of the rural deprived tells us that. Perhaps, also too much has somehow come to be expected of it, both because academic commentators have repeatedly accepted at face value the idealistic objectives of statutory plans, and because planners have been slow to recognise and face up to the opposition towards any efforts to respond effectively to rural social disadvantage.

REFERENCES

Blacksell, M. and Gilg, A. 1981. *The countryside: planning and change.* (Allen and Unwin, London).

Blowers, A. 1980. *The limits of power.* (Pergamon, Oxford).

Cambridgeshire County Council, 1966. A rural planning policy and its implementation. *Official Architecture and Planning,* 29, 1126-41.

Castells, M. 1978. *City, class and power.* (Macmillan, London).

Cockburn, C. 1977. *The local state.* (Pluto, London).

Davies, J.G. 1972. *The evangelistic bureaucrat.* (Tavistock, London).

Dunleavy, P. 1980. Social and political theory and the issues in central-local relations. in *New approaches to the study of central-local relationships,* Jones, G. (ed), (Gower, Hampshire).

Flynn, R. 1981. Managing consensus: strategies and rationales in policy-making. in *New perspectives in urban change and conflict,* Harloe, M. (ed), (Heinemann, London).

Gilder, I.M. 1979. Rural planning policies: an economic appraisal. *Progress in Planning,* 11, 213-71.

Glyn-Jones, A. 1979. *Rural recovery: has it begun?* (Devon County Council and Exeter University, Exeter).

Greenwood, R., Walsh, K., Hinings, C.R. and Ranson, S. 1980. *Patterns of management in local government.* (Martin Robertson, Oxford).

Gyford, J. 1976. *Local politics in Britain.* (Croom Helm, London).

Hereford and Worcester County Council, 1978. *Rural community development report.* (The Council, Worcester).

Lambert, J., Paris, C. and Blackaby, B. 1978. *Housing policy and the state.* (Macmillan, London).

Leach, S.N. 1980. Organisational interests and inter-organisational behaviour in town planning. *Town Planning Review,* 51, 286-99.

Leach, S.N. and Moore, N. 1979. County/district relations in shire and metropolitan counties in the field of town and country planning. *Policy and Politics*, 7, 165-79.

Miliband, R. 1969. *The state in capitalist society*. (Weidenfeld and Nicholson, London).

Moseley, M.J. 1979. *Accessibility: the rural challenge*. (Methuen, London).

Murie, A. 1980. The housing service. *Town Planning Review*, 51, 309-315.

Newby, H., Bell, C., Rose, D. and Saunders, P. 1978. *Property, paternalism and power*. (Hutchinson, London).

Newton, K. 1976. *Second city politics*. (University Press, Oxford).

Pahl, R.E. 1970. *Whose city?* (Longman, London).

Poulantzas, N. 1973. *Political power and social classes*. (New Left Books, London).

Rhodes, R.A.W. 1980. Some myths in central-local relations. *Town Planning Review*, 51, 270-85.

Saunders, P. 1980. *Urban politics*. (Penguin, Harmondsworth).

Saunders, P. 1981. Community power, urban managerialism and the local state. in *New perspectives in urban change and conflict*, Harloe, M. (ed), (Heinemann, London).

Shaw, J.M. (ed) 1979. *Rural deprivation and planning*. (Geo Abstracts, Norwich).

Walker, A. (ed) 1978. *Rural poverty*. (Child Poverty Action Group, London).

Working Party on Rural Settlement Policies, 1979. *A future for the village*. (H.M.S.O., London).

SECTION FOUR:

POLITICAL INFLUENCES IN THE COUNTRYSIDE

Chapter 24

Geography and nationalism:
the case of rural Wales

Gareth Edwards

INTRODUCTION

The emergence of nationalist forces within Wales since the Second
World War has led to increasing conflict between the Principality of
Wales and the British state. This has been particularly so since
the political recognition gained by Plaid Cymru (the Welsh Nationalist Party) in the General Election of 1974. This led to a review of
government policies toward the region and the granting of a number
of concessions to its people (Nairn, 1977). This process of conflict
and concession has been evident in the case of rural Wales, where
resources have been allocated and utilised against a background of
political and cultural struggle. Consequently, the human geography
of that area has been substantially affected by the confrontation
between the nationalist movement and central government at Westminster. The purpose of this paper is therefore to discuss the
following factors:
1) the impact of a capitalist system on a peripheral rural area;
2) the power of a regional nationalist movement to modify central
   policy;
3) the socialist alternative for rural Wales advocated by the
   nationalist party;
4) the implication of the above for geographical research in rural
   Wales.

BENEFITS OF THE CONFLICT FOR RURAL WALES

The pressure from nationalist movements within Wales during the
1950s and early 1960s had only limited success in influencing central
government decisions. However, Griffiths (1979) pointed out that,
'This situation was transformed by the Carmarthen by-election in
July 1966. Gwynfor Evans' victory was of great significance. Plaid
Cymru was beginning to modernise itself substantially... In the next
few years its progress in parliamentary elections was to be spec-
tacular. There were high polls at Rhondda West (1967), Caerphilly
(1968) and Merthyr (1972)'. This emergence of Plaid Cymru from the
political wilderness culminated in its winning three rural constit-
uencies in the 1974 General Election. The 'hung' parliament after
this election ensured that the Labour government needed to keep the
support of the three Plaid Cymru Members of Parliament in order to
remain in power. This in turn resulted in a series of measures of
benefit to Wales, including its rural areas. An immediate benefit
to Wales was the increase in public expenditure in the Principality

which saw its share of total expenditure in the United Kingdom rise from 3.85 per cent in 1973 to 4.65 per cent in 1977. Although this increase was small in terms of overall government spending, the rise in expenditure per head of the population between 1974-75 and 1978-79 was greater in Wales (85.1 per cent) than in either England (68.9 per cent) or Scotland (72.9 per cent). This increase in expenditure was particularly welcomed in rural Wales. For example, the Transport Supplementary Grant for the county of Powys rose from £8.30 per head of population in 1975-76 to £35.30 in 1978-79 whilst the grant for the counties of Dyfed and Gwynedd rose from £3.20 to £14.40 and from £4.50 to £15.30 respectively. During the same period the total Transport Supplementary Grant for England and Wales decreased from £6.00 per head of the population in 1975-76 to £5.60 in 1978-79. Although it is true to say that during the Labour government of 1974-79 public expenditure increased throughout Britain, it would appear that Wales in particular, including its rural areas, benefited as a result of the government's reliance on the support of the Nationalist M.P.s.

Of greater significance, however, is the major legislation of specific benefit to rural Wales introduced during the period 1974-79. The first major measure was the Welsh Development Agency Act (1975) which established an agency with a borrowing limit of £150 million to attract and generate industrial development in the Principality. A further step in the same direction took place the following year with the Development of Rural Wales Act which established the Development Board for Rural Wales (D.B.R.W.), which is now called Mid-Wales Development. This was followed in 1977 by the Water Charges Equalisation Act which sought to restrict the difference in the cost of water ('water rates') between the various regions of Britain. The impetus for this measure came mainly from within Wales where water rates were considered excessively high in comparison with those in the English conurbations which drew their water from reservoirs in rural Wales. Further legislation which was thought likely to be of direct benefit to Wales was the Community Land Act (1975) which established the Land Authority for Wales. Throughout this period, the ill-fated plans for the devolution of government to Wales were being drawn up which put the government's slender majority in Parliament in even greater danger and forced further dependence upon the support of Plaid Cymru. In 1979, this resulted in the payment of compensation to quarrymen in north Wales who suffered from the lung disease, silicosis: such compensation had been demanded by Plaid Cymru in return for their support of the government.

Thus, the rise of Plaid Cymru, as an expression of the wider nationalist movement in Wales, coupled with the favourable political conditions prevailing in a 'hung' parliament, resulted in a number of apparently favourable commitments to the Principality during the period 1975-79. Unfortunately, the rejection of the government's devolution proposals in the referendum of 1979 brought these advantageous conditions to an end. Even after this defeat and the return of a substantial Conservative majority in the present parliament, however, Welsh nationalists have continued to gain concessions from Westminster by extra-parliamentary activity. These concessions have primarily been directed towards enhancing the cultural life of Wales: for example, the placing of Welsh words above English ones on road signs (this was originally refused on grounds of safety), and the setting up of a fourth television channel using the Welsh language, which was rejected at first by the government on economic grounds. Currently, the conflict is taking on a more political and economic tone with the present campaign of non-payment of water rates. More militant signs of the conflict are seen in the burning down of holiday homes and the placing of incendiary devices at public buildings in both Wales and England.

The last decade demonstrates that pressure from nationalist sources has brought concessions from central government which have been of some benefit to the inhabitants of rural Wales. However, it can be argued that such concessions have only been cosmetic in nature and have done little to solve the underlying problems of rural Wales which are rooted in its relationship with the British capitalist state.

RURAL WALES: THE PROBLEM REMAINS

The 'nullification' of nationalist forces (Nairn, 1977) and the strengthening of 'regionalist consensus' (Rees and Lambert, 1981) through the use of the above concessions has been of only limited relevance to the underlying problems of rural Wales. Williams (1980) has defined these problems as being the result of '... the control of resources by externally based powers (i.e. from outside rural Wales) which denies the indigenous population the ability to sustain a locally based pattern of balanced economic development. The process whereby core areas organise the production of the periphery has the effect of concentrating resources at the core and creates a progressive economic and political dependence of the periphery on the core'. Thus politically, economically and socially the development of rural Wales is still determined not according to the needs and aspirations of the local community but according to the dictates of the 'core' capitalist interests which dominate central government. This may be illustrated by the following examples.

The present government with its large majority can effectively ignore political pressure from Welsh nationalists and consequently the problems of Wales receive less attention than previously. For example, during 1980-81 an average of only five minutes a day was spend in discussing Welsh affairs in Parliament. In contrast to the 1974-79 'Welsh Parliament', Wales has been returned again to the political periphery where the problems of rural Wales can no longer attract the political will necessary for their solution. This has been recently demonstrated by the government's withdrawal of Assisted Area status from the majority of mid-Wales, a decision which effectively undermines the working of the D.B.R.W. Although its ability to build advance factories and to provide financial inducements to industrialists has been enhanced, the loss of Assisted Area status has reduced its ability to attract new industries. Thus, Iain Skewis, Chief Executive of the D.B.R.W., recently stated that, 'The flag we have been flying to attract industrialists has been pulled down. We have to find a new one to pin to the flagpole' (*Western Mail, 28.7.82*).

However, even at the outset of the Board's life, its ability to achieve its aims was limited by the fact that it was the servant of the 'core' economy working in the periphery. As the Board was required to concentrate almost exclusively on manufacturing industries, many of which have little history in the area, its success has been mainly in attracting companies into the area rather than in helping to establish new, indigenous developments (D.B.R.W., 1981). This has resulted in the industry of mid-Wales being dependent upon the vagaries of the 'core' economy. For example, Newtown, the Board's show-piece, recently lost 500 jobs when two companies, B.D.R. (a subsidiary of G.K.N.) and Dowty Engineering closed as a result of national circumstances rather than because of difficulties with their Newtown factories. Although the Board

245

claims to have created 6350 'job opportunities' between 1977 and 1981 through its factory-building programme, unemployment in the area rose from 8.4 per cent in March 1977 to 11.7 per cent in March 1981. At the present time the towns of Lampeter, Cardigan and Tywyn, all of which are in the Board's area, are amongst those with the highest unemployment figures in Wales with one in every five workers being unemployed. Furthermore, the 'job opportunities' created by the D.B.R.W. are not necessarily new jobs as there are many instances of existing businesses in the area moving to its industrial estates to take advantage of preferable rent arrangements on newer premises. Wenger (1982) has pointed out that there is often a conflict between the D.B.R.W. and local communities regarding the type of development needed in rural Wales. The duties of the D.B.R.W. mean that it can only provide limited benefits for rural Wales. It is able to counter neither the uneven development of its area nor the apathy expressed towards it by the local communities.

A further example of the core exploiting this peripheral area of rural Wales is the decision of the government to revoke the Water Equalisation Act, which allows the core region of the English Midlands to extract water from Welsh reservoirs and to charge its consumers half the price of water in Wales. Thus, water rates in Birmingham, which is served by reservoirs in mid-Wales, are 15p in the pound compared with 34p in the pound for Welsh rural communities within sight of those same reservoirs (*Y Ddraig Goch,* May 1982). In 1983 the Severn-Trent Water Authority paid £1.5 million for the 360 million litres of water a day it extracted from mid-Wales reservoirs but the Welsh Water Authority was trying to charge the Severn-Trent Water Authority a further £3 million a year. That authority, however, refused to meet this increase in charges and the government supported their position. Plaid Cymru is calling for a total payment of £40 million a year from both the Severn-Trent and North West (England) Water Authorities for the water they extract from Wales. This situation is further aggravated by the fact that the North West Water Authority pays the Severn-Trent Water Authority £400 000 a year for water extracted from Lake Vyrnwy on the Powys-Gwynedd border which is under the latter's control. Thus Liverpool pays Birmingham for water collected in a drowned Welsh valley.

Even when developments do take place, they often do so for the benefit of the 'core' rather than for the benefit of rural Wales itself. Williams (1980) cites the Dinorwic hydro-electric power station as a development which provides long-term benefits for the core economy in the form of cheaper electricity from this 'pumped-storage' scheme, while providing only short-term employment in the local area during the construction phase. He concluded that, 'In fact, projects such as these may have a positively de-stabilising effect on the local labour market. Despite being among the lowest paid of the workers employed during the construction stages, local people can still earn more than previously. Accordingly, there is a reluctance to exchange new-found 'affluence' for the lower rewards available locally after construction is completed. Hence, among the local work-force at Dinorwic, 67 per cent of those under 30 years of age expressed the intention of leaving the area when they were made redundant. Thus, a project which was presented as one which, in part, was intended to ameliorate levels of unemployment and associated out-migration may well serve to stimulate the latter'. Thus, the concessions made at Westminster bring only limited benefits to rural Wales and leave untouched the basic problems of its uneven development as a politically and economically peripheral area.

246

AN ALTERNATIVE FOR RURAL WALES

The problems of rural Wales are consequently not unique and exhibit many of the characteristics present throughout the peripheral rural areas of north-west Europe. However, these problems in rural Wales are given added weight by the area's cultural distinctiveness (Williams, 1980). The presence of the Welsh language and the existence of unique cultural institutions such as eisteddfodau (cultural festivals) in rural Wales exacerbate the conflict with Westminster. Thus, the issue of second homes, a symptom of uneven development in peripheral areas, is given added significance in rural Wales because of the detrimental impact on the Welsh language of English-speaking in-migrants. This combination of uneven development and cultural distinctiveness has sustained the call that the communities of rural Wales be more directly involved 'in the change and future direction of the society' (Wenger, 1980).

The nationalist alternative for rural Wales is embodied in this principle of self-determination for its peoples. Consequently, Plaid Cymru (1970) stated that, 'We believe that self-government is a prerequisite to Welsh economic growth, not because Welshmen are any more able than Englishmen or Scotsmen in these matters, but because only a government serving Wales, and Wales alone, can give the unqualified commitment and unwavering attention to the problems of Wales that is necessary to solve our problems'. Thus the nationalist view is that while concessions won at Westminster may bring limited benefits, it is only by divorcing Wales from the control of 'core' capitalist interests in London that the economic and concomitant social development of rural Wales may take place. The basis of this development would lie in the control of the resources and money within Wales by the people through industrial democracy, workers' co-operatives and a Welsh Parliament. Plaid Cymru (1978) stated that, 'We believe in 'co-operative democracy' where the workers elect a board of directors to manage their factories and workshops in conjunction with representatives from the local community, the consumers and central government'. Such policies are enshrined in the party's acceptance of the aims of 'decentralised socialism' at its 1981 Conference. The purpose of the party is not simply to replace government from Westminster by government from Cardiff, but to give local elected bodies and local workers a greater control over their affairs. An example of how such policies could be implemented is seen in the town co-operative system of Mondragón in the Spanish Basque country. The co-operative, which was founded in 1956, has created nearly 20 000 jobs, and it has its own bank (the People's Savings Bank) and industrial research centre. At a time when unemployment has risen to around 20 per cent in the Basque country, not one Mondragón worker has lost his job. It must also be noted that the co-operative's aims are not solely eocnomic but are also closely linked to the maintenance of the local community and culture (Reeves, 1980).

Elements of this approach are already evident in the prominent position held by members of Plaid Cymru in such community co-operatives as Antur Aelhaearn and Cymdeithas De Gwynedd in north Wales. The emphasis within the nationalist plan for rural Wales is therefore upon the development of self-reliant community and workers' co-operatives to provide economic and social stability. In contrast Plaid Cymru (1978) noted that, 'The London government will never be able to save rural Wales. The rural areas have already been treated unfairly under the present systems of Rate Support Grants, Transport Supplementary Grants, heavy tolls on petrol etc. whilst artificial new towns or growth towns will do nothing towards sustaining natural rural communities'.

There are a number of implications for research in human geography in rural Wales which arise out of the above discussion. First, although rural Wales exhibits many of the same characteristics as other rural areas in Britain, it has a uniqueness which derives from its peculiar cultural and political institutions. Consequently, modes of analysis developed in other rural areas are not always applicable to the case of rural Wales, a point which has not always been appreciated by rural geographers and their research has suffered accordingly. Second, the researcher in rural Wales can only gain an understanding of the area's development and the failure of past policies by referring to the political and cultural conflict between Welsh nationalism and Westminster. For example, the creation of the D.B.R.W. as a result of pressure from within Wales should have been an important development in the area, but its limited success can be seen as the result of the limited powers granted to it by Parliament in the interest of 'core' capitalism. Finally, it is clear that policy recommendations by rural geographers to counteract uneven development in rural areas must be more radical. The provision of key settlements, the creation of development boards, the introduction of integrated transport systems and other such policies are at best palliatives designed to counter the symptons of rural decline without attacking the root of the problem in the political and economic system that encourages the predominantly urban 'core' to exploit the resources of the rural 'periphery'. The call to geographers working in rural Wales is therefore to move away from solely providing palliative measures whereby central government can nullify local aspirations. Instead, geographers should support radical political and economic change, similar to that proposed by the nationalists, whereby the population can become the authors of their own destiny.

CONCLUSION

The pattern of government intervention in rural Wales has evidently been substantially influenced by the forces of nationalism within the Principality. However, those concessions of apparent benefit to rural Wales which were achieved by Plaid Cymru and others, have not answered the area's problems as they have not removed the exploitation of the area by the interests of 'core' capital present at Westminster. Consequently, an alternative political and economic system is needed for the area if it is to solve its problems, and it is clear that the nationalist movement with its call for decentralised socialism represents a vehicle through which this could be achieved. In conclusion, Williams (1980) has stated that, 'I have emphasised the dependent nature of economic development in this peripheral area (i.e. rural Wales); development is controlled by a 'foreign' bourgeoisie, aided and abetted by the state; in consequence this development is produced in the primary interests of core capitalism, rather than in the interests of the local population... it is in these conditions of economic change, allied to the cultural distinctiveness of rural Wales, that the explanation of political conflict must be sought and that it is in the latter that a major avenue of change in material conditions lies'. The challenge to rural geographers is therefore not to shore up the interests of core capitalism in rural Wales, but to seek the introduction of those changes advocated by Welsh nationalists.

REFERENCES

Development Board for Rural Wales, 1981. *Annual report*. (The Board, Newtown).

Griffiths, T. 1979. Wales nationhood and socialism. *Broad Left Journal*.

Nairn, T. 1977. *The break-up of Britain*. (New Left Books, London).

Plaid Cymru, 1970. *An economic plan for Wales*. (Plaid Cymru, Cardiff).

Plaid Cymru, 1978. *Gwlad ein plant*. (Plaid Cymru, Cardiff).

Plaid Cymru, 1981. *An energy policy for Wales*. (Plaid Cymru, Cardiff).

Rees, G. and Lambert, J. 1981. Nationalism as legitimation? Notes towards a political economy of regional development in South Wales. in *New perspectives in urban change and conflict*, Halse, M. (ed), (Heinemann, London).

Reeves, R. 1980. Can a Mandragón work here? *Arcade, 6*.

Wenger, G.C. 1980. *Mid-Wales: deprivation or development*. (Social Science Monographs no.5), (Board of Celtic Studies, University of Wales).

Wenger, G.C. 1982. The problem of perspective in development policy. *Sociologia Ruralis, 22*, 5–16.

Williams, G. 1980. Industrialisation, inequality and deprivation in rural Wales. in *Poverty and social inequality in Wales*, Rees, G.R. and Rees, T.L. (eds), (Croom Helm, London).

Chapter 25

Politics and the countryside:

the British example

Andrew Gilg

Planning is not a value-free activity and to a large extent it
reflects the norms and mores of the society it operates within.
Similarly, the government of the day largely reflects the views of
the society it governs, although in Britain very few governments
have been elected on a true majority vote. Many post-war governments have indeed been elected with less than 50 per cent of the
vote or with even fewer votes than the main opposition party.
Nevertheless, for most of the post-war period the gap between the
two main parties has been fairly narrow and between the mid-1950s
and 1970s a consensus type of politics emerged. The reintroduction
of sharp divisions between a newly socialist Labour Party and a
sharply right-of-centre Conservative Government has however removed
this consensus, unless the vacuum that is left is filled by the
newly formed Social Democrat-Liberal Alliance.

It is the main thesis of this paper that the politics of the
countryside have been dominated by four main periods of socio-
economic thought, which have both been reflected in and by the
government of the day. In other words the government not only seeks
to control but is at the same time controlled by the wider environment outside Westminster. A good example of this is provided by the
Heath government of 1970-74 which began with a policy of non-
intervention, but was forced by both internal and external circumstances to do a 'U-turn' and introduce interventionist and semi-
socialist measures in its last two years of office, thus returning
to the social democratic consensus years of the 1950s and 1960s.
However, either side of these consensus years, two quite different
periods occurred as outlined in Table 25.1. The rest of this paper
uses this table as a model for a description of the relationship
between politics and the countryside.

THE 1945-51 LABOUR GOVERNMENT

At the end of the Second World War the British people were
conditioned to planning and the belief that 'a brave, new world'
could only be provided by the state. They remembered too the high
unemployment of the pre-war Conservative years. Accordingly they
gave a massive majority to a Labour Goverment committed to fundamental reforms, namely the nationalisation of all public utilities,
services and staple industries, and state control of prices
including post-war food rationing. In agriculture and forestry,
the need to continue the expansion of the war years was paramount

251

Table 25.1 Trends in planning

GOVERNMENT	SOCIO-POLITICAL PERIOD	AGRICULTURE & FORESTRY	SETTLEMENT	CONSERVATION & RECREATION
1945-51 Labour (socialist)	Nationalisation of major public services, state intervention and post-war recovery	*Agriculture Act 1947* Expansion at all costs. Price guarantees and grants. In forestry, dedication agreements and felling licences	*Town and Country Planning Act 1947* The state to act as developer, the production of end-state plans	*National Parks Act 1949* Agriculture the natural conservator of the landscape and wildlife. Recreation, health outdoor type The designation of National Parks and Nature Reserves
1951-64 Conservative (one-nation consensus)	Gradual relaxation of state control, increasing prosperity and growth. 'Never had it so good'	*Agriculture Act 1957* Limitation of change in guaranteed prices to ± 3%p.a. Quotas and standard quantities. Forestry as rural employer	*Town and Country Planning Acts 1953, 1954 and 1959* Allowed free market in land and the re-entry of the private developer	
1964-74 Labour and Conservative (social democrat)	The flowering of economic and social freedom. Everything thought possible, but limits perceived by the early 1970s	Accession to the EEC, 1972. The problems of overproduction begin to appear. But expansion still the goal. Economic return on forestry queried	*Town and Country Planning Acts 1968 and 1971* Replacement of end-state, land-use plans by flexible structure plans. Re-organisation of local government into two tiers	*Countryside Act 1968* Growth of recreation leads to positive provision, e.g. Country Parks
1974-82 Labour and Conservative (Centre and right)	The oil crisis and world recession. Strict financial controls forced onto Labour, gladly espoused by new Tory right	Surpluses appear in a number of products but overall the UK not self-sufficient. Expansion leads to landscape damage. Re-assertion of forestry expansion	The production of structure plans, against a background of counter-urbanisation, urban decay and rural growth	*Wildlife and Countryside Act 1981* The conservation decade ends with a voluntary process for conservation, backed by compensation payments

and the Agriculture Act 1947 gave farmers the stability and confidence needed for investment. It provided price guarantees for most
products so that if the market price fell below a certain limit the
state would make up the difference. Generous grants were also pay-
able for the modernisation of agricultural production. In the
longer term the Act allowed the Government to invest in both basic
and applied research via research stations and experimental husbandry
farms so that the long-term productivity of agriculture could be
raised. Interestingly this was all to be achieved largely through
the private sector, the Labour Party recognising that farming could
not be nationalised for practical and economic reasons. However,
the 1947 Act did impose a good deal of price control, even if in
the long term this has been favourable to the farming community. In
forestry, the 1947 Forestry Act allowed private owners to dedicate
land to forestry in return for cash grants, and the 1951 Forestry Act
introduced legislation to restrict the felling of most trees without
the grant of a felling licence from the Forestry Commission, a
government agency set up in 1919.

In settlement planning the 1947 Town and Country Planning Act
was a far more socialist piece of legislation. It effectively
nationalised development land by imposing a 100 per cent tax on land
sold for development. Furthermore it envisaged that almost all new
settlements would be purpose-built local authority estates provided
with basic services, such as schools, shops and health centres
(Cherry, 1974). So that the authorities would know where to build
their new estates, the Act introduced the concept of 'development
plans'. These were to be detailed land-use plans showing the
expected land use of an area in twenty years' time. Their main aim
was to show the development intentions of the local authority,
hence their title, 'development plans'.

In conservation and recreation, the Government encountered the
same problems as in agriculture, namely, that the nationalisation
of land for these purposes would be too expensive, especially for a
nation so crippled by the war effort. Accordingly, they had to
compromise by legislating for National Parks, in name only, in the
National Parks and Access to the Countryside Act 1949. Another
problem was that the proposed National Park areas were already
inhabited and extensively farmed. However, this was not seen as too
much of a problem at the time, since agriculture was not considered
a threat to the landscape as it is today in some quarters. Instead,
it was viewed as its natural conservator. In Scotland the pressures
on the countryside were far less severe and no National Parks were
designated. In some special areas the need for extra conservation
was provided by the power to designate 'nature reserves'. The other
main concern of the 1947 Act was access to the countryside, which
had been constantly denied by landowners in some parts of Great
Britain in the 1930s. The Act allowed local authorities to negotiate
'access agreements' with private landowners, and this was not
expected to be too difficult since most of the demand was expected
to come from the healthy, outdoor-type of rambler who knew and
respected the countryside.

In conclusion the 1945-51 period was one of contrast. In the
field of settlement planning the powers were draconian and in the
long term excessive, in the area of agriculture and forestry they
pitched the balance between free enterprise and state control just
about correctly, while in the field of conservation and recreation
they fell woefully short of the sort of positive action expected
from a left-of-centre party.

THE 1951-64 CONSERVATIVE GOVERNMENT

The Labour Government was defeated in 1951 at a general election by an incoming Conservative administration who had used the slogan 'set the people free' as part of their election propaganda. This set the tenor for the next thirteen years of Conservative government, one of the longest periods of one-party rule in Britain. The Conservatives set about retaining the best features of the Labour reforms, for example, the National Health Service and the Agriculture Act, but partially dismantled less acceptable forms of state control, for example, the effective nationalisation of land and property development in the Town and Country Planning Act 1947. This was a period of freedom with responsibility, and a one-nation consensus. With increasing prosperity new concepts developed, for example, a one-class, property-owning democracy, and this new wealth distributed among all classes and almost all regions allowed Harold Macmillan to proclaim that the British people had 'never had it so good'.

This certainly applied to agriculture and forestry where farmers entered a period of prosperity unparalleled for over a century. The basis for this was the Agriculture Act 1947 which provided a firm base for farm prices and although many prices fell throughout the 1950s, increases in productivity and yields allowed farm incomes to rise dramatically. Further stability was introduced by the 1957 Agriculture Act which limited the allowable change in farm support prices to ±3 per cent per year. To counter over-production as yields rose, quotas and standard quantities were set for an increasing number of products, for example, potatoes. Very few people questioned the expansion of production, notably, the increase of arable production, although Wibberley did point out the greater productivity of land given over to industry (Wibberley, 1959). However, the main thesis remained that agricultural expansion was sacrosanct even when food rationing ended in the early 1950s. However, in forestry the policy emphasis was changed in 1958 from being entirely based on building up a strategic reserve to one of also providing employment in rural areas where jobs were scarce.

In settlement planning, most planners saw their primary role as the protection of farmland from urban sprawl. The central role of the planners, however, changed dramatically in the 1950s as the Conservatives dismantled controls and taxes on land development and not only allowed but actively encouraged the private building sector to take over as the main vehicle for home construction. The role of the planner thus changed from actively promoting planned development to controlling the development programmes of private house builders. With increasing affluence more people could afford cars and commuting from villages to towns and cities became an increasingly popular activity. Naturally, house builders wished to cater for this trend and one of the main planning issues of this period was the battle between planners trying to preserve villages and the open countryside and developers wanting to build where people wished to live (Hall, 1973). To counter this trend the government extended the concept of 'green belts' from London (where it was applied in the 1930s) to many British cities (see the papers by Munton and Sharp in this book). Planning was thus relegated to the role of guiding and modifying change instead of directing it.

This applied even more so in the field of conservation and recreation where the National Parks Commission and the Nature Conservancy set up by the National Parks Act 1949 were only able to designate National Parks and Nature Reserves. Little real work could be done and as Conservative philosophies of free enterprise began to take root, the idea of extending planning controls over large areas of open countryside like the National Parks became less acceptable.

THE 1964-74 LABOUR AND CONSERVATIVE GOVERNMENTS

Controls of any kind became less acceptable during the 1960s as the libertarian movement swept not only the United Kingdom but Europe and the U.S.A. In the U.K., reaction to an increasingly out-of-touch Conservative Government and the attraction of a new brand of socialism offering a loosely planned but free society, allowed the election of the first Labour Government for thirteen years. This Government espoused pragmatism as its central credo, and managed the country in a benignly libertarian manner. Its successor, the Conservative Government of 1970-74, in spite of some stubborn policies which cost it the general election of 1974, also broadly espoused a centrist position. Generally, the period was one of extreme optimism and rates of growth of 5 per cent in Labour's National Plan of 1965, and a 'dash for growth' in the Conservative budget of 1972 were indicative of the belief that rapid expansion was the key to political success. Nonetheless, growth rates failed to meet expectations and warning voices were raised that post-war growth could not continue for ever (Meadows, 1972).

One area where growth continued was in agriculture where productivity and yields continued to grow. Support for agriculture was further reinforced when, after ten years of negotiation, Britain joined the E.E.C. in 1972. At that time, C.A.P. prices were markedly higher than those in the U.K. and this gave a further boost to U.K. agricultural production, although the five-year transition period and the adoption of a green pound rate below the normal currency rate offset this advantage to some degree (Marsh and Swanney, 1980). However, U.K. production could still not meet home demand except in a few products, such as liquid milk, and so the problem of surpluses was still seen as a European rather than a U.K. matter and expansion was still officially encouraged. However, in forestry, the policy of expansion was seriously questioned by a report on *Forestry Policy* (Agriculture, 1972) which argued that the estimated rate of return of around 1 per cent on new forestry planting did not justify new planting on economic grounds. It could only be justified in terms of employment, the balance of payments or the provision of recreational opportunities.

Expansion continued for a while in settlement planning and the high birth rate of the 1960s forced planners to prepare for a U.K. population of 70 million by 2001. Several new towns and cities in the British countryside were envisaged, especially along the estuaries of the Severn, Humber and Tay. This was also the period when the motor car was rampant and the advent of systems planning placed great emphasis on modern communications, notably motorways, which opened up huge areas for commuting. The broad-brush systems approach was also reflected in a new type of plan introduced by the 1968 Town and Country Planning Act, the structure plan. Structure plans were intended to be broad strategies for unitary city-regions, but unfortunately plans to reorganise local government along city-region lines were abandoned by the Conservatives, and instead a two-tier system based on the county with subsidiary districts was introduced in 1974 (Roberts, 1976). In Scotland the system was revised more realistically with regional authorities producing structure plans and district authorities, local plans.

Renewed expansion of the economy was reflected in more affluence, cars and leisure time which stimulated countryside recreation in the

1960s. This allowed recreation to become the more important component in the conservation and recreation equation, and the Countryside Act 1968 recognised this by making positive provision for increased countryside recreation, mainly in the form of country parks, whose main task was to deflect recreational use away from sensitive or over-used sites towards purpose-built areas. Legislation for countryside conservation and recreation was first enacted in Scotland with the passing of the Countryside (Scotland) Act 1967 which set up the Countryside Commission for Scotland. Leisure was the great new industry of the 1960s and only a few questioned whether the phenomenal economic growth of the post-war period could be continued, or queried the effects of this growth on the environment.

THE 1974-82 LABOUR AND CONSERVATIVE GOVERNMENTS

The post-war period was brought abruptly to a halt by the oil crisis of 1973 and, although most economies continued to grow throughout the 1970s, the engine of growth had begun to falter. By the 1980s the U.K. had entered a severe recession. At first the Labour Government tried to spend its way out of trouble by applying traditional demand-led policies but as inflation soared to 25 per cent and the pound fell to below 1.60 dollars, severe limits on monetary expansion had to be introduced, out of necessity rather than belief. In Japan these monetary policies were espoused more readily and their apparent escape from the recession encouraged the Conservative Party to introduce severe monetary policies allied to a relaxation of government controls when they were elected in 1979. Although the Labour Party has now rejected this approach, there is little doubt that over most of the 1974-82 period both parties agreed in restraining the growth of the money supply and relaxing planning controls in order to encourage economic growth.

In the area of agriculture, over-production not only in the E.E.C. but increasingly so in the U.K. too, led to calls for a slowdown in the rate of agricultural expansion and even for a withdrawal from the E.E.C. Growing criticism came from the newly established conservation lobby, who objected not only to the cost and over-supply of farm products but also to the effects of modern farming on the landscape (Shoard, 1980). The lobby pointed to the loss of landscape features, elimination of habitat and destruction of wildlife because of the removal of hedgerows, the ploughing of meadows, downland and upland and the drainage of wetlands. Often only government grants made this financially feasible. Nonetheless, two government policy statements endorsed continued expansion. They recognised the side-effects of expansion but believed that agricultural improvement was compatible with conservation. In forestry new planting was encouraged by a number of reports which argued for a doubling of the afforested area (Centre for Agricultural Strategy, 1980). In spite of economic and conservation arguments against increased afforestation, the government accepted an increased target rate for planting of 22 000 hectares per annum.

The rise of an anti-farming lobby was partly due to a new type of rural dweller, the exurbanite commuter or retiree. The move of population to the countryside, first noted in the 1930s, had gathered momentum throughout the post-war period and by the 1970s nearly all rural areas were gaining population while nearly all cities were declining (Fielding, 1982). Against this background of counter-urbanisation, the first generation of broad-brush structure plans was produced and most of them attempted to distribute rural growth among a small number of selected towns and villages so that the cost of public services could be minimised (Martin and

Voorhees, 1982). The net result of these policies would be a more even population distribution throughout most of lowland Britain, but with the more remote villages, which had not been selected for growth, subject to decline and an increasing degree of deprivation as public facilities were moved elsewhere (Gilder, 1979). In response to the recession, however, the government has relaxed planning controls and suggested that growth would be acceptable in more places than before, though still not in areas of protected landscape, green belts or the open countryside (Department of the Environment, 1980).

Fears about the open countryside grew to such an extent in the late 1970s that demands developed for some more stringent legislation in the field of conservation and recreation. The crux of the problem was the fact that the British countryside is a man-made creation, largely the result of farming practice and that it can only be conserved by farmers. Accordingly, negative powers alone will not be sufficient except in the shortest of terms. Both political parties agreed that positive measures were essential but they differed on the degree of sanction to be employed. The Labour Party wanted to buy out the right to convert moorland to agricultural use and to acquire the power to buy land if voluntary agreements to manage it for conservation purposes were not adhered to. However, the Conservative Government believed that voluntary agreements backed up by compensation would be sufficient. The Wildlife and Countryside Act 1981 introduced 'management agreements' which allow local authorities to negotiate a management regime with a farmer for the purpose of conservation, backed up with compensation if necessary (MacEwen, 1982). In the field of species protection, more flora and fauna were placed under direct legal protection, but of course if habitats continue to be destroyed these devices will only be hollow measures.

In conclusion, the post-war period has demonstrated that planning policies evolve as the economy and society change (Booth, 1982). This means that the first failure of countryside planning has been that it has been too reactive and responded to changing events too readily. The main areas where this has not been the case are, of course, agriculture and forestry where strongly expansionist policies have been continued long past the time when they were really needed. The over-dominant position of agriculture in rural planning leads directly to the second failure of rural planning, the lack of integration between policies. For example, some government departments attempt to conserve moorland scenery while others are actively discouraging its alternative use for farmland or defence training. One reason for this lack of integration relates to the third failure of rural planning, the inability to formulate a general view about the role of the countryside, and thus to provide a focus for a less reactive and more stable planning system.

Nonetheless there have been continuities in policy and attempts to provide integration. By and large, agricultural expansion and the containment of urban growth have been continuing aims, and the overall goals of land-use planning remain little altered, although the means of achieving these goals have evolved. Attempts to provide integration have been provided by a number of measures, for example, Section 11 of the Countryside Act 1968 asked all government agencies to have due regard to the conservation of the countryside and the provision of access. The National Park plans introduced by the Local Government Act of 1972 attempted to integrate socio-economic and physical objectives in the National Parks (Hookway, 1978). However, real integration has yet to be achieved although most parties pay lip-service to it. A parliamentary sub-committee on National Parks and the countryside (Expenditure Committee, 1976)

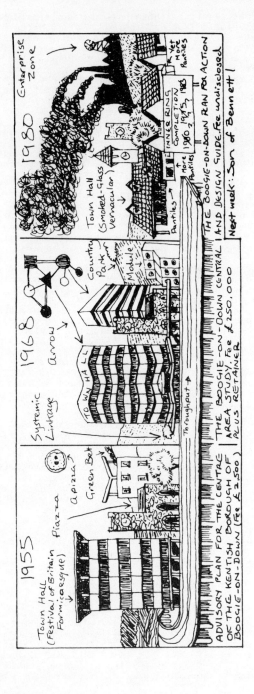

Figure 25.1 Changes in planning fashion

Source: *Planning*, 409, 1981; available weekly from College Green, Gloucester, GL1 2LX

recommended that: '(a) The government should give early consideration to the need for the streamlining of the statutory machinery as
it affects the countryside. (b) The government should attempt to
formulate an overall strategy for land use in the countryside.'
However, these ideals have only once been put into practice, when in
1972 a Standing Committee on Rural Land Use was set up in Scotland
to consider general questions of land use (Coppock, 1979). Unfortunately this Committee was disbanded in 1980 (Gilg, 1982).

It would be unfair to end however on a negative note (Figure
25.1). The post-war system of rural planning has produced an
agriculture which provides a high proportion of Britain's dietary
requirements at a reasonable price, many villages have been
attractively developed or conserved, large areas of open countryside
remain scenically beautiful, much bad development has been prevented,
and much pollution of the air and water has been cleaned up
(Blacksell and Gilg, 1981). The only sadness is that planning,
having achieved so much, has failed to achieve even more (Davidson
and Wibberley, 1977; Gilg, 1978; Cherry, 1976; Gilg, 1980-82).

REFERENCES

Agriculture, Fisheries and Food, Ministry of, 1972. *Forestry policy,*
(H.M.S.O., London).

Blacksell, M. and Gilg, A. 1981. *The countryside: planning and
change.* (Allen and Unwin, London).

Booth, G. 1982. Fashions in environmental planning. *The Planner,*
68, 68.

Centre for Agricultural Strategy, 1980. *Strategy for the UK forest
industry.* (The Centre, Reading).

Cherry, G. (ed) 1976. *Rural planning problems.* (Leonard Hill,
London).

Cherry, G. 1974. The evolution of the Planning Acts: 1909, 1919,
1932, 1947 and 1968. *The Planner,* 60, 675-703.

Coppock, J. 1979. Rural land management in Scotland. *Town and
Country Planning,* 48, 47-8.

Davidson, J. and Wibberley, G. 1977. *Planning and the rural
environment.* (Pergamon, Oxford).

Department of the Environment, 1980. *Circular 22/80: development
control: policy and practice.* (H.M.S.O., London).

Expenditure Committee of the House of Commons, 1976. *National parks
and the countryside.* H.C.433(75-76), (H.M.S.O., London).

Fielding, A. 1982. Counterurbanisation in Western Europe. *Progress
in Planning,* 17, 1-52.

Gilder, I. 1979. *Rural planning policies: an economic appraisal.*
(Pergamon, Oxford).

Gilg, A. 1978. *Countryside planning.* (David and Charles, Newton
Abbot).

Gilg, A. (ed) 1980-82. *Countryside planning yearbook,* vols 1-3,
(Geo Books, Norwich).

Gilg, A. 1982. Legislative review. *Countryside planning yearbook,*
3, 114-15.

Hall, P., Drewett, R., Gracey, H. and Thomas, R. 1973. *The containment of urban England*. (Allen and Unwin, London).

Hookway, R. 1978. National park plans. *The Planner*, 64, 20-3.

MacEwen, A. and M. 1982. The Wildlife and Countryside Act. *The Planner*, 68, 69-71.

Marsh, J. and Swanney, P. 1980. *Agriculture and the European Community*. (Allen and Unwin, London).

Martin and Voorhees Associates, 1982. *Review of rural settlement policies*. (Department of the Environment, Bristol).

Meadows, D. 1972. *The limits to growth*. (Earth Island, New York).

Roberts, N. 1976. *The reform of planning law*. (Macmillan, London).

Shoard, M. 1980. *The theft of the countryside*. (Maurice Temple and Smith, London).

Wibberley, G. 1959. *Agriculture and urban growth*. (Michael Joseph, London).

Chapter 26

The implementation of land-use policy in the urban fringe: the North Middlesex Green Belt estates 1920-1950

Elizabeth Sharp

In many environments studied by geographers a multiplicity of public organizations influence, directly and indirectly, the social processes responsible for land-use change. This paper examines certain aspects of the conflict between urban and suburban interests in the implementation of land-use policy, and the effects of the conflict on the land-use pattern. The particular policy in question concerns the public purchase of land in the periphery of London in order to control the outward spread of London's suburbs.

In peri-urban areas, different local authorities often have separate and conflicting responsibilities for suburban and urban areas. The literature contains some examples, including those of land-use zoning in Denver, U.S.A. (Johnston, 1981), social and political power in Croydon, south London (Saunders, 1979), and public housing policy in the London Borough of Bromley (Young and Kramer, 1978). More generally, however, inter-authority relations and their effect on public policy, as an area of enquiry, have been neglected by geographers, particularly where land-use planning conflicts are concerned.

The case to be examined here is that of the 'green belt' estates on the periphery of London. These properties were acquired by local authorities during the period 1920-80 and are not to be confused with the post-1947 Metropolitan Green Belt (M.G.B.) statutory planning designation, although most of the properties are de facto within this area (Thomas, 1970; SCLSERP, 1976; Ferguson and Munton, 1978). These estates, covering over 300 properties and 20 000 hectares, have had an important bearing on the land-use and ownership pattern of London's periphery. They include much open land in agriculture and recreational uses important to the local amenities of past and present residents living in the area.

An understanding of the role of inter-authority relations is critical to any analysis of the location and land use of these estates. This is because of the extensive participation by means of a grant-aid scheme of the London County Council (L.C.C.) in their acquisition, although the properties themselves lay beyond the jurisdiction and control of the Council. Despite the often quoted importance of the L.C.C. grant-aid scheme for the pattern of land acquisition (Hall et al., 1973) this study will suggest that whilst the role of the L.C.C. was to influence the volume of acquisition, as regards fine detail of location and specifically the type of land use to which the properties were devoted, the local authorities on the periphery of London played the major part. Furthermore, the

activities of the L.C.C. were concentrated into a relatively short
period.

LOCAL GOVERNMENT STRUCTURE IN ENGLAND AND WALES, 1920-50

In England and Wales there are four main levels of government
administration. They are, in decreasing spatial scale: central,
county, district and parish. This description, however, over-
simplifies the structure, and for the purposes of this paper more
detail will only be provided on the county and district levels.
With the exception of authorities known as County Boroughs, which
were unitary authorities arising from nineteenth-century development (Byrne, 1981), the main local government division of concern
here is the county. Outside the area of the L.C.C.,counties were
subdivided into various district authorities of roughly equivalent
functions - urban district councils, rural district councils and
'non-county' borough councils. Members of county and district
councils were elected separately. Each raised their own revenues
through a property tax known as 'rates', and each had different
responsibilities.

County and district authorities were the main public bodies
responsible for land-use planning, education, housing and highways.
The exact division and extent of responsibilities varied during the
study period, but in the case of land-use planning and open-space
provision, it was the county and county borough authorities which
were responsible for most activity, although the district
authorities retained some power. Until 1947 the district authorities
were the primary land-use planning bodies, but in some cases they
transferred their powers and associated financial responsibility to
the counties as early as 1932. In short, the situation is complex,
arising out of the pragmatic origins of English local government.
For the purposes of this study, it simplifies matters to say that
the L.C.C. dealt largely with the County Councils of Essex, Hertfordshire, Middlesex, Buckinghamshire, Surrey and Kent, and the
County Borough Councils of East Ham, West Ham and Croydon, while
the district authorities retained certain land-use planning powers
over, and interests in, the same properties.

THE GREEN BELT ESTATES: BACKGROUND

The emergence of the green-belt concept and the expansion of London
is amply documented elsewhere (Thomas, 1970; Munton, 1983; Hall et
al., 1973). Briefly it is worth recalling that the green belt, in
its modern form, was conceived as a linear park providing a break
in the outward expansion of London in the late-nineteenth century.
By the mid-1920s, following the widespread publication and discussion
of the Garden City concept, the green-belt idea had become rather
more complex and ambiguous. This confusion was reinforced in the
1920s by the publication of a number of regional advisory plans.
These were non-statutory plans prepared by professional planners on
behalf of groups of district authorities (e.g. Adams and Thompson,
1924; Adams, Thompson and Fry, 1928; Davidge, 1927). These early
planners were heavily influenced by the ideas of Ebenezer Howard
and his disciples, and some had worked on the development of
Welwyn and Letchworth Garden Cities.

More practically, these regional plans revealed the actual
extent and rate of expansion of London in the 1920s, this period
being noted for its absence of revised Ordnance Survey maps (Sheail,
1981, 142-4). Suburban growth had been accelerated by changes in
transport technology, extensions to the public transportation

262

network, and by changes in public and private housing finance
(Jackson, 1973; Swenarton, 1981; Bowley, 1944). The regional
advisory plans universally recommended some form of 'regional open
space' or 'green belt' on the periphery of existing development, in
order to 'preserve amenity', and to provide open space and 'lungs
for the city'. Sir Raymond Unwin's reports, prepared for the Greater
London Regional Planning Committee (G.L.R.P.C., 1929; 1933), unified
these sub-regional proposals into a proposal for a 'green belt' or
'green girdle' running around the whole of London. The proposal
emphasised the green belt's dual purpose, namely recreational
provision and the prevention of further development.

Actual powers to bring such proposals into being were limited.
Until 1932, planning powers (effectively land-use zoning) were
clumsy and slow to implement (Cherry, 1981; Ashworth, 1954; Sheail,
1981). Furthermore, an owner objecting to any proposed zoning, even
after 1932, could usually claim compensation from the local planning
authority for loss of development profits at full market value.
Given the large differences in value between agricultural land and
land 'ripe for development' in the 1930s (Vallis, 1972), local
district authorities faced the prospect of heavy claims for compensation each time they refused planning permission for development.
In such circumstances the reaction of those local authorities which
supported the proposals for a green belt was to purchase and manage
the land themselves.

This decision was heavily influenced by the separation of local
political and administrative control between the suburbs and rural
areas on the one hand and the urbanised area of London on the other.
The boundaries of the County of London had not been revised since
1889, and much of the expansion of London in the twentieth century
had taken place in counties and county boroughs adjacent to the
capital. This had the twofold effect of creating resistance to
further spread of the urban area among the existing rural establishment, and the wish of the new suburbs also to see themselves as
organisationally, if not physically, independent of the metropolis
(Jackson, 1973; Young and Garside, 1982). This wish for separate
control was heightened by the extensive house-building activities
of the L.C.C. during the 1920s. The L.C.C. built large satellite
housing estates, such as at Becontree in Essex, without prior
consultation with the peripheral local authorities which were then
left to provide such basic services as education and health care
for its new population (Young and Garside, 1982). From the late
1920s, many of the authorities adjacent to London, and especially
Surrey and Middlesex County Councils, began a slow process of
acquiring that land recommended as green belt in the regional
advisory plans, and zoning it in town planning schemes as open
space.

A major transformation occurred in 1935 when the L.C.C. decided
to participate in the acquisition of green-belt properties. The
location of these properties was well outside the boundaries of the
L.C.C. and, politically, the Council was held in low esteem by its
adjacent local authorities. This was due to the above-mentioned
housing schemes and to the election in 1934 of its first socialist
council, which put it in political opposition to the Conservative-
and Independent-controlled counties. The aim of the green-belt
scheme announced by the L.C.C. in 1935, however, apparently accorded
with that of the adjacent local authorities. This aim was to
improve actual and potential levels of recreational provision in the
London area as a whole by preserving a green belt around London.
The green belt would compensate for the low levels of provision of
the remainder of the built-up area (Morrison, 1935; Dalton, 1939).
The L.C.C. made £2 million available to six counties and three

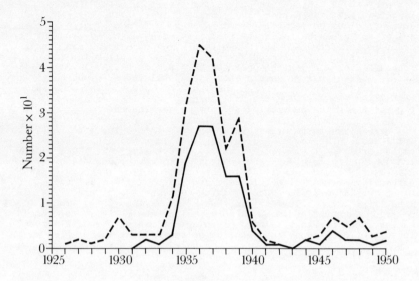

Figure 26.1(a)　Number of properties approved for acquistion as green belt in the Greater London area as a whole, in total and with L.C.C. grant aid only

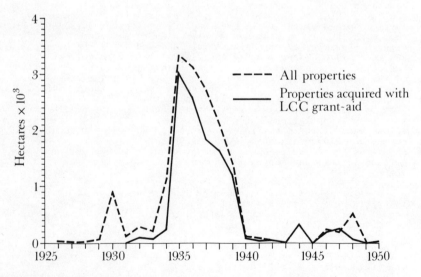

Figure 26.1(b)　Area approved for aquistion as green belt in the Greater London area as a whole, in total and with L.C.C. grant aid only

county borough councils to aid them in their purchase of green-belt. Applications from district authorities had to be supported and sponsored by the relevant county authority. The finance consisted of grants worth up to 50 per cent of the cost of the properties, providing certain conditions of use and public access to the properties were adhered to. Initially the scheme was introduced without any change in legislation; but problems with the technical operation of the scheme, notably the enforcement of restrictive covenants,led to the enactment of the Green Belt (London and Home Counties) Bill in 1938.

The inception of the scheme resulted in a dramatic expansion of acquisition (Figures 26.1a and 26.1b). Between 1935 and 1950, the total number of properties purchased rose from 33 to 243, and the area covered increased from 2742.9 ha to 14 097 ha. Superficially these figures suggest that the L.C.C. played a highly significant role in the acquisition of the green-belt estates, but detailed analysis of the changing levels of acquisition and of the nature of the properties acquired, reveals a number of inconsistencies. Firstly, properties acquired with L.C.C. aid were large. They had an average size of 89.81 ha compared with an area of 20.81 ha for all the other properties. They were also mainly in agricultural use at the time of their purchase, and they have remained so subsequently, whereas those properties rejected for grant aid were frequently in, or about to be put into, recreational use. Secondly, by 1950, only £1 million of the original fund had been expended, although participant authorities continued to buy properties for green belt and open space without applying for L.C.C. grant aid. This indicates that either there were major external changes which mitigated against the full participation of the L.C.C. in the scheme in the way it originally intended, or that the aims of the authorities were in practice less than congruent. The rest of this paper will examine in more detail the operation of the scheme in North Middlesex, an area characterised by high acquisition activity and severe development pressures (Jackson, 1973), and for which excellent records survive[1].

INTER-AUTHORITY RELATIONS

From work such as that of Friend (1976; Friend and Jessop, 1969; Friend et al., 1974) and Blowers (1980), it can be argued that where a number of authorities have an interest in a particular policy and its implementation, those authorities normally try to achieve a consensus. This is preferred to a state of conflict over the control of resources (Pondy, 1967; Leach and Moore, 1979; Thompson, 1967). Negotiation is likely to lead to compromise between the interested parties. These studies pay little attention, however, to situations where the degree of spatial control over the object of concern is not equally divided among authorities - as in the case of the green-belt estates - and it is argued here that where authorities are of roughly equal administrative, financial and political stature, the authority in which the resource is located will be the dominant one.

The relations between Middlesex County Council and the L.C.C. are examined with this position in mind. By looking also at the North Middlesex area, consisting of Potters Bar Urban District, Enfield

---

[1] References to official minutes and papers are prefixed G.L.R.O. (Greater London Record Office). Records held at County record offices such as the Greater London Record Office may include county council minutes, council committee and sub-committee minutes, technical papers and reports. Sharp (1980) gives a résumé of those records available on green belt issues in the Greater London area.

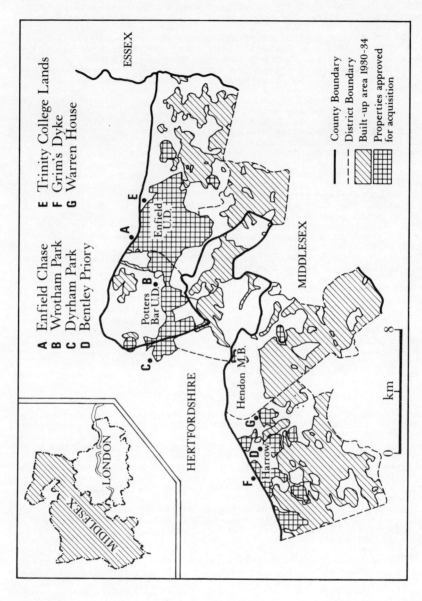

A Enfield Chase   **E** Trinity College Lands
B Wrotham Park   **F** Grim's Dyke
C Dyrham Park   **G** Warren House
D Bentley Priory

ESSEX

HERTFORDSHIRE

MIDDLESEX

Enfield U.D.

Potters Bar U.D.

Hendon M.B.

Harrow

MIDDLESEX

LONDON

— County Boundary
- - - District Boundary
Built-up area 1930-34
Properties approved for acquisition

km

0   8

Figure 26.2   The North Middlesex study area

Urban District and Harrow Urban District (Figure 26.2), a comparison can be made with the relations between Middlesex County Council and these district authorities.

The apparent aim of acquisition was to provide recreational open space for the urban population of Middlesex and London - an aim publicly espoused by both Middlesex C.C. and the L.C.C. Examination of the Middlesex C.C. records shows, however, that there were other reasons for the adoption of a policy of land acquisition, most notably the prevention of further residential development in general and, specifically, L.C.C. housing schemes. This intention became more explicit in the Middlesex C.C. minutes as the 1930s progressed. These aims contrast with those of the L.C.C. which specifically wished to see land acquired for open-space purposes, and which made public access a stringent condition of its grant aid (G.L.R.O.(2): M.C.C./C/E&H/508/1). Failure to reach a compromise on this critical difference in policy, and the threat the L.C.C. grant conditions posed to Middlesex's control over the property within the area of its jurisdiction help to explain the limited timespan of the LCC's activity in Middlesex.

THE NORTH MIDDLESEX GREEN BELT SCHEME

Local authority records show that Middlesex C.C. had a long history of active interest in open-space provision. This stemmed from the late-nineteenth century and before the passing of its own legislation in 1898, which was itself in advance of the national open-space legislation enacted in 1906[2]. Until the early 1930s, the mechanism by which Middlesex C.C. acquired open space was through the agency of its constituency district authorities to which it gave grants. The district authorities were responsible for the identification of the properties and the negotiations with their owners regarding purchase.

After the publication of the north and west Middlesex regional planning schemes (Adams and Thompson, 1924; Adams, Thompson and Fry, 1928), Middlesex C.C.'s policy toward 'green belt' developed a three-fold emphasis - prevention or control of urban sprawl, protection of 'amenity', and provision of open space for recreation. The physical implementation of the scheme drew heavily on the regional advisory schemes. In 1934, two important developments occurred. firstly, a clear target of 8335 hectares of green-belt open space was set by the county for acquisition and protection. The land was to include all the appropriately located, private golf courses and woodland (G.L.R.O.: M.C.C. Development Committee minutes; 21st November 1934). Due to the extent of urban development in the south and east of the county, this area was sought in the northern and western fringes of the county - those areas designated as 'green belt' in the advisory plans of the 1920s. Secondly, Middlesex C.C. took over the initiative for the identification of, and negotiation for, properties from the district authorities. The announcement of the L.C.C. scheme in 1935 was enthusiastically received by Middlesex C.C. and the proposed grant aid led it to enter into serious negotiation for many of the larger properties in the county.

The change in responsibility for the acquisition of properties, combined with the additional finance proffered by the L.C.C. scheme increased the level of purchasing considerably (Figures 26.3a and 26.3b). In North Middlesex, for example, Enfield Chase (837 ha), Wrotham Park (487 ha) and Dyrham Park (460 ha) became part of the

---

[2] Middlesex County Council Act, 1898; Open Space Act, 1906

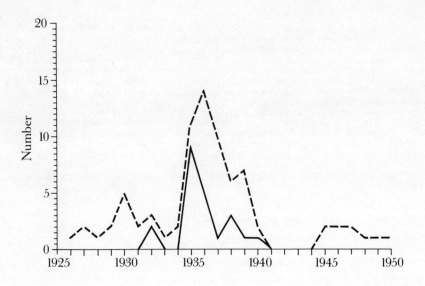

Figure 26.3(a)   Number of properties approved for acquistion as
green belt in Middlesex, in total and with L.C.C.
grant aid only

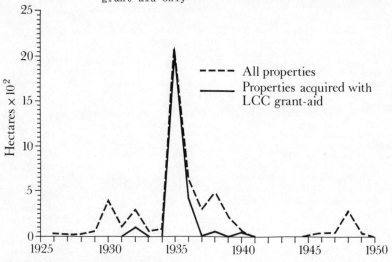

Figure 26.3(b)   Area approved for acquistion as green belt in
Middlesex, in total and with L.C.C. grant aid only

green-belt estate (Figure 26.2). Yet by the end of 1938, the rate
of acquisition had slackened appreciably. The reason for this
particular pattern of acquisition will now be examined in more detail,
by looking at the relations between, first, the L.C.C. and Middlesex
C.C. and, second, between the local district authorities and
Middlesex C.C.

1. Relations with the L.C.C.

By 1934 Middlesex had already started compiling an inventory of sites
suitable for purchase. The majority of Middlesex C.C.'s applications
for grant aid from the L.C.C. from 1935 onwards were successful

Table 26.1   North Middlesex: Properties approved for acquisition,
              1928-48 by Middlesex County Council

Year	Number of properties	Area (ha)	Receiving L.C.C. grant aid		Rejected by L.C.C. for aid	
			Number	Area (ha)	Number	Area (ha)
1928	1	18.3	0	0	0	0
1929	2	52.08	0	0	0	0
1930	3	265.4	0	0	0	0
1931	1	6.6	0	0	0	0
1932	3	122.08	2*	99.1	1*	22.9
1933	0	0	0	0	0	0
1934	0	0	0	0	0	0
1935	6	1287.9	5	1277.5	0	0
1936	8	458.75	5	386.35	3	72.5
1937	4	115.0	0	0	2	82.08
1938	4	381.25	3	373.3	0	0
1939	0	0	0	0	0	0
1940	1	43.75	1	43.75	0	0
1941-7	0	0	0	0	0	0
1948	1	287.08	0	0	1	287.08
TOTALS	34	3038.19	16	2179.90	7	464.56

* These properties were considered by the L.C.C. for grant aid after 1935,
  although they were approved for acquisition by Middlesex C.C. before that date

(Table 26.1), but the L.C.C. were inconsistent in awarding grants for
the acquisition of golf courses, and in no case did the L.C.C. offer
anything approaching the 50 per cent the county council had interpreted the original L.C.C. resolution to be offering. The grants
offered for properties in North Middlesex ranged between 20 and 35
per cent of the net acquisition cost - that is the gross purchase
price less revenue from the property capitalised over 20 years. In
the case of Enfield Chase, for example, a proffered grant of 33 per
cent amounted to only 25.6 per cent of the gross acquisition cost
through the application of this formula.

The insistence of the politicians and officals of the L.C.C. on
the provision of public access to properties to which they were
making financial contributions proved to be the critical aspect of

the acceptability of L.C.C. participation in the North Middlesex area, although this only became a major issue after a large number of acquisitions had been agreed. Middlesex C.C.'s overriding interest in making purchases at this time was to forestall the sale of land to developers and it was felt by their officers and members that the conditions attached to the L.C.C.'s grants formed an excessive intrusion into their autonomy (see G.L.R.O: M.C.C./C/E&H/ 524/3; M.C.C., Open Spaces Sub-committee, 7th October 1937; 4th November 1937; 8th June 1939). Furthermore, the practicalities of managing this sudden increase in local authority land holdings, and its concomitant financial commitment, encouraged them to retain the land in some form of revenue-generating use (for example, agriculture rather than public open space) as this helped to offset maintenance costs and meet loan charges.

The crucial case was that of Bentley Priory (see Figure 26.2), an estate of 67.9 ha, approved for purchase by Middlesex C.C. in 1936. This formed one of the major properties in the second phase of purchase of Harrow green-belt properties, along with others such as the Grove estate and Grims Dyke. The L.C.C. insisted that public access to this property should be made available before grant aid was allocated (G.L.R.O.: M.C.C./C/E&H/524/3 and /4). Negotiations between the two authorities continued until 1964, almost 30 years. Objections to the imposition of public access provisions came not only from Middlesex C.C., but also from Harrow Urban District, under whose control the property would eventually come. The L.C.C. had not previously attempted to enforce seriously the public access provisions entailed in the grant-aid provisions. Throughout the War, however, although Middlesex had proceeded with the acquisition of the major part of the property, the L.C.C. insisted on minimum standards of improvement of public access (increasing the public footpath network) within five years of the cessation of hostilities. Such a time constraint was objected to both by Middlesex C.C. and the Harrow U.D.C. As a result, although the offer of grant aid for a further 70 hectares elsewhere was outstanding after the War, Middlesex and Harrow proceeded to buy the property on their own account, with the exception of small portions amounting to some 17 hectares, purchase of which was abandoned. The financial assistance offered by the L.C.C. was considered to be considerably outweighed by the infringements of the authority and jurisdiction of both Middlesex and Harrow (G.L.R.O.: M.C.C., Estates and Housing Committee minutes, 9th February, 1950).

It would appear that although the L.C.C. achieved some of its objectives in terms of land acquisition, its achievements in land use and public access were only a qualified success because of the obstruction of the Middlesex C.C. Furthermore, the fine detail of location of the properties could only be marginally influenced by the L.C.C. since it could only react to the applications from the counties and, in the case of Middlesex, that authority continued to buy other properties, including those rejected for grant aid, on its own account.

2. Relations with the district councils

The objections of Harrow U.D.C. to the conditions attached to grant aid from the L.C.C. have already been briefly mentioned. The nature of the relationship between the district councils (Potters Bar, Enfield and Harrow) and the Middlesex C.C. differs from the relationship between the L.C.C. and Middlesex C.C. Between 1920 and 1950, responsibilities for land-use planning and open-space provision lay primarily with the district authorities, but they were in the process of being moved up the local authority hierarchy to the county level. Frequently, responsibilities for public utilities were

not allocated consistently to separate levels of local government –
the precise distribution of power depended on local needs and power
structures. However, the district had sufficient legislative independence to ensure that the county authority had to respect the
wishes of the district to some extent, in order to ensure cooperation
in other related policy fields such as public health, education and
housing.

Analysis of the estates in the study area shows that the relationship between the Middlesex C.C. and the district authority was
more critical in determining the form and land use of the estate
than that with the L.C.C. From 1929 onwards, Middlesex C.C. had
contributed 75 per cent or more of the cost of acquiring properties
it considered to form open space of a 'regional or green belt'
nature. If the county contributed more than 50 per cent of the
purchase money, the land belonged to the county, but the district
authorities were still expected at some time after the acquisition
to take over the management of the properties for recreational purposes. Hence the agreement of the district authorities to acquisition was important as this indicated their acceptance of future
responsibility for the property. The importance of this aspect of
county-district relations can be illustrated by examining Enfield
U.D.C. and Harrow U.D.C. In 1937, both these authorities raised
objections to the acquisition of further green belt and open space
within their bounds, although different reasons accounted for their
objections. Table 26.2 illustrates the importance of these decisions
for the rate of acquisition in these districts.

Table 26.2   North Middlesex: Properties approved for acquisition
              in Harrow and Enfield Urban Districts

Year	Harrow Number of properties	Area (ha)	Enfield Number of properties	Area (ha)
1929	2	52.08	0	0
1930	3	160.83	1	104.58
1931	1	6.6	0	0
1932	3	122.08	0	0
1933	0	0	0	0
1934	0	0	0	0
1935	0	0	1	837.08
1936	7	214.16	1	244.58
1937	0	0	2	84.58
1938	0	0	0	0
1939	0	0	0	0
1940	1	43.75	0	0
1941-7	0	0	0	0
1948	0	0	1	287.08
TOTALS	17	599.50	6	1557.9

In Harrow, a large amount of open space - some 274.3 ha - had been acquired by 1936. When measured against the size of the local population this gave a standard of provision well above the 3 ha per thousand head of population which was the nationally recommended level. Proposals were then advanced by Middlesex in 1936 for the purchase of a further 600 ha in the north of the district. These proposals had been under discussion for some years, but never actively promoted. This proposed increase of over 200 per cent in open space, the area to be ultimately managed by the district authority, was considered excessive by Harrow U.D.C., especially as existing provision met the required standard in terms of the district's population. Eventually a compromise was negotiated which was critical to the land-use pattern in this part of the green belt. Middlesex C.C. agreed that Harrow U.D.C. should only take over the management of further properties as and when the district needed additional local open space. Meanwhile, Middlesex C.C. were to maintain all the land acquired in some form of remunerative (non-recreational) use. This normally meant farming or an institutional use. (G.L.R.O.: M.C.C. Open Spaces Sub Committee, 12th July 1937).

In Enfield U.D.C. a large amount (933.28 ha) of land had been purchased by 1937 to prevent further development. However, this conflicted with that authority's need to strengthen its local financial base. The housing stock in Enfield was predominantly old and of low rateable value, with the result that the district had one of the lowest rateable values per capita in Middlesex (£8.10 compared with a county average of £9.38 for 1935-35), but the highest rate in the pound (65p compared with a county average of 51p) (London Statistics, 40, 1935-37). In 1936, Middlesex C.C. promoted the acquisition of lands owned by Trinity College, Cambridge, adjacent to Enfield Chase (Figure 26.2). This would have increased the area of green belt by a further 245 ha. By early 1937, this property was under threat of sale to housing developers (G.L.R.O.: M.C.C./C/E&H/68) and Enfield U.D.C. had still not resolved to support the acquisition of the whole area. In June 1937 Enfield finally decided to support only the acquisition of the portion of the property in use as a golf course. In order to finance the acquisition of the whole property, Middlesex C.C. had to make provision for the re-sale of part of the land for development purposes, which also met the demands of Enfield U.D.C.

As can be seen from Table 26.2, 1937 formed an abrupt threshold between very active acquisition and very low levels, and between 1937 and 1950 Middlesex C.C. discontinued negotiations for 16 properties totalling 176.5 ha in North Middlesex. Much of this can be attributed to the stand made by the local district authorities who had strong objections to the acquisition of further open space. While the availability of L.C.C. grant aid contributed to the overall level and pattern of acquisition, the separate aims of the district councils and Middlesex C.C. were crucial in the latter half of the study period in determining the detail of acquisition. The county balanced its existing land holdings and its green-belt policy against future relations with the districts over other issues such as housing. The implementation of green-belt policy has to be seen against a wider range of local government considerations. The results of this were threefold. Although Middlesex C.C. had taken the initiative since 1934 in identifying and acquiring properties, the initiative returned to the district authorities after 1937 (G.L.R.O.: M.C.C. Open Spaces Sub-committee, 6th January, 1938; 10th March 1938). Secondly, lands surplus to local open-space requirements remained in agricultural or institutional use, and except for marginal improvements in footpath access were not released for recreational use as required by green-belt policy. Thirdly, the original purchasing intentions of Middlesex C.C. were trimmed, and some properties intended as further links or infill for the green-belt girdle had to be abandoned.

CONCLUSIONS

This study shows that an understanding of the nature of inter-authority relations is important to an understanding of the processes responsible for shaping the form, area and land use of the London green-belt estates. A county dependent on the cooperation of its districts for the implementation of the policy and for cooperation in other policy areas will negotiate and bargain. It will concede on certain issues once the greater part of its policy objectives have been achieved. In the case of the relationship between Middlesex C.C. and the L.C.C. however, the county authority does not need to rely on the compliance and cooperation of the L.C.C. in its other local government functions. Thus once most of its own objectives were reached, it chose to proceed without L.C.C. financial assistance. In contrast, the L.C.C. is dependent wholly on its adjacent suburban authorities to carry out its particular green-belt policy. It must exchange financial assistance in return for some say in the use of physical resources under the spatial control of another local authority. Given the reaction of Middlesex and the districts to the conditions of the L.C.C.'s grant aid, it is questionable whether the L.C.C. in fact gave enough consideration to the types of restriction on use and access it attached to its grant aid, since these authorities considered that the restrictions outweighed the relatively low levels of grant that actually materialised.

The availability of L.C.C. aid certainly allowed the authorities to buy considerably more land, and more rapidly, than would otherwise have been possible. Actual location and use of the purchased properties was determined largely by the proposals outlined in the regional plans - the L.C.C. taking no positive role in identification - and by the wishes of the district authorities. In the case of Harrow, for example, the ideal of acquiring further land in 1937 was quashed by the practicalities of managing such land - practicalities which had not been considered at the outset of the green-belt scheme.

For example, less than 25 per cent of the land acquired by Middlesex C.C. with L.C.C. grant aid in the north of the county was put into open-space use, the rest remaining in agricultural or other non-recreational uses. Furthermore, other aspects, such as accessibility from the centre of the urban area had also received relatively low priority in selecting the sites or allocating grant aid. The specific implications of this for the urban fringe land-use pattern around London were that the additional land released for open-space uses met only local rather than regional needs. These local people could hardly be described as being among the most deprived members of the community, and yet their additional benefits were heavily subsidised by the county council and also often by the L.C.C.

Putting the implications of this particular policy into a wider context, it is interesting to note the early use in Great Britain of land acquisition to achieve recreational and urban planning objectives. Moreover, these were achieved with the co-operation of authorities predominantly under Conservative political control. These findings, given the peculiar arrangement of British local government, will give some interesting comparisons with other locations where, for example, land-use planning activities are controlled at central government level, or where there is no equivalent of a county level of government. In such situations relationships may be quite different and have equally different but interesting impacts on urban fringe land-use patterns. Hanrahan and Cloke's paper in this book suggests that the findings of the inter-war study are still valid for inter-organisational relations today.

ACKNOWLEDGEMENTS

I am indebted to Dr R.J.C. Munton for comments on this and earlier drafts of the paper, also to Dr H.C. Prince and Dr P.L. Garside.

REFERENCES

Adams, T. and Thompson, L. 1924. *West Middlesex Regional Planning Scheme*.

Adams, T., Thompson, L. and Fry, M. 1928. *North Middlesex Regional Planning Scheme*.

Ashworth, W. 1954. *The genesis of modern British town planning*. (Routledge and Kegan Paul, London).

Blowers, A. 1980. *The limits of power*. (Pergamon, Oxford).

Bowley, M. 1944. *Housing and the state, 1919-1944*. (George Allen and Unwin, London).

Byrne, T. 1981. *Local government in Britain: everyone's guide to how it all works*. (Penguin, Harmondsworth).

Cherry, G. 1981. *Pioneers in British planning*. (Architectural Press, London).

Dalton, H. 1939. The green belt round London. *Journal of the London Society*, 255, 68-76.

Davidge, W.R. 1927. *Hertfordshire Regional Planning Report*.

Ferguson, M. and Munton, R.J.C. 1978. Informal recreation in the urban fringe: The provision and management of sites in London's green belt. *Land for Informal Recreation: a Geographical Analysis of Provision and Management Priorities*, Working Paper 2, (Department of Geography, University College, London).

Friend, J.K. 1976. Planners, policies and organisational boundaries: some recent developments in Britain. *Policy and Politics*, 5, 25-46.

Friend, J.K. and Jessup, W.N. 1969. *Local government and strategic choice*. (Tavistock, London).

Friend, J.K., Power, J.K. and Yewlett, C. 1974. *Public planning: the intercorporate dimension*. (Tavistock, London).

G.L.R.P.C. 1929. *First report of the Greater London Regional Planning Committee*.

G.L.R.P.C. 1933. *Second report of the Greater London Regional Planning Committee*.

Hall, P. *et al.* 1973. *The containment of urban England,* vols 1 and 2. (P.E.P., London).

Jackson, A.A. 1973. *Semi-detached London*. (George Allen and Unwin, London).

Johnston, R.L. 1981. The political element in suburbia: a key influence on the urban geography of the United States. *Geography*, 66, 286-296.

Leach, S. and Moore, N. 1979. County/district relations in shire and metropolitan counties in the field of town and country planning: a comparison. *Policy and Politics,* 7(2), 165-179.

Morrison, H.S. 1935. All about the L.C.C.'s new plan. *G.L.R.O.: CL/PK/1/29,* (script for BBC radio broadcast).

Munton, R.J.C. 1983. *London's green belt: containment in practice.* (George Allen and Unwin, London).

Pondy, L.R. 1967. Organizational conflict: concepts and models. *Administrative Science Quarterly,* 12, 296-320.

Saunders, P. 1979. *Urban politics: a sociological interpretation.* (Penguin Educational, Harmondsworth).

Sharp, E.G. 1980. The London County Council green belt scheme - a note on some primary sources. *Planning History Bulletin,* 2(2), 12-16.

Sheail, J. 1981. *Rural conservation in inter-war Britain.* (Oxford University Press, Oxford).

S.C.L.S.E.R.P. 1976. *The improvement of London's green belt.* SC 620 (London).

Swenarton, M. 1981. *Homes fit for heroes - the politics and architecture of early state housing in Britain.* (Heinemann Educational, London).

Thomas, D. 1970. *London's green belt.* (Faber, London).

Thompson, J.D. 1967. *Organizations in action.* (McGraw Hill, New York).

Vallis, E.A. 1972. Urban land and building prices, 1892-1969. *Estates Gazette,* I-IV, 1015-1019, 1209-1213, 1406-1407, 1604-1605.

Young, K. and Garside, P.L. 1982. *Metropolitan London - politics and urban change, 1837-1981.* Studies in Urban History, 6, (Edward Arnold, London).

Young, K. and Kramer, J.K. 1978. *Strategy and conflict in Metropolitan housing: suburbia versus the Greater London Council, 1965-1975.* (Heinemann for C.E.S., London).

Chapter 27

Small rural communities and the reorganisation of Dutch local government districts

Jacob Groot

INTRODUCTION

At first sight small rural communities and the reorganisation of local government districts have nothing to do with each other. Small rural communities can be described as population concentrations of limited size, having only a few services, and wanting to distinguish themselves from other socio-spatial units. Reorganisation of local government districts merges small municipalities into larger districts of local government. What consequences of any importance could such a reorganisation have for the social life of the small communities concerned? Most often a reorganisation of local government only results in the closure of some municipal halls. Even then it may be possible that the building will be put to a use which will meet a real need in the community.

Nevertheless, the executive committee of the Dutch National Association of Small Communities decided to devote their annual congress in 1982 to the reorganisation of local government districts. Founded in 1974, this association has as its main objective looking after small communities' interests especially against all kinds of governmental measures which could be detrimental for either all of the small communities or some of them. The membership of the association consists to a large extent of administrators of rural municipalities. As many of these municipalities are threatened by a reorganisation, it can easily be understood why the Dutch National Association of Small Communities devoted their annual congress to this subject. It was to this congress that the present author gave an address in which he tried to show how social life in small villages is linked to municipal re-division.

Looking for a starting point for how to approach the present theme, one can advance the thesis that local government must be considered as an essential element in the social life of the whole community. Early this century Max Weber launched his incisive analysis of the way bureaucratisation was penetrating social life. This means that doing research into local society, as well as teaching about it, one has always to pay attention to the way the government is acting within this local society. Equally, government activity can never be considered apart from what is happening in the local society. To a large extent Dutch rural society is a society of small villages. This means that a sociological approach to abolishing small municipalities cannot be discussed without considering social life within the small communities.

SMALL VILLAGES AND RURAL MUNICIPALITIES

In the Netherlands there is hardly a rural municipality without at least one small village. Even some urban municipalities, such as Amsterdam and Groningen, have a number of small villages within their boundaries. Although the total number of municipalities has decreased from about 1000 to about 800 during the last thirty years, there are still many small municipalities which consist of small villages. Some small municipalities, especially those in the provinces of Groningen, Gelderland, Noord-Holland and Zuid-Holland, comprise either one small village or a few small villages and hamlets. There are also quite a number of rural municipalities, judged by central government as too small, that encompass up to twenty or more small villages.

The municipality is the lowest level of Dutch local government. This means that apart from where a municipality comprises only one village, Dutch villages have not experienced public administration comparable with the parish in Britain. In this way reorganisation of local government districts means a change in the direct relations between citizens and local government officials.

All small rural communities are characterised by a strong physical and social identity. Physical identity encompassses the location and appearance of the houses, the location and the shape of the church and particularly the church tower or its absence - in short, everything that determines the appearance of the village. Social identity expresses itself in a specific place name which has an important symbolic value for most of the villagers. Social identity also finds its expression in having a church, school, its corporate life, a village hall and so on. All these are elements by which the inhabitants distinguish themselves from other village societies. They are conscious of this because in present Dutch society (as well as in international society) these villagers have a need to separate themselves as a local society from outsiders. Because of this need for social identity, it is important to consider the regional reorganisations of Dutch local government districts which are either under way or are proposed in the near future.

As a consequence of the small scale of social life within a rural municipality, social distances between governors and governed are mostly short. Burgomaster, eldermen, council members and usually also the civil servants working at the municipal hall all come into contact with the municipality's inhabitants in social roles which are different from those they occupy as municipal officers. He is a fellow churchgoer or father or mother of a classmate of your child, he is a fellow member of a club or partner of a person who is participating in the same discussion group, in the evening his dog sometimes meets yours or he is your neighbour. This multitude of connecting social roles in which the governors and the governed in a small municipality come into contact with each other is characteristic of the social life in rural municipalities. In small municipalities many governors and governed are acquainted with each other personally. From an investigation in Baden-Württemberg (West Germany) Jauch (1975) found that in small municipalities which after the 'Verwaltungsreform' remained independent, 45 per cent of those interviewed were acquainted personally with all or nearly all the members of the municipal council. On the other hand, in the former small municipalities which had been amalgamated only 3 per cent of those interviewed were acquainted with their council members.

Social life in a rural municipality is characterised by a rather extended connection of role patterns as a consequence of which governor-governed relations have a different character from those in larger municipalities with a few thousand inhabitants. The general patterns of both norms and values are no less highly differentiated in many small municipalities than in larger ones, and this also influences the way governors and governed deal with each other. For example, values such as tolerance and helpfulness in personal relations generally have a higher priority in rural communities than in urban ones. Therefore the relations between governors and governed, as well as those between governors and civil servants and those inside the groupings of governors and civil servants are different from those in urban societies, where frankness and liberty are generally valued more highly.

THE PROBLEM

The only way to assess the effect of local government reorganisation is to consider what happens in the municipality as part of what happens in the whole of local society. Following Jauch's (1975) work, we shall look in two ways at the relations between municipal government and the local/regional society. First, we shall consider what changes will occur in the relations between citizens and officials at the municipal hall when a reorganisation of districts is carried through. In doing so, we can distinguish between, first, the social relations which the citizens, following their individual or group interests, have with the municipal officials and, second, the involvement which citizens may have in municipal affairs.

We shall also consider the local/regional society and we shall ask what changes may be expected to occur in it as a consequence of a reorganisation. In doing so, we shall pay particular attention to (a) community attachment, (b) small village services, (c) networks of social relations both within the individual rural communities and within the regional society, and (d) the promotion of farmers' interests.

CITIZEN-MUNICIPAL GOVERNMENT RELATIONSHIPS

As to the change in social relations between citizens and municipal government, the hypothesis is self-evident that, after an enlargement of local government districts, these relations will decrease in intensity. For many inhabitants the physical distance to the municipal hall will have increased. Some of the inhabitants will have to use public transport in order to visit the hall, or they will have to ask for help from either relatives or acquaintances, whereas they could pay their own way before. It can also be suggested that after reorganisation the social distance between inhabitants and local government officials will have increased considerably.

Jauch (1975) found that in the municipalities which had remained independent only 17 per cent of those interviewed had not visited the municipal hall in the preceding year. In the newly formed municipalities, on the contrary, about 60 per cent of interviewees had not visited the hall during the preceding year. From an investigation in the newly formed municipality of Tholen (the Netherlands), ten Berge (1975) found that the number of people attending the public meetings of the municipal council varied greatly according to their home village. The number of attenders who were living in either the village where the council meetings were held or in two nearby villages appeared to be 30 per cent higher than the number of attenders from other villages in the municipality. Another finding from the Tholen investigation is that the inhabitants of the chief village, where the municipal hall is situated, as well as those from an adjacent village, availed themselves of the visiting hours held by both burgomaster and eldermen to a much greater extent than did the inhabitants of other villages.

There are clear indications that after a municipal re-division those with a low socio-economic status in particular will be in a disadvantaged position for contacting local government officials. After a re-division, citizens have to contact their officials more by writing and telephoning and less in a personal and direct way, which disadvantages people of low socio-economic status. It may be noted that many inhabitants of small villages, especially those in remote areas, are of a low socio-economic status.

Not only will many of the inhabitants of small villages be in a disadvantaged position after a municipal re-division, but also most of these villages themselves will suffer. The promotion of small villages' interests will be a much more difficult task for council members than before. More often than not, fewer members in the new municipal council will be from a small village. The national political parties may also have more influence, so a council member will probably have to show a greater loyalty to his or her political party than to his or her local constituency, i.e. his or her own village. It is not improbable that party and village interests will not always coincide.

CITIZEN INVOLVEMENT IN MUNICIPAL AFFAIRS

Citizen involvement in municipal matters means that the average citizen is well informed about what is going on in the municipal domain and further that, if necessary, he or she is ready to participate in public activities like council advisory committees, council elections, the rebuilding of a village hall and the voluntary fire-brigade.

The hypothesis may be advanced that after a reorganisation of local government districts, citizen involvement in municipal affairs will decrease. If the involvement of people living in small villages did decrease, this would be caused very probably by a change in the communication structure concerning municipal affairs. After reorganisation the inhabitants of small villages are informed about municipal matters more by means of newspapers and other written material. Probably on this point too people with a low socio-economic status may find themselves in a more disadvantaged position than persons with higher education. Since proportionally more people with a low socio-economic status live in small rural communities, this means that these communities particularly will experience the adverse effects of the change in the communication structure caused by a municipal re-division. The probable consequences of such a re-division may be that low-ranked people and in particular many inhabitants of small villages may feel themselves less involved in public activities at the local level. Additionally, the complexity of issues under discussion at the municipal level will have increased. The discussion will have been prepared by better qualified and more specialised civil servants than formerly.

On the other hand it may be that the greater professionalism of both the municipal politicians and civil servants (the main purpose of a reorganisation of local government districts) may lead to a situation in which the decline in information for the average people living in small villages will not be as severe as might be imagined. In the new municipal hall an information department will probably be created which will supply the necessary municipal information to local and regional newspapers. In view of the great interest many low-status Dutch country people nowadays seem to have in illegal 'pirate' radio stations, it would be useful if the municipal inform-ation officers would also deliver their information to these stations - assuming that this would be legal!

FROM LOCAL SOCIETY TOWARDS LOCAL GOVERNMENT

Out of the numerous possible social changes caused by a reorganisa-tion of local government districts, the topics we have selected to study in detail can be set out as four questions. Does community attachment decrease after reorganisation? Will a re-division cause a decline in the services available in small villages? Do networks of social relations both within the rural community and within the regional society experience changes as a consequence of a reorganisa-tion? Do farmers' interests run the risk of being promoted less forcefully after reorganisation?

Before answering the first question, community attachment should be defined as the emotional alliance community members have with their own settlement as well as their own community combined with their need to separate themselves from the outside world. It can-not be denied that the municipality's name and its coat of arms have a symbolic meaning which may contribute to community attachment. Without any doubt this will be the case for a municipality which coincides with a small village. On the basis of this evidence it may be expected that, by being merged into a larger municipality (with a new name, new arms and a new flag) community attachment will sometimes be reduced. Moreover, one may suppose that a certain disintegration may appear in the social life of the small villages concerned.

However, on the basis of his findings, Jauch (1975) came to the conclusion that after the 'Verwaltungsreform' a diminution of village attachment had not taken place in the small villages he investigated. On the contrary, his investigation indicated that in some cases the municipal re-division had strengthened village attachment. We are very curious to know if this strengthening will also happen in the Dutch small villages which will be involved in a reorganisation of local government districts to be executed in the near future.

Concerning the second question, the possible deterioration in village services as a consequence of a municipal redivision, some Dutchmen who are familiar with recent public administration in newly formed rural municipalities, have revealed such amazing facts that thorough research into this question is required. This kind of research has not been done so far in the Netherlands.

As to the question of possible changes in social relations both within the rural community and within the regional society, we only mention some of the findings by Jauch. From his investigation it appeared that the patterns of social relations had not been disturbed through the 'Gemeindereform'. This appeared to be the case particularly for the structure and function of local and regional associations. However, it did appear that after reorganisa-tion the associations became more important for the exchange of information on municipal matters as well as for the purpose of achieving consensus regarding solutions to difficult problems. On the other hand, an increase in the influence of the local associa-tions as pressure groups acting upon municipal policy could not be detected. Furthermore, it turned out that the political parties had been playing a more significant part in municipal policy-making.

The final question concerns the promotion of farmers' interests after a municipal re-division. Since a number of rural municipalities are being united with a market-town municipality, we may expect farmers' influence upon municipal government policy to have decreased after reorganisation. The findings from Jauch's investigation may reassure the farmers. His hypothesis that after the 'Verwaltungsreform' the farmers would have lost influence over municipal policy was not confirmed. However, Jauch's investigation yielded indications that some of the seats on the councils of the newly formed municipalities will cease to be held by farmers in the near future. An obvious trend was noted for farmers to retire from municipal policy. Very often the increasing burden of modern farming was mentioned as a reason for them retiring.

This paper has attempted to show that before a scheme for the reorganisation of Dutch local government districts is executed, much more sociological research is needed and, we think, geographical research too.

REFERENCES

Berge, J.B.J.M. ten, 1975. Het platteland: proeftuin voor gedecentraliseerd bestuur? (The countryside: experimental garden for decentralised public administration?). *Tijdschrift voor openbaar bestuur,* 1, 167-169, 189-192, 213-216.

Jauch, D. 1975. *Auswirkungen der Verwaltungsreform in ländlichen Gemeinden.* (Effects of government reorganisation in rural municipalities). Hohenheimer Arbeiten 82, (Ulmer, Stuttgart).

Chapter 28

A key to settlement growth in rural areas: local administrators, their powers and the size of their territories

Jan Groenendijk[1]

That rural settlement patterns are bestowed upon us by history is more than a cliché. Planners and politicians ought to be encouraged by this fact to evaluate the meaning and function of settlements as environments for living. Today, the responsibility of governments covers these issues. Policies for housing, planning and economic growth, and also less spatially explicit actions are taken by governments in a number of fields, the outcomes of which may form specific patterns in space. These patterns (e.g. of population growth in settlements) should be viewed as the outcome of decision-making processes that are influenced by distinct political and administrative structures (the ordering of roles in a political system) and by political cultures (the patterns of attitudes and orientation towards politics among the members of a political culture system (Muir and Paddison, 1981)).

We cannot possibly hope to explain fully differential population growth processes and the patterns they form solely on the basis of regularities in decision making. One step towards describing relationships between political institutions and spatial processes of growth can be taken by comparing the outcomes of different political and administrative systems.

As we are interested in the way these systems may influence patterns of population growth, the administrative guidance for the location of new housing will form the core of this study. We shall compare rural population growth in the northern Netherlands with that in Norfolk as outcomes of different administrative structures. Again, many factors will be responsible for differences in population growth in specific locations. Here, we investigate the effects of administrative guidance on the growth of settlements. This comparison is made as part of a more general investigation into population growth patterns as spatial outcomes of municipal political processes in the rural areas of the Netherlands.

ADMINISTRATIVE SYSTEMS OF LOCAL GOVERNMENT

The second half of the nineteenth century was a crucial period in the formation of British and Dutch local government structures. By

[1] I am indebted to John Ayton and Mike Terry (Norfolk County Council), and to Malcolm Moseley and John Packman (University of East Anglia) who introduced me to the Norfolk data.

1851 Dutch local government had been formed into democratic jurisdictions, referred to here as municipalities. Territorially they
were a direct inheritance from a civil parish structure. The
implementation of the unitary 'gemeente' system implied a territorial enlargement for a great number of these jurisdictions.
However, extension of suffrage for central government elections
made the bourgeoisie determined to hold on to their position in the
power structure of the smaller rural municipalities (Brasz, 1961).

Meanwhile, British local government was re-structured on a wider
territorial scale of districts and counties in order to provide
modern services. So while in the Netherlands the parish level
survived, in England it was almost abandoned. Why and how could it
survive in the Netherlands?

In the first place, a large number of rural municipalities had
larger populations than their British counterparts. Until the end
of the nineteenth century, some were in thinly populated areas with
very wide, civil-parish territories. They attracted much population
when agrarian production was intensified after the advent of
fertilisers. Reclaimed land was made into larger new municipalities.

Secondly, when the tasks of modern local government were widened,
urban municipalities rather than central government took the lead
in the provision of water, gas, electricity and housing, and rural
municipalities were frequently able to join these systems. In this
way municipalities became an important level for local government.
Later initiatives, planned and financed by the state, were also
given to the municipalities. The provinces were only given supervisory power over the performance of municipalities and no executive
power or substantial budgets. Precisely the fact that the execution
of so many tasks - from schools and sporting facilities to housing
and sewerage - is at one level gives the municipality real power.
In all but the smallest village hall, one will find a staff of at
least ten to fifteen civil servants under the management of an
appointed burgomaster. This burgomaster in rural areas is often a
career man who has to prove his worth. Achievements which will be
applauded by the locals and further his career include the introduction of modern urban-style services.

Growth of population will serve as a means to several ends. It
brings the municipality closer to servicing thresholds, it pays off
handsomely in the budget, it expands the local administration and
will further staff careers. Having only limited powers of taxation,
municipalities derive their income from the central government,
population and, since 1983, built-up area being important yard-
sticks for funding municipalities.

The territories of rural municipalities mostly contain a number
of villages and hamlets. Municipal expansion schemes show a strong
tendency to favour the largest village within the territory. In
that way a local population threshold might be attained for higher
order, commercial services. Business may also be 'stolen' from a
(competing) neighbouring municipality. Smaller villages within the
territories show much less growth, if any. A provincial planning
policy that precludes growth of some municipalities' largest
villages will meet with bitter opposition and will not have much
chance of being implemented by those municipalities.

In rural England, administrative territorial organisation
before 1974 provided no stimulus for the rural districts to concentrate population growth. Most market towns - the concentration
poles - were contained in tiny urban districts, jurisdictions
territorially separated from the rural districts which were

divorced from their more obvious nodes for concentration. This lay-
out of territories hindered selection for growth among the villages
of rural districts. A growth selection policy by the county, the
need for which is made clear by Ayton (1976), was not acceptable to
these districts when housing demand with purchasing power is
spatially diffuse. Apart from that, the wording of the 'Interim
Settlement Policy' adopted by the County Council in 1972 was rather
flexible. It was permissive rather than stimulating towards
selected villages. The county might not be able to raise a stern
'no' to prospective builders in inappropriate villages.

GOVERNMENTAL INSTRUMENTS IN THE HOUSEBUILDING PROCESS

Planning and implementation

What powers do local governments have to control the spatial
distribution of new housing? There is the 'planning permission'
given by the county or, when delegated, by the district. On the
Dutch side there is the 'building permission' given by the municipality following its statutory plan which has to be approved by the
province. This outlines the legal framework, but for the
behavioural scientist other variables should come to the forefront
to explain where and when building of houses is to be expected under
some public guidance.

The building process can be seen as organised by a coalition of
actors (Burie, 1977). These actors, among whom is local government,
participate in a coalition for their own benefit. In order to
produce houses, the coalition has to get hold of a number of
resources, such as finance, the capacity to build, the land and
different sorts of governmental permission.

The building process will take effect when all the resources
have been obtained. Participants in the coalition obtain one or
more of these resources and offer them to the coalition; these are
the roles they play. Many different coalitions can be formed. In
our age of great companies, many roles can be performed by just one
company. A building company may act beyond the building resources
it controls by financing housing and selling them for a profit.
The same company may have gained control over the land long before.
As long as civil servants perform as they should, builders will
never control the granting of planning permission. So local
government always has to play at least one crucial part in the
coalition.

Local government may sometimes have a virtual monopoly in the
coalition. It finances public housing, it gives the permission, it
can own the land and it could, in exceptional circumstances, own a
building company. Governmental influences in these coalitions can
vary markedly.

In the formation of these coalitions to produce rural housing,
there are two important differences between Britain and the Netherlands. Both make the position of local government in the Netherlands much stronger especially in rural regions. First, public
housing in the Netherlands is more equally distributed between urban
and rural regions. Evidence will be given below when comparing
Norfolk to the northern Netherlands. Second, almost all building
land in the Netherlands goes through public ownership. Municipalities buy land (expropriate it when needed), service it (sewerage,
etc.) and sell it to prospective builders on their own conditions.
This means that no building can take place until the basic physical

services have been provided.   These investments are paid for by the prospective builder when he buys the land. .

Britain has seen three post-war attempts to obtain governmental control of building land.   However, in a two-party political system, this highly disputed issue will probably never be settled for the considerable time which is needed to make it work.   Besides this, the bodies that provide services like sewerage are certainly not 'local'.

Public housing

Apart from the spatial guidance of private housing investments through planning, there is also public investment in housing, the location of which is a governmental decision.   Potentially, this makes public housing an instrument for organising the spatial distribution of housing in which government has a powerful position. However, public housing in the countryside is usually less important than it is in cities.   Even then public housing is bound to be over-represented outside the smallest villages.   For one thing the farmers' dwellings will form a larger part of the housing stock in small villages.   On the other hand, only in somewhat larger villages can the public body as 'landlord' be sure that the houses will be required on a permanent basis.   In this way rural municipalities in the Netherlands show in the modal case an over-representation by 50 per cent of public housing in the largest villages.   This figure is influenced as well by a tendency to concentration in the largest village of a municipality which is a feature of the Netherlands.

Table 28.1   Public and private housing in western parts of Norfolk
              and the North Netherlands

	Houses built 1960/1-1970	% Public housing
Breckland	8869	40
Thetford	3218	79
East Dereham	1031	34
other settlements < 5000	4620	14
West Norfolk	9126	35
King's Lynn	2833	76
other settlements < 5000	6293	16
Groningen	46 360	49
settlements > 5000	34 070	51
settlements < 5000	12 290	43
Friesland	46 875	51
settlements > 5000	26 710	59
settlements < 5000	20 165	41
Drente	38 295	47
settlements > 5000	21 685	55
settlements < 5000	16 610	35

We can compare the part played by public housing in the building processes between 1960-1 and 1971. We can differentiate between settlements with more and less than 5000 inhabitants in 1960-1 to distinguish rural settlements from small urban centres (Table 28.1). Although public housing forms a larger part of the total new housing stock in the Dutch provinces (which have more than the Dutch average), a striking difference is to be noted between the rural settlements' share in Groningen, Friesland and Drente compared to those of West Norfolk and Breckland. Clearly, rural Norfolk is accommodating different people than its cities (Moseley, 1982).

System-biased effects

It is obvious that the power local governments potentially have in the Netherlands in the formation of house-building coalitions places them in a good position to select within their territory which settlements to develop. The scale of local government units may have made its own contribution. For Norfolk and the Dutch northern provinces the following mean areas illustrate this: municipalities 70 km^2, parishes 10 km^2, former rural districts 700 km^2. One might think that local government on a large scale (the county level) is in a better position to concentrate housing in selected settlements. This may be true for the higher ranks of a settlement hierarchy. For the lower ranks, however, the Dutch decentralised system may give more opportunity for selection. We have already seen that, in contrast to the rural municipality, the former rural district had not much to gain by concentration, the real centre being in another, 'urban', district. Apart from that, the decisive power of the county to grant planning permissions leaves this statutory power at the end of information and decision lines rather than in the centre. This is particularly relevant to the avalanche of demands for building individual houses and small estates that reach the county. These demands will be underpinned by local arguments which the county cannot easily counter.

After 1974 the new districts were in a better position, encompassing nodal regions *in toto*. The effects of the change in administrative structure, however, cannot accurately be observed due to the change in economic climate that took place at the same time, and the number of pending planning permissions.

GROWTH OF RURAL SETTLEMENTS: MUNICIPAL SELECTION IN THE NETHERLANDS

Research on the selection of settlements for growth in the Netherlands (Groenendijk, 1983) has made clear to what degree non-urban municipalities concentrate new housing through their planning powers. In more than 80 per cent of them the percentage population growth of the largest ('chief') settlement of the municipality exceeds the percentage growth for the population of the other settlements.

That larger settlements grow rather than smaller ones is furthered by the aspirations that municipalities have as providers of services (mainly to be realised in their largest settlement) and the selective planning policy they have as a consequence. A complementary effect of this policy is that relatively few planning permissions will be distributed to the other settlements in a municipality, leading to a loss of population in some of them. So, although only a small number of municipalities in the Netherlands are actually declining in population, many settlements are in this situation.

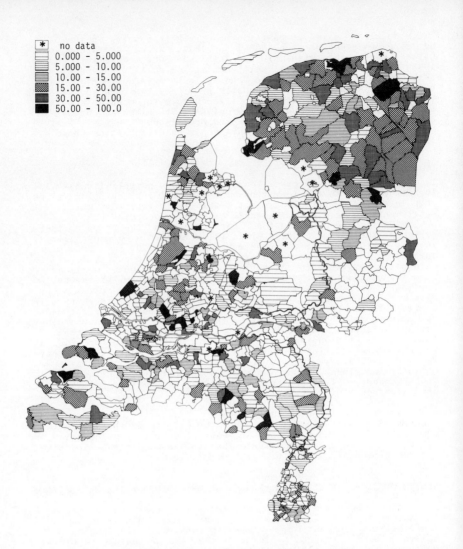

Figure 28.1   Population living in declining settlements (population
index 1960-71 ≤ 90) as a percentage of the total population in the municipalities' settlements

Owing to this selective policy there is hardly any relationship
between the population growth of municipalities and the population
growth of their settlements. As can be seen from Figure 28.1,
people living in declining settlements may form a considerable part
of the municipal population throughout the Netherlands, even in peri-
urban areas with booming population. Northern parts of the Netherlands show the highest percentage of the population living in declining settlements. Even there the relationship with population
growth is limited; the relatively quickly growing province of Drente
has as high percentages of population living in declining settlements as elsewhere in the north of the Netherlands.

Larger settlements have a greater chance to be selected for growth than smaller ones. Size is only taken into account with respect to the other settlements within the same municipal territory. This is to say that being the largest ('chief') village of a municipality is more conducive to selection for growth than size in itself (Table 28.2).

Table 28.2 Percentage growth 1960-71 of size groups of settlements, 'chief' settlements and other settlements

Population	<1000		1000 - 1999		2000 - 3999		4000 - 9999		>10 000	
	'Chief'	Other	'Chief'	Other	'Chief'	Other	'Chief'	Other	'Chief'	Other
Groningen	(8)	(230)	(21)	(27)	(10)	(2)	(6)	(3)	(5)	(0)
	5.4	-7.3	15.3	0.17	31.8	53.9	41.2	-2.2	16.3	-
Friesland	(10)	(325)	(9)	(35)	(14)	(2)	(6)	(0)	(5)	(0)
	13.2	1.6	16.5	13.3	25.0	44.4	25.8	-	23.9	-
Drente	(6)	(229)	(15)	(13)	(6)	(8)	(2)	(2)	(4)	(0)
	49.7	-0.8	43.9	16.6	59.2	4.3	43.5	15.3	62.2	-

(..) = number of settlements

From this table it is clear that settlements of widely different sizes can be 'chief' ones. Small settlements occasionally are selected because they are part of a gemeente which has little territory and/or little population. Whims of history are responsible for the differentiation in territorial sizes. Analysis of variance (Groenendijk, 1983) confirms that in all but the pressured Randstad provinces the status of 'chief' village rather than size differentiates growth figures between rural settlements.

GROWTH OF RURAL SETTLEMENTS: NORFOLK AND NORTHERN

PARTS OF THE NETHERLANDS

Comparisons of Norfolk and the northern Netherlands are not uncommon (Buursink, 1971). These are clearly rural parts of both countries: growth and density of population are similar (Table 28.3).

Table 28.3 Growth and density of population in Dutch northern provinces and Norfolk

	population 1960/1	density/km²	growth % 1960/1 -71
Groningen	475 462	190	9.3
Friesland	478 931	124	9.8
Drente	312 176	116	19.3
Norfolk	561 071	106	10.1

Both regions know the problems common in extreme rural areas such as relative deprivation of modern facilities and few job opportunities. Major differences are to be noted, however, in the social history of the regions. Landlordism has not played any significant role in the north of the Netherlands. Obvious class distinctions did however

289

Table 28.4 Percentage of settlements declining (index 90 or less), stable (index 91-110) or growing (index ≥ 110) per size group 1960/1-1971

	1 <1000	2 1000-1999	3 2000-4999	4 5000-9999	5 10 000-19 999	6 20 000-50 000	7 >50 000
Oost Groningen	72/15/13	25/56/19	-/40/60	33/33/33	-/-/100		
Delfzijl	48/37/15	-/100/-	-/100/-	-/-/100	-/-/100		
Centraal Groningen	49/28/23	21/45/34	-/25/75	-/33/67	-/-/100		-/100/-
N. Friesland	25/38/37	15/45/40	9/9/82	-/50/50	-/100/-		-/100/-
Z.W. Friesland	40/32/28	11/33/56	-/67/33	-/-/100		-/-/100	
Z.O. Friesland	51/28/22	33/25/42	-/33/67	-/-/100	-/-/100		
N. Drente	48/37/15	13/13/75	-/17/83		-/-/100	-/-/100	
Emmen	57/19/24	-/33/67	33/50/17	-/33/67	-/-/100		
Z.W. Drente	49/28/23	-/-/100	-/-/100		-/-/100		
Norfolk							
East Central	22/42/36	-/22/78	-/-/100	-/-/100	-/50/50		-/100/-
North	37/44/19	29/43/29	17/50/33	-/-/100			
Breckland	40/44/16	25/-/75	-/-/100				
West	47/29/24	22/56/22	29/14/57				-/100/-

exist between the prosperous farmers of Groningen and their labourers. Socialism has taken root firmly in the councils. The rural parts of the region differ from other rural parts of the Netherlands since socialists play a considerable part in the executive and legislative power of municipalities and provinces. While farm labour was ebbing away, some success has been seen from efforts to replace it 'on the spot'. Eastern parts of Friesland and Drente in particular have seen the development of some industrial centres in the countryside. Nevertheless, the region as a whole is the most rural in the Netherlands and has the largest share of people living in small settlements. In this respect there is little divergence from Norfolk.

What population development is taking place with respect to the small settlements in the regions we selected for our study? Can differences that may show up be explained as effects of differences in the administrative systems on either side of the North Sea?

Table 28.4 shows how many settlements are declining in population, stable or growing. The smallest size group in the northern Netherlands has half or more of its settlements declining. Norfolk shows fewer declining settlements in this group especially in the east and central parts.

The few settlements with over 1000 people in the north of the Netherlands that are growing are mostly 'chief' villages. In Norfolk the growing settlements of this size group are found in the faster growing region, whereas in the north of the Netherlands the faster growing Drente regions do not distinguish themselves from the rest. On the contrary, the regions of Friesland that grew the least have more settlements in the growing category.

This arouses our curiosity as to what difference there is between Norfolk and the North of the Netherlands in the distribution of growth and decline of population in settlements (Figure 28.2).

The pattern in Norfolk clearly shows a tendency to clustering. Clusters of growing settlements are found towards Norwich, King's Lynn, Thetford, Wymondham and the coast. Large areas without any growing settlements lie in between. Friesland and Groningen have their peripheral disadvantaged areas and one-settlement municipalities that could not select any settlement for 10 per cent growth. Municipalities in a broad arc from West Friesland via North Friesland, North Groningen, East Groningen and part of East Drente did not manage to get 10 per cent growth. The great majority of them, however, were able to select at least one settlement for more than 10 per cent growth. So did all the municipalities of Drente.

Figures 28.1 and 28.2 are really complementary since the selection of some settlements for growth is also the selection of others for decline. This is as true in the faster growing province of Drente and the suburban fringes of the cities of Groningen and Leeuwarden. Even there, the municipalities manage to make in-migration converge on selected villages.

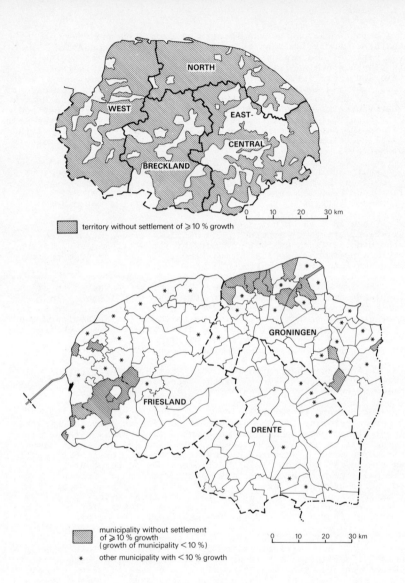

territory without settlement of ⩾10 % growth

municipality without settlement
of ⩾10 % growth
(growth of municipality <10 %)

*   other municipality with <10 % growth

Figure 28.2 Areas without settlements with population index > 110
              between 1960/61 and 1971
              Norfolk   - parishes, urban districts and boroughs
              Netherlands - municipalities and settlements as part
                             of municipalities

CONCLUSION

The relative uniformity in population growth in the northern provinces of the Netherlands is in striking contrast with the clustering in Norfolk. The Dutch administrative framework places local
government in a position to select settlements for growth and
effectively reshape a settlement pattern, seen as obsolete from a
modern servicing point of view.

The fact that municipalities in the north of the Netherlands
together with higher authorities ad ere to and implement this policy
can be ascribed in part as well to different political ideas preva-
lent in the north of the Netherlands and Norfolk and the different
spatial scales of local government territories.

REFERENCES

Ayton, J.B. 1976. Rural settlement policy, problems and conflict.
in *Regional and rural development,* Drury, P.J. (ed), (Alpha
Academic, Chalfont St Giles).

Brasz, H.A. 1961. *Veranderingen in het nederlands communalisme.*
(Van Gorcum, Assen).

Burie, J.B. 1977. De structuur van het bouwproces. *Handboek bouwen
en wonen,* Aflevering 10.

Buursink, J. 1971. *Centraliteit en hiërarchie.* (Van Gorcum, Assen).

Groenendijk, J.G. 1983. *Centrumdorpen 1947-1971, beleid op gemeente-
lijke schaal.* (Vrije Universiteit, Amsterdam).

Moseley, M.J. 1982. *Power, planning and people in rural East Anglia.*
(Centre of East Anglian Studies, University of East Anglia,
Norwich).

Muir, R. and Paddison, R. 1981. *Politics, geography and behaviour.*
(Methuen, London).

SECTION FIVE:

RURAL HOUSING

Chapter 29

Housing in small villages:

a classification of contexts

Frans Thissen

INTRODUCTION

In the past decade physical planning and research with respect to small villages in the Netherlands have witnessed an important development. The attention paid to small villages has greatly increased and changed character. The number of publications about small villages grew rapidly, and the small village soon became a recognised part of national, provincial and municipal physical planning. Apart from this the character of research and policy has changed. Social developments such as a continuing increase in the scale of society, suburbanisation, the growing influence of government and an emancipation of citizens are important in this connection. However, research which affects policy is essential. The publications by Groot have had a particularly great influence on the character of many local studies and on national, provincial and municipal policies (Groot, 1981).

Groot (1980) considers the problem of small villages to be social, centred round the evaluation of the living conditions for the inhabitants of small villages. Essentially these problems are the result of the fact that these villages are small, and because some of the inhabitants wish to distinguish themselves from other local communities. This problem is also geographically relevant for various aspects of the evaluation of living conditions (for example, service provision and the physical conditions of living) and because deficiencies in village life can only be removed at relatively high cost.

When the evaluation of living conditions is placed in a central position (a more social view) a strong integration of different sectors and more attention to the local situation have to be essential elements of policy for small villages. At all levels of administration these elements can be found to an increasing degree. Nevertheless, the increase or decrease of population related to the scale of housing provision will remain an important keystone in policy for peri-urban rural areas as well as remoter ones.

The policy of central government with respect to small villages is based on the principle that the housing needs of those who have social or economic ties with the village should be met, although in certain areas there may be restrictions concerning growth and dispersal of population (M.V.R.O., 1979). The provinces decide on the policy with respect to housing provision in villages and regions, whereas the municipalities - who are important as initiators of

housing - are increasingly made responsible for housing policy,
which includes more than just housing provision (M.V.R.O., 1981).
Local village plans are mostly fitted into regional village plans.

Apart from an excessive concern for restricting housing in small
villages in national and provincial policies, the availability of
houses for those who have social or economic ties with the village
appears to be one of the most important aspects of the evaluation of
living conditions (Streekorgaan, 1981). Limited housing provision
in small villages (or none at all) implies in a quantitative and
qualitative sense that local housing need cannot be satisfied. This
may become particularly problematic if there is strong interest in
the local housing stock from outside the village and if the public
authorities lack the means to house those people who have social or
economic ties with the village. Besides, a limitation on housing
provision in small villages may create a feeling among the inhabitants of being 'out of the picture', so that a potential local
demand may not become manifest.

Research into the housing situation in small villages shows that
the native inhabitants of small villages have only a limited chance
of finding a house in their own village, should they want one. In
addition to the quantitative and qualitative shortages, the influence of outsiders and the limited possibilities of housing allocation, the financing of small housing projects and the irregular
vacancy of houses are problematic (Bolsius *et al.*, 1981).

In this chapter the housing situation in small villages in the
Netherlands is approached by a number of classifications of COROP
regions (see Figure 1.1) or provinces with respect to three relevant
characteristics.
1. The population development of small villages in different size
   categories in COROP regions. This classification indicates what
   kind of intra-regional processes of differentiation are important
   in different parts of the Netherlands. These processes have
   important consequences for the characteristics of the population
   and the housing stock of small villages.
2. The nature of the housing stock of COROP regions. The relevance
   of a description of the housing stock of COROP regions lies
   especially in the significance of the housing stock of the region
   for the differences between settlements and neighbourhoods in the
   region in terms of the nature of the housing stock and characteristics of the population.
3. National and provincial policies with respect to small villages,
   described at the provincial level. The national and the
   provincial policies - related to the kind of problem definition -
   are important as contexts for the small villages.

The locational patterns identified can be related to the distribution of the degree of urbanisation and rurality (see the papers by
Ostendorf and Hauer in this book).

CLASSIFICATION OF REGIONS WITH RESPECT TO SOME RELEVANT VARIABLES

The population development of small villages

The concentration of population, housing, employment and facilities
in the larger settlements with a central position has been an
important aspect of physical planning in the Netherlands. The process of concentration occurs especially within the framework of the
municipality (see the paper by Groenendijk in this book). Near the
cities, concentration is inter-woven with suburbanisation, with
respect to the policy of clustered decentralisation as well as to

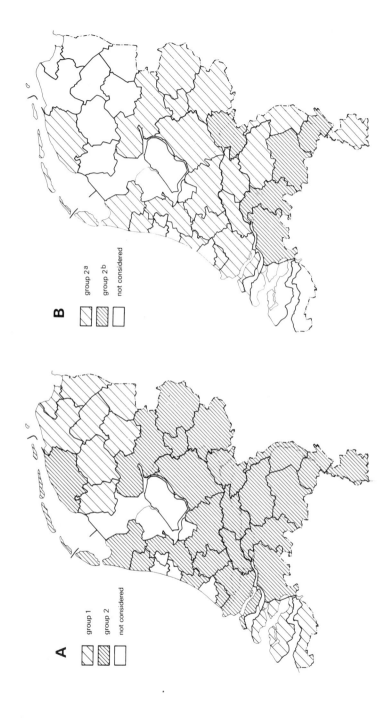

Figure 29.1   Growth/size types of village by COROP region (see Tables 29.1 and 29.2 and footnote 1)

Table 29.1   Distribution for 32 growth/size categories (population 1960, population growth 1960-71) of settlements with less than 4000 inhabitants in two groups of COROP regions (group 1 right hand top corner; group 2 bottom left; see Figure 29.1a)

Growth index 1960-71 (1960 = 100)

Number of inhabitants 1960	<66	66-80	81-95	96-110	111-125	126-140	141-155	>155
< 500	8.4 .9 / 3.1 -.4	14.1 1.3 / 5.6 -.6	20.2 .8 / 12.3 -.4	9.9 -.2 / 11.3 .1	4.5 -.6 / 7.0 .3	2.5 -.2 / 3.1 .1	1.3 -.2 / 1.8 .1	3.5 -.5 / 6.1 .2
500-999	1.2 .1	1.8 .2 / 1.3 -.2	4.3 .4 / 3.2 -.2	4.7 -.2 / 5.3 .0	3.2 -.3 / 4.4 .1	1.6 -.4 / 2.8 .1	1.3 .2	2.6 .4
1000-1999			2.0 .4 / 1.5 -.3	4.6 .4 / 3.0 -.2	3.4 .1 / 3.2 .0	1.0 -.7 / 3.2 .3	1.5 .1	1.3 -.7 / 3.6 .3
2000-3999				2.1 .4	1.1 -.2 / 1.5 .1	2.1 .3	1.7 .3	2.8 .3

The first number is the percentage of settlements in the relevant growth/size category, the second number is the deviation with respect to the distribution for the 34 analysed COROP regions in standard deviation units.

Scores < 1.0% are not considered

```
100% = 970
       settlements

100% = 2266
       settlements
```

Table 29.2 Distribution for 32 growth/size categories (population 1960, population growth 1960–71) of settlements with less than 4000 inhabitants in two groups of COROP regions (group 2 b right-hand top corner; group 2 a bottom-left; see Figure 29.1b)

Each cell lists the group 2 b figures (upper-right) and the group 2 a figures (lower-left); each figure pair is "percentage / deviation".

Number of inhabitants 1960	Growth index 1960-1971 (1960 = 100)							
	<66	66-80	81-95	96-110	111-125	126-140	141-155	>155
< 500	2b: 3.6 / -.3 2a: 3.1 / .4	2b: 6.4 / -.4 2a: 5.4 / -.6	2b: 9.7 / -.7 2a: 13.0 / -.3	2b: 8.2 / -.7 2a: 12.1 / .3	2b: 4.3 / -.7 2a: 7.7 / .5	2b: 2.8 / -.1 2a: 3.2 / .2	2b: .6 / -.6 2a: 2.1 / .3	2b: 4.7 / -.2 2a: 6.5 / .3
500-999	2b: 1.1 / .1 2a: 1.3 / .2	2b: 1.5 / .0 2a: 1.3 / -.2	2b: 4.1 / .3 2a: 3.0 / -.3	2b: 6.0 / .3 2a: 5.2 / .0	2b: 5.2 / .4 2a: 4.2 / .1	2b: 2.1 / -.2 2a: 3.0 / .2	2b: 1.7 / .5 2a: 1.2 / .1	2b: 2.8 / .5 2a: 2.5 / .3
1000-1999			2b: 3.2 / 1.9 2a: 1.1 / -.8	2b: 3.6 / .0 2a: 2.9 / -.2	2b: 4.9 / .8 2a: 2.7 / -.2	2b: 4.3 / .8 2a: 2.9 / .1	2b: 2.1 / .5 2a: 1.4 / .1	2b: 4.7 / .8 2a: 3.3 / .2
2000-3999				2a: 1.0 / -.2	2b: 2.4 / .8 2a: 1.3 / -.1	2b: 2.4 / .5 2a: 2.1 / .3	2b: 2.8 / 1. 2a: 1.4 / .1	2b: 3.0 / .3 2a: 2.8 / .3

100% = 466 settlements

100% = 1800 settlements

The first number is the percentage of settlements in the relevant growth/size category, the second number is the deviation with respect to the distribution for the 24 analysed COROP regions in standard deviation units.

Scores < 1.0% are not considered

actual development (Engelsdorp Gastelaars and Ostendorf, 1974; Harts, 1978). A classification of COROP regions with respect to the occurrence of small settlements in 32 growth/size categories (Table 29.1) produces an initial division into two groups[1] (Figure 29.1a). Concentration in the three nothern provinces and in Zeeland has taken place in the period 1960-71. Very small, declining villages are amply represented and fast-growing settlements are almost absent. In the remaining part of the Netherlands, fast-growing settlements can be found in nearly all size categories, but very small villages with declining populations are also found. In addition to growth in small villages through suburbanisation, concentration plays an important role here in the population development of small villages. In most of the peri-urban rural areas in the Netherlands, suburbanisation *and* concentration are both important processes of differentiation at an intra-regional scale with consequences for the characteristics of the housing stock and population in small villages.

A second analysis directed at the differences between COROP regions within group 2 (Figure 29.1b) supplies another division into two groups. Some COROP regions in the southern part of the country (especially in the province of Noord-Brabant) have considerably fewer settlements with less than 500 inhabitants and these are less often growing settlements. Apart from the relatively large number of very small settlements which has existed for a long time in Noord-Holland (COROP regions 18 and 23), it seems that villages with less than 500 inhabitants are still recognised as valid settlements in provincial physical planning in eastern and western Netherlands. In these parts of the country the growth of population has also been of greater significance for very small settlements: the greater pressure of suburbanisation in the west and the greater dispersal of the settlement pattern in the east may supply an explanation here. In the south - where urban and industrial development took place - suburbanisation has had fewer consequences for the smallest villages.

In addition to suburbanisation and concentration, the provincial policy with respect to villages of less than 500 inhabitants is important for their population development, among other things principally through the definition of the 'small village'. This policy, however, is influenced by the pressure on small villages and the significance of those villages in the regional spatial structure.

Characteristics of the housing stock

In the study by Engelsdorp Gastelaars, Ostendorf and De Vos (1980), the characteristics of the local housing stock outside the central cities appear to be relevant for the socio-morphological structure of the region. Physical-morphological criteria with respect to the housing stock of municipalities show a statistical relationship with important socio-morphological dimensions. The suburbanisation process, which is characteristic of the urbanised regions of the western Netherlands, is the most important motor behind this intra-regional differentiation. The physical-morphological characteristics of the housing stock which are important for the socio-morphological structure are the proportion of recently completed dwellings and the proportion of more expensive dwellings (see the paper by Ostendorf in this book).

---

[1] The groups are distinguished by a cluster analysis: Ward's hierarchical method is used with squared Euclidean distance as the similarity measure (Figure 29.4, binary data). The variables are standardised and a relocation procedure is applied to the results (Berkouwer, 1978).

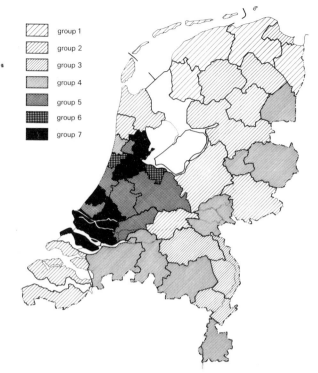

group 1	
group 2	
group 3	
group 4	
group 5	
group 6	
group 7	

Figure 29.2   Types of COROP region by housing stock, 1971 (see footnote 1)

It is clear that concentration also leads to a certain socio-morphological structure.   Great differences in the nature of the housing stock and related characteristics of the population may be expected, especially between fast-growing and declining villages (Groot, 1980; Sociaal en Cultureel Planbureau, 1980).   The occurrence of recently completed dwellings in peripheral rural areas is also connected with important characteristics of the population (Engelsdorp Gastelaars *et al.*, 1980, table 11). Further industrialisation outside the central region leads to a specific socio-morphological structure in rural areas, which partly relates to the occurrence of public housing (for instance in COROP region 8, see Engelsdorp Gastelaars *et al.*, 1980, 47-8).   Finally, suburbanisation which is not intra-regional leads to a specific socio-morphological structure which is connected with the occurrence of recently completed dwellings, expensive dwellings and possibly other characteristics of the housing stock and the material spatial structure.

The housing stock in central and peripheral regions varies greatly in terms of house size, tenure and rates of turnover.   The proportion of owner-occupied houses (generally with a low rate of turnover) varies widely between central and peripheral regions as does the proportion of multiple-occupied housing where turnover rates are usually high.

303

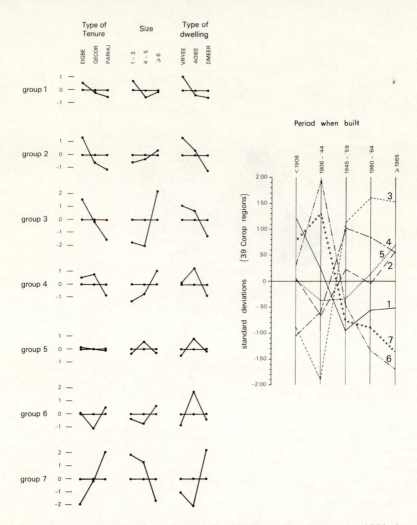

Figure 29.3   Aspects of housing in COROP regional types, 1971 (see
Figure 29.2 and Table 29.3).   Units are standard
deviations

   In   Figures 29.2 and 29.3, seven groups of COROP regions are
distinguished on the basis of 14 housing variables; type or tenure,
size, type of dwelling and construction period (see Table 29.3).
Three dimensions are apparent here.   The differences are clear
between central (group 7), intermediate and peripheral regions
(group 1): central and peripheral regions are similar with respect to
the age of the housing stock and the occurrence of small dwellings,
central regions distinguish themselves by the occurrence of private
rented housing.   A second dimension reflects the demographic differences between the southern part of the Netherlands (groups 3 and 4)

304

Table 29.3   Variables describing the housing stock, used in the
              cluster analysis of 39 COROP regions

A. Type of tenure:

EIGBE	%	owner-occupied
GECOR	%	public rented housing
PARHU	%	other rented housing

B. Size:

1-3	%	dwellings with less than 4 rooms
4-5	%	dwellings with 4 or 5 rooms
> 6	%	dwellings with 6 or more rooms

C. Type of dwelling:

VRYEE	%	single-family, detached
AOIEE	%	single-family, terraced
DMEER	%	part of a multiple-occupied structure

D. Period when built:

< 1906	%	dwellings constructed before 1906
1906-1944	%	dwellings constructed 1906-1944
1945-1959	%	dwellings constructed 1945-1959
1960-1964	%	dwellings constructed 1960-1964
> 1965	%	dwellings constructed in 1965 or after

and the remaining regions; a high proportion of large dwellings is
characteristic. Finally, there are distinctive COROP regions which
have experienced urban-industrial development (group 4); their most
important characteristic is a large proportion of public rented
housing.

National and provincial policy for small villages

The provinces are classified on the basis of the six criteria shown
in Table 29.4. The analysis allows a division into both two and
five groups (Figure 29.4).

In group I the most important rural settlement areas are represented. An integrated policy with respect to small villages is
characteristic of the northern provinces, 'liveability' is an
important goal in the whole group. In Drente and Friesland the
improvement of the hierarchy of services is considered a major goal.
In the eastern provinces of Overijssel and Gelderland and in Zeeland,
the definition of the 'small village' as a settlement with minimal
services is crucial. Policy here deals with the problem of the
spacing of the settlement pattern. This is of special significance
since the distances between settlements in these provinces are
larger than elsewhere in the country, which has a major impact on
the availability of essential facilities. In addition this sub-
group is characterised by the development of long-stay recreation
and the occurrence of interregional suburbanisation (for instance,
immigration of the elderly from the Randstad).

In the western and southern parts of the Netherlands, rural areas
are peri-urban in almost every respect (group II). Especially in
the central, physical-morphological urbanised provinces (Noord-
Holland, Zuid-Holland and Utrecht), suburbanisation has been of
great significance. The partial failure of the policy of clustered
decentralisation in the early seventies in these provinces has
strongly affected the policy for small villages. In Zuid-Holland
and Noord-Brabant, 'liveability' is more important as a goal than
in the northern part of the Randstad.

Figure 29.4.   Small village policies in the provinces (see Table
29.4 and footnote 1)

CONCLUSION

The locational patterns described here give a picure of the diversity of contexts in the Netherlands for small villages.   Such a
description can be a starting point for further research on the
position of small villages and their inhabitants within a specific
regional context.   This research can be directed at three themes.
1.   We may conclude that the population characteristics of small
     villages are part of a regional socio-morphological structure
     and result from several processes of differentiation. Important
     for the housing situation in small villages are aspects of the
     material spatial structure such as the characteristics of the
     local housing stock, the accessibility of activity centres and
     other physical-morphological qualities of the village.
2.   Future research should also consider the migration pattern in a
     region and the relative position of small villages and different
     groups of migrants in the regional migration system.   The func-
     tion of small villages as a part of the regional housing market
     can be described, and the effects of policies such as the con-
     centration of housing production in key settlements may be
     evaluated.

Table 29.4   Frequency ratios* of groups of provinces with a
similar national and provincial policy with
respect to small villages (see Figure 29.4)

	1	2	3	4	5	I	II
Restrictions with regard to growth and decentralisation of population (national policy)	0	0	.61	1.83	1.83	.31	1.83
Definition of 'small village': minimal service provision is one of the criteria (provincial policy)	1.57	0	1.57	1.05	.79	1.05	.94
Important goal of provincial policy: hierarchy of services	0	2.20	1.47	.73	0	1.47	.44
Important goal of provincial policy: 'liveability'	1.38	1.38	1.38	0	1.38	1.38	.55
Important goal of provincial policy: restriction of suburbanisation	1.38	0	0	1.83	1.83	.31	1.83
Integral small settlement policy (provincial policy)	5.50	2.75	0	0	0	1.83	0

Sources: M.V.R.O., 1979; P.P.O., 1979; Niessen, 1979

* Frequency ratio = % in the group / % in the population

3. The migration of households to and from small villages can also
   be studied. A relevant question is to what extent the housing
   supply in small villages is important for households migrating to
   and from small villages. Hypotheses derived from correlations at
   an aggregated level can then be tested at an individual level.

REFERENCES

Berkouwer, Y. 1978. *Programmabeschrijving van clustan 1-c.* (Centrum
voor data-analyse, Faculteit der Sociale Wetenschappen, Rijksuniversiteit Utrecht, Utrecht).

Bolsius, E. *et al.* 1981. *Leefbaarheid kleine kernen.* Eindrapportage
van de projectgroep vastgesteld door de provinciale onderzoekers-
vergadering, ('s-Gravenhage).

Engelsdorp Gastelaars, R. van and Ostendorf, W. 1974. De groei van
plattelandskernen in Nederland. *Tijdschrift voor Economische
en Sociale Geografie,* 65, 162-173.

Engelsdorp Gastelaars, R. van, Ostendorf, W. and de Vos, S. 1980. *Typologieën van Nederlandse gemeenten naar stedelijkheidsgraad.* Monografieën Volkstelling 1971, 15b ('s-Gravenhage).

Groot, J.P. 1980. *Groeiende en kwijnende plattelandskernen.* Monografieën Volkstelling 1971, 14 ('s-Gravenhage).

Groot, J.P. 1981. *Dutch small rural communities in research and in public policy.* (Helsinki/Wageningen).

Harts, J.J. 1978. Kleine gemeenten in een stadsgewest: een empirische studie van suburbanisatie van bevolking rond Utrecht. *Geografisch Tijdschrift,* XII, 4, 351-361.

M.V.R.O. 1979. *Nota Landelijke Gebieden.* Derde nota over de rumitelijke ordening, deel 3d Regeringsbeslissing met nota van toelichting, ('s-Gravenhage).

M.V.R.O. 1981. *Structuurschema Volkshuisvesting.* Deel 3d Regeringsbeslissing, ('s-Gravenhage).

Niessen, L.C. 1979. *Summier overzicht van de benadering van de kleine dorpenproblematiek in de verschillende provincies.*

P.P.O. 1979. *Overwegingen bij de afbakening van het begrip 'kleine kern' mede in het licht van de bij de Volkstelling te hanteren wijk- en buurtindeling.* ('s-Gravenhage).

Sociaal en Cultureel Planbureau 1980. *Sociale achterstand in wijken en gemeenten.* SCP-cahier, no.14, ('s-Gravenhage).

Streekorgaan Gewest Helmond, 1981. *Dorpenplan.* (Helmond).

Chapter 30

Changing migration to rural

areas in the Netherlands

Oedzge Atzema

INTRODUCTION

Living conditions in rural areas are strongly affected by population changes in which migration is the most dynamic component. Selective migration flows can initiate significant structural changes within the community involved (Lewis and Maund, 1976). These structural changes along with changes in people's values alter the behavioural pattern of rural communities.

In the sixties and early seventies a large part of the rural areas in the Netherlands showed a growth in population largely as a consequence of a net in-migration of people coming out of the cities - residential suburbanisation (Van Ginkel, 1979). The volume and direction of migration between the regions of the Netherlands has changed remarkably in the course of the seventies. The regional pattern of movement in Great Britain seems to have altered in the same way (Ogilvy, 1982). One of the most outstanding changes in the Dutch migration pattern is the declining migration towards rural areas. This suggests that the process of suburbanisation is waning.

In this article the main changes in the Dutch migration pattern will be described first, particularly migration to rural areas. The second part of the paper gives a brief enumeration of the possible causes of these changes. There will not be any discussion about the effects of the changes on rural communities. The changes in migration are too recent to make any empirical observations on this.

THE CHANGING DUTCH MIGRATION PATTERN

Unlike the situation in Great Britain, Dutch researchers have easy access to annual figures on migration. The data are collected by the Central Bureau for Statistics; the smallest administrative unit is the municipality. Total mobility is measured by the total number of persons changing residence between municipalities. In the second half of the sixties total mobility increased (Ter Heide and Eichperger, 1974) and the growth in mobility continued in the early seventies. In 1973 the peak was reached with 716 500 migrants (53.3 per thousand of the Dutch population). Since 1973 there has been a considerable decrease in total mobility. Mobility in 1979 (530 500 migrants, 37.8 per thousand) sank to a level never approached since World War II (Verhoef, 1981). In 1980 and 1981 there is a slight increase in mobility.

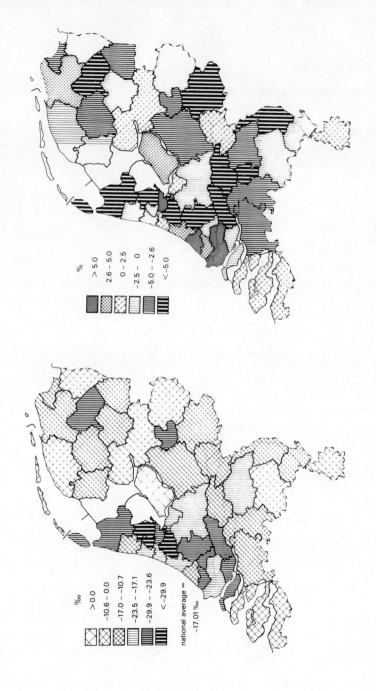

%
> 5.0
2.6 – 5.0
0 – 2.5
-2.5 – 0
-5.0 – -2.6
< -5.0

Figure 30.2   Change in proportion of in-migrants
in regional mobility, 1973–79

‰
> 0.0
-10.6 – 0.0
-17.0 – -10.7
-23.5 – -17.1
-29.9 – -23.6
< -29.9

national average =
-17.01 ‰

Figure 30.1   Decrease in regional mobility (per
thousand of the average population),
1973–79

Because of the trend in total mobility, a comparision will be made of the regional mobility figures for 1973 and 1979. Regional mobility means the sum of the persons who left a specific region and those who settled in that region. As regional units we use COROP regions (see Figure 1.1).

As with total mobility, regional mobility decreased by nearly 26 per cent between 1973 and 1979. This percentage decline is almost equal to the decline of migration within the COROP regions (26.3 per cent). The quantitative similarity indicates that the mobility decline contains equal proportions of interregional migration and intra-regional migration. This development contrasts with the growth of mobility in the sixties and early seventies. According to Van Ginkel and Harts (1979) the mobility growth in that period can mainly be attributed to intra-regional suburbanisation.

The greatest decline in regional mobility took place in the western part of the Netherlands. Both urban areas (Amsterdam and the surrounding region) and rural areas (the Green Heart region) show a significant decline in mobility (Figure 30.1). However, there is a clear difference between rural and urban regions in this respect. Figure 30.2 shows an increase of in-migration into the highly urbanised regions (Amsterdam, Rotterdam and Den Haag) and a relative decrease in more rural regions (the northern part of Noord-Holland and the Green Heart region). The mobility decline in urban regions is mostly related to a decrease in people leaving the city region, while the decline in mobility in rural areas is largely related to fewer in-migrants. The urbanised rural regions near large cities have the highest decrease in in-migrants. In other words more people have remained in urban areas during the past few years.

Table 30.1   Arrivals and departures (absolute and per thousand of the population) for groups of municipalities with different degrees of urbanisation, 1973 and 1979

	1973				1979			
	arrival		departure		arrival		departure	
	absolute	‰	absolute	‰	absolute	‰	absolute	‰
Rural municipalities	86 400	56.9	65 600	43.2	67 700	41.1	59 500	36.1
Industrialised rural municipalities	166 000	58.3	121 400	42.6	107 500	34.7	101 600	32.8
Commuter municipalities	144 800	80.0	117 300	64.8	97 400	49.3	87 000	44.0
Small towns	89 300	67.1	72 000	54.1	67 200	44.3	59 600	39.3
Medium-sized towns	110 500	50.2	129 900	59.0	82 500	37.0	86 000	38.6
Large cities	118 800	31.9	209 500	56.3	107 600	30.2	136 300	38.2
Total	715 800	53.3	715 700	53.3	529 900	37.7	530 000	37.7

Source: C.B.S.

311

More detailed information is given by migration between municipalities. A categorisation of municipalities by the Central Bureau
of Statistics has been made on the basis of their degree of urbanisation (Table 30.1). There are large decreases in the movements
out of the major cities (municipalities of over 100 000 inhabitants)
and in the movement towards industrialised rural municipalities and
commuter municipalities.

All flows between categories of municipalities decreased between
1973 and 1979. Nevertheless, there are important differences in the
degree of decline. By calculating the percentage each of the 36
flows forms of the total migration between 1973 and 1979 and then
subtracting the percentages (Table 30.2), it is obvious that the
proportion of migration flows towards the group of municipalities
with a higher degree of urbanisation has increased. The most extreme
decrease occurs in the migration flows from large cities towards
industrialised rural municipalities. The reduction in this flow
(49 per cent) is almost double the reduction in total mobility (26
per cent). On the other hand, there is an increase in the proportion
of migration towards large cities.

Table 30.2   The percentage difference in the migration flows between
groups of municipalities with different degrees of
urbanisation, 1973 and 1979

Origin \ Destination	(a)	(b)	(c)	(d)	(e)	(f)
Rural municipalities (a)	0.47	0.32	0.22	0.33	0.28	0.46
Industrialised rural mun. (b)	0.33	-0.72	0.10	0.39	0.38	0.72
Commuter municipalities (c)	-0.04	-0.66	-0.29	-0.06	0.08	0.97
Small towns (d)	0.26	-0.03	0.16	0.13	0.19	0.49
Medium sized towns (e)	-0.41	-0.98	-0.52	-0.09	-0.25	0.31
Large cities (f)	0.06	-1.84	-1.51	-0.50	-0.55	0.77

Source: C.B.S.

These changes in migration were predicted by Van Ginkel (1979).
He distinguished four stages in the migration from large cities.
The last stage is characterised by a considerable reduction in net
out-migration from cities as a consequence of a heavy decrease in
gross out-migration. The reduction in net out-migration from large
cities is also partly due to the changing scale of external in-
migration which is concentrated in the large cities. Van Ginkel
(1979) called this last stage 'concentrated suburbanisation' because
the reduced in-migration to rural areas results in (and maybe from)
a concentration of the arrivals. This means that only a few rural
municipalities receive the small number of arrivals from cities.
The other rural municipalities receive many fewer suburbanites.
If the number of departures from these rural municipalities is
stable or falls, we may expect an increase in the number of rural
municipalities with net out-migration. By comparing the first and
second halves of the seventies, the number of municipalities with
net out-migration has grown by 26 per cent (258 municipalities
compared with 205).

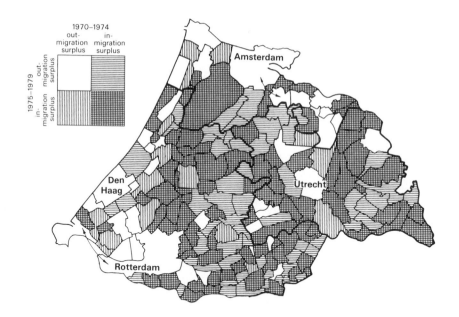

Figure 30.3   Migration surpluses of municipalities in the Randstad
region, 1970-74 and 1975-79

Figure 30.3 shows these developments for 1970-74 and 1975-79 in
one of the most urbanised parts of the Netherlands - the Green Heart
region between Amsterdam, Den Haag, Rotterdam and Utrecht.

The large cities still had net out-migration in the second half
of the seventies, and almost 45 per cent of all municipalities had
net in-migration in both periods. Nevertheless, it is clear that
many rural municipalities, which had net in-migration in the first
half of the seventies, had become municipalities with net out-
migration in the second half of the seventies (32 per cent of the
municipalities in this region). Only 9 per cent of the municipalities changed from net out-migration to net in-migration. On the
basis of these figures we can assume that the suburbanisation
process is slackening.

We are able to demonstrate this development in more detail for
another part of the Netherlands, the Midden-Brabant region (parts of
COROP regions 34, 35 and 36, see Figure 1.1). In this region there
are three central cities, Eindhoven (195 000 population), Tilburg
(152 000) and 's-Hertogenbosch (88 000). Within the regional housing
markets surrounding these cities we can classify municipalities on
the basis of planning criteria. There are *urbanised municipalities*
within the city regions with important new housing developments, and
more *rural municipalities* which, through planning, are kept as much
as possible outside the urbanisation process and which are to be
preserved for their rural character. Looking at the migration
between the three cities and both categories of municipalities
(Table 30.3), we see that the migration from the central cities has
proportionately decreased the most. Most of the rural

Table 30.3   Intra-regional migration between the central cities,
           Eindhoven, Tilburg and 's-Hertogenbosch and the other
           municipalities within their regional housing markets,
           1971-79

Other municipalities	1971 - 73			1974 - 76			1977 - 79		
	arr.	dep.	sur.	arr.	dep.	sur.	arr.	dep.	sur.
Urbanised	19 800	10 800	9 000	17 900	10 700	7 200	13 700	9 000	4 700
(index: 1971-73 =100)	*100*	*100*		*90*	*99*		*69*	*83*	
Rural	9 800	4 900	4 900	7 400	5 100	2 300	5 400	4 400	1 000
(index: 1971-73=100)	*100*	*100*		*75*	*104*		*55*	*90*	

arr. = arrival from central cities
dep. = departure to central cities
sur. = surplus

Source: C.B.S.

municipalities show a remarkable decline in urban arrivals and have
a net out-migration with the central cities. The universal out-
migration at the beginning of the seventies has now changed to
several inter-municipal in-migration surpluses in the larger rural
municipalities.

SOME EXPLANATIONS

The decline in mobility in rural areas is caused by a reduced number
of people arriving from the cities. In some rural municipalities,
the volume of urban arrivals has fallen faster than in other rural
municipalities. This seems to reflect the population of the rural
municipality and its physical planning policy for housing.

    The central question is 'What are the causes of the decreased
migration towards rural areas?' This is particularly important for
the evaluation of physical planning activities. In order to understand why people are migrants or non-migrants, it is necessary to
look at their motives and possibilities for migration. The
decision-making process of individual households is linked to the
social and housing systems (Short, 1977). Cortie and Ostendorf
(1977) pleaded for migration research where attention is paid both to
the constraints of the residential environment and the social
position of the household (which are both conditions for freedom of
choice in the decision-making process), and also to the individually
bounded preferences of the household. The mass migration flow
towards rural areas at the beginning of the seventies can be
explained by three factors.
1. Preference: The availability of good quality and, in the short
   term, financially attractive, houses formed the main incentive
   for urban-rural migration.
2. Housing-system conditions: There are relatively few possibilities for obtaining a dwelling in urban areas in comparison with
   rural areas. At that time there was a growing housing shortage
   in the cities leading to a 'forced' out-migration from the cities.
   In many rural villages relatively extensive house-building
   activities were being undertaken.
3. Social-system conditions: In the sixties the Netherlands passed
   through a period of great economic growth. Wage increases led to
   an increase in income and a growth of personal prosperity. This

meant that more households could realise their housing preferences and buy a car to bridge the distance to the city where
most of the employment was still located. Thus middle-class
people were able to live in the countryside.

Consequently, migration from the cities towards rural areas was
selective. Middle and higher status people with young children
moved to subsidised and private housing in the form of single-
family terraced housing (Ottens, 1976).

In an analogous fashion we can explain the subsequent decrease
in urban-rural migration on the basis of possible changes in these
preferences and conditions.
1. Changing preference: It may be that urban people now have a
reduced preference for the types of houses found in rural areas and
for the countryside itself. There is, however, little evidence for
such a change, although some publications do suggest a re-evaluation
of city life (gentrification).
2. Changing housing-market conditions: The planning authorities
have tried to locate the recent growth in the regional housing stock
in or near existing urban areas. These house-building activities
reduce the need for searching for a dwelling in the countryside.
The planning authorities have also tried to close the housing market
in rural areas to urban settlers. This policy did not succeed for
a long time, but is now becoming more effective since the regional
authorities have put a stop to the growth of rural villages
(Glasbergen and Simonis, 1979). This policy has caused a differentiated planning of house-building for the rural villages (in
small villages less than in larger villages) and a municipal housing
policy which works on the basis of settlement condition and housing
allocation. We can presume that a restrictive settlement policy
decreases the chances of urban people obtaining a house in the
countryside.
3. Changing social-system conditions: The social system has been
changed by economic recession and demographic shifts. The relationship between economic recession and the decreased urban-rural
migration is based on both general and specific conditions. There
are economic variables which slow down the propensity to migrate in
general terms, such as reduced growth of income and the uncertainty
about future income and continuation of employment. Other economic
variables are linked to the working of the housing market, especially for private housing, the lengthening of the period during which
a house is empty before being sold and the widening gap between
asking prices and settling prices.

Besides these general relationships there are also economic
variables which work more specifically in terms of urban-rural
migration, for example differences between average house prices in
urban and rural areas and the growth of travel costs by car.
Economic variables work not only on the demand side of the housing
market, but also on the supply side. In recent years there has been
a decrease in house-building in the Netherlands. One of the most
important reasons is that building costs are outstripping selling
prices. The free housing market, one of the important mechanisms
for suburbanisation, has collapsed. The supply of dwellings subsidised by the government is dominant. These new dwellings are
mostly sited according to the policy of the planning authorities.
Also, building costs are high in rural areas because one can only
build a few houses at a time.

Demographic shifts in the cities have equally resulted in changes
in urban-rural migration. Because of the selectivity of suburbanisation, some groups of urban inhabitants were over-represented in urban-
rural migration. Van Gestel (1981) has calculated that a major part

of the future decrease in the out-migration from the northern fringe
of the Randstad is due to demographic processes in the urban region.
Fewer families prone to suburbanisation are present in the central
cities.

CONCLUSION

This brief discussion makes it clear that there are rather complex
relationships causing the reduction in urban-rural migration. The
causes can be grouped as follows:
a) possible changes in the preferences of upper- and middle-class
   people for living in the cities;
b) the planning policy conducted with reference to suburbanisation
   (the house-building policy in urban and rural areas and the
   public housing policy of rural municipalities);
c) the economic recession (the effects of decreasing prosperity on
   housing construction and migration behaviour);
d) demographic shifts in the cities (changing population size, age
   structure, marriage pattern and family size).

These possible causes have to be translated into operational
variables in models which give explanations at both the aggregate
level and the individual level for the changes in urban-rural
migration. At the present time, our section in the Geographic
Institute at Nijmegen is working on a research project to develop
such explanatory models.

REFERENCES

Cortie, C. and Ostendorf, W. 1977. Bevolkingsmigratie en de dynamiek
van het stedelijk systeem. *Geografisch Tijdschrift,* 11, 295-306.

Gestel, P.W.C. van 1981. *Overloopperspectieven in de Noord-Vleugel
van de Randstad tot 1990 in het licht van de demografische
ontwikkelingen.* (Amsterdam).

Ginkel, J.A. van 1979. *Suburbanisatie en recente woonmilieu's.*
2 dln., (GIRUU, Utrecht).

Ginkel, J.A. van and Harts, J.J. 1979. Bevolkingsspreiding en mobili-
teit. in *Nederland op weg naar een post-industriele samenleving,*
Kwee, G.L. *et al.* (eds)(Van Gorcum, Assen), 203-217.

Glasbergen, P. and Simonis, J.B.D. 1979. *Ruimtelijk beleid in de
verzorgingsstaat.* (Amsterdam).

Heide, H. ter and Eichperger, Ch.L. 1974. De interne migratie. in
*Van nu tot nul; bevolkingsgroei en bevolkingspolitiek in Neder-
land,* Heeren, H.J. and Praag, P.H. van (eds), (Utrecht), 222-243.

Lewis, G.J. and Maund, D.J. 1976. The urbanization of the countryside: a framework for analysis. *Geografiska Annaler,* 58, 17-27.

Ogilvy, A.A. 1982. Population migration between the regions of
Great Britain 1971-1979. *Regional Studies,* 65-73.

Ottens, H. 1976. *Het Groene Hart binnen de Randstad: een beeld van
de suburbanisatie in West-Nederland.* (Van Gorcum, Assen).

Short, J.R. 1977. The intra-urban migration process: comments and
empirical findings. *Tijdschrift voor Economische en Sociale
Geografie,* 68, 362-370.

Verhoef, R. 1981. Demografie van Nederland 1980. *Centraal Bureau
voor Statistiek maandschrift,* 6, 463-478.

Chapter 31

Housing and conservation in the Lake District:
a study in ambiguity

Gordon Clark

INTRODUCTION

In 1949, a sustained campaign for the creation of national parks in
the United Kingdom met with partial success when the National Parks
and Access to the Countryside Act was passed by the British Parliament (Cherry, 1975; Sheail, 1975; Sandbach, 1978). During the next
eight years, ten national parks were designated in England and Wales.
The largest and most populous national park is in north-west.
England and covers the area known as the Lake District (Figure 31.1).

English and Welsh national parks are unusual by international
standards since they are arguably neither parks nor national.
They are not 'parks' because most of the land is devoted to farming
and forestry and not to recreation. Neither are they 'national'
since only one per cent of the land in national parks is owned by
the national park authorities (Gilg, 1978). In the Lake District,
the largest and most populous national park, the dominance of
private ownership is more limited than usual in the central area.
Nonetheless, like all the national parks, it contains a sizeable
rural population (46 000 people live in the Lake District) whose
needs must be safeguarded. There are also visitors and holiday
makers in large numbers to be catered for and from whose pressures
the area may need to be protected. It has been estimated that more
tourists visit the Lake District Park than any other in England and
Wales.

What is now called the Lake District Special Planning Board is
charged with the following statutory duties which are laid on all
the national park authorities - 'the preservation and enhancement of
natural beauty in England and Wales' and 'encouraging the provision
or improvement, for persons resorting to the National Parks, of
facilities for the enjoyment thereof..'. They also have a statutory
responsibility to promote the social and economic well-being of the
Park.

The national park authorities are required to strike a balance
between conservation (which is the national interest in the parks)
and development for local people and tourists. The Sandford Report
in 1974 examined the way in which that balance had been struck since
1949 and it concluded that greater emphasis should be given in the
future to conservation and the protection of the character of the
landscape and less weight should be given to development (Department

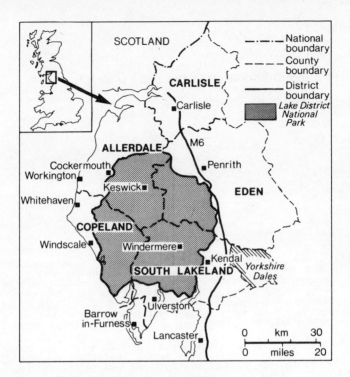

Figure 31.1   The English Lake District

of the Environment, 1974). This re-ordering of priorities was given
official sanction in Circular 4/76 from the Department of the
Environment (1976).

    The central conflict in national parks was characterised as that
between conservation and development and it is within this framework
that the national parks have conventionally been assessed (MacEwen,
1982; T.R.R.U., 1981). In recent years, however, a rather different
set of conflicts has been assuming renewed importance. This is the
question of the balance to be struck between local interests (both
public and private) on the one hand, and national public and private
interests on the other. These conflicts are no less difficult to
resolve than those between conservation and development with which
they interact. It is clear from the National Parks Act that the
conservationist element inherent in national park policy was not
intended to affect housing policy, but in the last ten years the
Lake District Special Planning Board has attempted to implement a
housing policy which is consistent with the principles of national
parks in the post-Sandford period. This has highlighted the conflict
between local and national interests and between development and
conservation. The history of housing policy in the Lake District in
particular is worthy of study because it raises several important
questions. First, what is to be the relationship between equity and
conservation? Second, under what conditions can a system designed
for land-use planning make a deliberate contribution to the planning
of socio-economic conditions? Third, what is to be the relationship
between planning for local needs and the national interest? Fourth,
what is the function of broad concepts such as 'natural beauty' and
'local need' in British rural planning?

Prior to 1974, the policy was to give planning permission for a relatively large number of new houses in the Lake District. There was a 45 per cent increase in the total number of houses in the National Park between 1951 (when the Park was established) and 1976. The reasons for this expansionist policy were fourfold. First, the population of the National Park was redistributing itself. Rural areas were continuing to lose population - 6.8 per cent between 1951 and 1971 - while the towns were growing; Windermere increased its population by 35 per cent between 1951 and 1971. The existing stock of houses was, therefore, poorly distributed in relation to where people wanted to live. Second, the area shared a national trend towards smaller households. The average size of a household in Cumbria declined by 12 per cent between 1961 and 1971 from 3.27 persons per household to 2.88 with a further decline to 2.60 anticipated by 1981. Such a trend requires more houses to cater for the same total population. Third, new houses were needed to replace sub-standard ones which were no longer fit to live in and needed to be demolished. Fourth, there was a pervasive concern among rural planners in the 1950s and 1960s over depopulation. This loss of people was ascribed in large part to the loss of employment from traditional rural industries. The Lake District lost 32 per cent of its agricultural workers and 52 per cent of those in quarrying in the 1960s, for example. Depopulation, it was believed, could only be stopped by attracting new industries - the same solution as was advocated within regional policy at that period - and new firms would only be attracted if there were sufficient houses for managers and key workers. Housing policy in the Lake District reflected a trend not only in rural planning but more widely in national policies towards housing and economic regeneration.

However, by the early 1970s this expansionist policy was subject to attack on four grounds. First, the policy of increasing the supply of houses had had no discernible effect in reducing the unusually high cost of housing in the Lake District. One survey noted that new houses of comparable standard were 12 to 17 per cent dearer inside the Park than outside it and another survey has put the differential much higher (Shucksmith, 1981). Second, concern was increasingly expressed over the extent to which houses of all types, including new houses, were being converted for use as holiday homes and second homes. This concern was intensified after a report in 1976 estimated that the proportion of holiday and second homes in Lakeland was 11 per cent of the housing stock and was over 30 per cent in some parishes (Bennett, 1976). It was argued that second-home ownership had gone beyond the conversion of redundant agricultural property in remote areas and had become so intense that local people were unable to obtain houses. This was felt to be harmful to local employment causing out-migration from the Park and destroying the 'community spirit' of villages. Third, there was concern at the visual impact of the new house-building. Conservation in the English and Welsh national parks has always been strongly concerned with the visual aspect of the countryside and nowhere more so than in the Lake District. The physical expansion of towns like Windermere was felt to be an indictment of the expansionist housing policy. Finally, the 45 per cent increase in the number of houses had not stopped rural depopulation. Sixty-one out of 82 parishes in the Lake District lost population between 1951 and 1971 while the Park as a whole lost only 1.4 per cent of its population (Lake District S.P.B., 1978).

The late 1970s witnessed a reversal of housing policy in the Lake District. The political background to this change has been described in some detail elsewhere (Clark, 1982). The rate of new house-building was to be reduced markedly under the new policy in order to preserve the landscape and conserve the stock of land suitable for house-building which was limited because of aesthetic and infra-structural constraints. The otherwise harmful effect this would have on the ability of local people to buy a house was to be offset by a novel use of a well-established planning power. Permission for building a new house would only be given if the house was sold to a local person who had been or who would be employed locally and who would live in the house for over six months of each year. This became known as the Section 52 policy since it is based on powers contained in Section 52 of the Town and Country Planning Act, 1971. In effect, this would reserve all new houses for local people and create a two-tier housing market - one group of houses open to all bidders and a second group open only to local people. Many planning reports in Great Britain have expressed a desire to help house local people. However, there is very little that planning authorities can do to give clear expression to this within the market for private houses. This is the first time that a British planning authority has tried to implement positive discrimination in favour of local people in such a way. The policy is now being used more or less openly by a number of districts in England and Wales.

The latest figures suggest that the rate of new house-building is rather higher than envisaged, while by November 1982 approval had been given for 264 houses under Section 52 powers with a further 115 houses approved under the similar powers of a local-person condition. Sixty-three Section 52 houses had been built or were under construction by mid-1980.

Housing policy in the Lake District mirrors the changing concerns of British rural planning in general. The earlier period was marked by attempts to stem the outflow of population and facilitate the raising of housing standards. The later period is marked by attempts both to reduce in-migration to the countryside without harming local interests and also to promote conservation more forcefully. In this sense, housing policy in the Lake District is a microcosm of many aspects of planning for the countryside.

Assessing any planning policy is not easy since there is the ever present uncertainty over the counterfactual - it is never clear what would have happened in the absence of the policy. How-ever, three critical aspects of the policy can be examined since they have a bearing on its social consequences and likely effective-ness. Consideration will be given to the effect of the policy on house prices, the extent to which different groups benefit from the policy, and finally, its legality and enforceability.

HOUSE PRICES

The effect of the policy on house prices is important because the policy was introduced partly in the belief that houses were being priced beyond the reach of local buyers because of the number of wealthy outsiders buying into the area for their retirement or holi-days. It was hoped that the exclusion of this external demand would allow Section 52 houses to be cheaper. There is no doubt that houses in the Lake District are more expensive than those in most other parts of northern England. However, this may be due partly to the Lake District not having the large number of small, cheap terraced houses which are so common a feature of most industrial towns in northern England. This is borne out by a survey of the prices of new houses constructed by the same builder and offered for sale simultaneously inside and outside the Park on housing estates of comparable size. The houses in the Park were between 12 and 17 per cent more expensive than houses outside (Clark, 1982). This differential is sufficiently small to be accounted for by higher standards of design in the National Park, by the higher construction costs found in most rural areas and, perhaps, a slightly higher rate of profit for the builder. In these circumstances the exclusion of potential second-home owners is unlikely to reduce prices. Indeed, it is arguable that the simultaneous reduction in the rate of house-building will raise prices since the few new houses which are built will be in such small numbers that the scale economies of the large housing estate cannot be tapped. On the other hand, while the cost of the few new houses may be higher than if they had been built on a large estate, prices might have been higher still without the new housing policy, given the overriding need to reduce the number of new houses on aesthetic and infrastructural grounds.

The new housing policy has two elements. One element will raise house prices since it seeks to prevent house-building so as to protect the visual quality of the Lake District and conserve the small stock of housing land which meets strict aesthetic and environmental standards. The other will attempt to moderate the inflationary consequences for the local population of the first element, but it is difficult to see its effect in this direction as anything other than limited. The National Park Officer for the Lake District noted, 'Section 52 has not brought home-owning any closer for the lower-paid Lakelander'.

THE DISTRIBUTION OF BENEFITS

The group for whom the policy is principally designed are the local people who cannot afford a house at current prices. The benefits of the policy for this group in terms of reduced house prices are not likely to be great. However, the costs and benefits of the policy for other groups are clearer. One group to benefit will be existing owners of houses who wish to sell and not buy another house in the Lake District because they are leaving the area or because of the death of a member of their family. The demand for Lake District properties will be confined to the almost static stock of existing houses and so gains can be expected here from sales or renting. Similarly, those who rely on the private rented sector can look to increased competition for houses and flats as some of the unsatisfied demand for houses to buy is transferred to the rented sector in the Lake District rather than to houses for purchase elsewhere. The Lake District is a unique area in England and it is likely that many outsiders will still seek to get into the area despite the restrictions on new house-building. Thus the owners of property to rent will gain from the policy and their tenants will lose.

The people eligible to buy a local-person house can be divided into two groups, those who already have a site for a house and those wishing to move into the area to work and who do not own land. The former group will find the benefits of the new policy easier to reap than the latter. Many of the early applications to build Section 52 houses have been from existing land-owners wishing to build another house on their own land. Potential in-migrants with jobs will find the new policy less helpful which is the opposite of the intention of the former policy of allowing rapid house building. This saw specific help for incoming workers as one of the objectives of housing policy.

321

The benefits of the new policy for the local marginal house-buyer are also uncertain because of the actions of building societies. These institutions provide a large, though diminishing proportion of the long-term finance for house purchase in Great Britain and a survey of their attitudes to Section 52 houses indicated that most would be willing to lend money for such property and several had already done so. However, there was a tendency to advance only 70-80 per cent of the purchase price compared with 85-95 per cent which is normal for most other houses. This would partly negate any benefit from a lower selling price particularly for the first-time buyers with limited savings.

The distribution of gains and losses from this policy may prove to be similar to that outlined above. If it is, various groups in society are being affected by the policy without any clear discussion of the merits of this. As with so many rural policies, no mechanism exists for estimating or monitoring the welfare consequences of Section 52 policy, nor for balancing the gains of various sizes for some groups against the losses for other groups. Such a mechanism, however rudimentary, would seem to be essential if rural planning is to acquire an overtly socio-economic purpose.

LEGALITY AND ENFORCEMENT

Any rural planning policy must be enforceable and legal. To be enforceable, there has to be a clear definition of what constitutes a 'local' person. If the definition is too restrictive, it may be very difficult to find anyone other than the current householder who would be eligible to buy the house when the householder wished to move to another house. This would unduly restrict migration and create a new form of tied cottage. Alternatively, too loose a definition would fail to meet the objective of keeping out people with tenuous local connections. The definition of 'local' for the Section 52 policy is set separately for each application to build a house. Most commonly, a 'local person' is one who will live and work in the local authority district in which the house will be built. This residential definition is clear and unambiguous and allows commuting for up to 30 km. In contrast, the definition of 'work' is unsatisfactory. It is not specified who in the household should work locally, nor for how long they should work each week nor is the period of employment laid down. This employment aspect of the definition is neither clear nor enforceable.

The legality of the policy is a less clear matter. It has been argued that Section 52 policy restricts the freedom of the individual to sell his house to anyone he chooses. A local authority have the right to rent or sell their council houses to anyone they wish, and discrimination in favour of local people is allowable. They cannot, it is argued, deny that freedom of sale to individuals. On the other hand, denying individuals the right to use or dispose of their land as they wish has been the essence of land-use planning, for over thirty years. Section 52 policy seeks to extend that right to the private housing market in order to achieve the same benefit for the public interest as has been sought by land-use planning. At the time of writing, the policy has been accepted by the Lake District Special Planning Board and it has been recommended by the Panel examining the National Park's Structure Plan (Cumbria C.C. and Lake District S.P.B., 1980). The Secretary of State for the Environment, who has ultimate responsibility for the Structure Plan, has stated that he wishes to see the policy deleted from the Structure Plan but strenuous representations to the Secretary of State by the Special Planning Board and local Members of Parliament have delayed a final judgement for an unusually long time. The policy, which was started in 1977, has already survived its fifth anniversary.

CONCLUSIONS

The study of housing policy in the Lake District is of more than local importance since it illustrates a number of dilemmas and themes during the last thirty years of British rural planning. The role of local autonomy in planning needs to be considered. Can a local area be allowed to adopt any planning policy it likes? Clearly not, since planning is a function derived from Acts of Parliament and it must be legal under the Planning Acts. However, the control of local initiatives by central Government extends far beyond ensuring legality into areas of detail. Two of the three inspectors at the Examination in Public of the Joint Structure Plan for the Lake District were officials of the Department of the Environment. The Department also appeared as a party at the Examination whose views were sought by the inspectors. The Secretary of State for the Environment has the final say on the contents of the Structure Plan. His department also sends out guidelines to local planning authorities in the form of circulars. Within this framework, there is little scope for local initiative in planning methods and policies even when local conditions are of a type or intensity which calls for a locally distinctive response. This leads to another feature of British planning, namely, that local planning must meet national needs as well as local ones. As the Scott Report observed in 1942, 'Every interest which in any respect transcends merely local interests should be regarded as a national interest'. Planning by British local authorities is closely, perhaps excessively, co-ordinated from London and Edinburgh.

One's attention is also drawn to the socio-economic content of rural planning. In Great Britain, planning originally meant land-use planning. More recently, there have been attempts to encourage local authorities to achieve wider socio-economic objectives than previously, most notably in the Sandford Report (D.O.E., 1974) and D.O.E. Circular 4/76 (D.O.E., 1976). Yet the Secretary of State for the Environment, in proposing to reject the Section 52 policy, noted that, 'planning is concerned with the manner of the use of land, not the identity or merits of the occupiers'.

There is clearly doubt in the mind of the Government as to whether socio-economic planning usurps the power of other departments of state and therefore is either a platitude or merely desirable in principle but unacceptable when put into practice. The current Government's inclination toward less planning implies a limited future for a socio-economic dimension to rural planning. This is as clear an example as one will find of two aspects common in British planning. First, the extent to which a policy is implemented is at least as important as the actual policy or Act of Parliament. For the academic researcher, this presents formidable problems when studying planning since decision-making is much more diffuse and less predictable in its outcome. Second, there is a pervasive fragmentation, both administrative and intellectual, in the way the British countryside is planned. Separate departments of central and local government and separate systems of physical and welfare planning create a balkanisation of the countryside and all its interdependent aspects. Also, there may be the corollary that land-use planning will never be able to do more than deal with geographical symptoms which can only be understood and, if necessary, controlled in the context of the socio-economic processes which generated them. A concern solely with land use suggests that the role of rural planning is that of palliative, modifying the

scale, speed and direction of trends without fundamentally altering
the course of events.

The need to monitor policies must also be studied. It is clear
that the most interesting effects of the successive housing policies
in the Lake District have been the indirect and unintended ones -
the effects on the private rented sector of the housing market, on
areas outside the National Park and on residential and occupational
mobility. If planning is to assume a responsibility for the socio-
economic welfare of areas, then it must develop mechanisms for fore-
casting and monitoring these indirect effects. Currently, rural
planning in Great Britain has very poorly developed procedures for
such monitoring. This is an area where improvements could be
beneficial and cost-effective.

This study raises important questions about the future of the
British countryside. For example, are retirement and second homes
so undesirable that they must be restricted? There has been no
local or national assessment of the net costs and benefits of, on
the one hand, second homes and, on the other, of a policy to restrict
them. One must also question whether current landscape preferences
are so important, unchanging and held with sufficient unanimity that
housing policy has to be altered to preserve landscape from extra
houses. It is not clear that extra houses of appropriate design
and siting have any bearing on the future quality of the 224 294
hectares of Lakeland landscape. Little thought seems to have been
given to defining precisely what is the legitimate national interest
in conserving National Parks, nor how conservation should be
pursued when it conflicts with the needs of sections of the local
population.

Finally, one notes the central importance in British rural
planning of under-defined concepts such as 'local need', 'natural
beauty' and 'community'. A 'local need' for housing undeniably
exists but its definition and scale are never made clear and the use
of the term in practical planning requires clarification. The Lake
District as a whole is certainly an area of natural beauty but there
are no criteria for assessing natural beauty which might serve as a
guide to the extent of conservation necessary in each part of the
Park. Similarly, the nature of the damage done to community spirit
by the retired and second-home owners is problematic. Changes in
the relative importance of such under-defined concepts are a major
factor altering the priorities within rural planning policy, yet the
meaning of the concepts, and hence the goals of policy, are
ambiguous. For the policy maker this fluidity is valuable because
it allows policy to evolve without the expense, delay and uncertainty of a new structure plan or a new Act of Parliament. Ambiguity
gives planners and politicians discretion over the scale and
direction of policy implementation and discretion with respect to
the priorities to be accorded to conflicting objectives. The
current debate on housing policy in the Lake District is a
fascinating struggle between central and local government to
determine which tier of authority should exercise that discretion.
However, since a multitude of different policies becomes justifiable
in these circumstances, arguing for one policy as the best is made
more difficult. Ambiguity of policy goals also makes assessing
past policy less clear-cut and less potentially embarrassing.
Consequently, ambiguity of purpose is a valuable tool for the
planner and politician faced with mutually incompatible goals and
opposing pressure groups in an ever-changing world.

In 1942, the Committee on Land Utilisation in Rural Areas (the
Scott Report) described the state of the English and Welsh

countryside and made many recommendations for its improvement
through a planning system. Major changes in the British countryside
are continuing, some the result of market forces and others the
intended or unintended results of the planning mechanism designed
to modify the effects of market forces. The most basic assumptions
of the Scott Report (for example, the universal benefits of a
prosperous agriculture) are now subject to critical appraisal. It
is perhaps time for another general review of the past and future
evolution of the countryside and the function and scope of rural
planning.

REFERENCES

Bennett, S. 1976. *Rural housing in the Lake District.* (Lancaster).

Cherry, G.E. 1975. *Environmental planning, vol.2 - National parks
and recreation in the countryside: an official peacetime
history.* (H.M.S.O., London)

Clark, G. 1982. Housing policy in the Lake District. *Transactions
of the Institute of British Geographers,* NS 7(1), 59-70.

Cumbria County Council and Lake District Special Planning Board,
1976. *Structure plan - report of survey.* (Carlisle and Kendal).

Cumbria County Council and Lake District Special Planning Board,
1980. *Cumbria and Lake District joint structure plan: written
statement.* (Carlisle and Kendal).

Department of the Environment, 1974. *Report of the National Park
Policies Review Committee* (The Sandford Report). (H.M.S.O.,
London).

Department of the Environment, 1976. *Report of the National Park
Policies Review Committee - statement of the conclusions of
the Secretaries of State for the Environment and Wales on the
report.* (Circular 4/76).

Gilg, A.W. 1978. *Countryside planning - the first three decades,
1945-76.* (Methuen, London).

Lake District Special Planning Board, 1978. *Lake District National
Park Plan.* (Kendal).

MacEwen, A. and M. 1982. *National parks: conservation or cosmetics?*
(Allen and Unwin, London).

Report of the committee on land utilisation in rural areas (The
Scott Report), 1942. Cmd. 6378, (H.M.S.O., London).

Sandbach, F.R. 1978. The early campaign for a national park in the
Lake District. *Transactions of the Institute of British
Geographers,* 3(4), 498-514.

Sheail, J. 1975. The concept of national parks in Great Britain,
1900-1950. *Transactions of the Institute of British
Geographers,* 52, 41-56.

Shucksmith, M. 1981. *No homes for locals?* (Gower, Farnborough).

Tourism and Recreation Research Unit (T.R.R.U.), 1981. *The economy
of rural communities in the national parks of England and
Wales.* Research Report no.47, (Edinburgh).

Chapter 32

Public-sector housing in
rural areas in England

David Phillips and Allan Williams

INTRODUCTION

The main keys to an understanding of living conditions in any area
are the local labour and housing markets, and the relationship
between these and larger-scale regional or national markets. There
have been important changes recently in both housing and employment
in the British countryside (Phillips and Williams, 1984). There is
evidence of employment gains in rural areas or, at least, outside
metropolitan areas (Spence et al., 1982) and the preliminary findings
of the 1981 census have identified substantial population shifts
from metropolitan to non-metropolitan areas (Randolph and Robert,
1981). This has increased the pressure on local housing markets in
rural areas at the same time as they have undergone important
changes. There has been a general decline in the supply of private
rented accommodation and an increase in owner-occupancy (Dunn et al.,
1981); in some areas the major impetus for the latter trend has been
the entry into rural housing markets of 'outsiders' such as retired
persons (Law and Warnes, 1976), tourists, second-home owners
(Bielckus et al., 1972) and urban commuters (Spence et al., 1982).
As entry to owner-occupation tends to be restricted to higher income
groups, lower income groups have tended to become increasingly
reliant on renting dwellings in the public sector as a means of
improving their housing standards (Phillips and Williams, 1982a).
However, there are only relatively small stocks of local authority
housing in rural areas and even these have been under threat from
both reduced building activity and sales to sitting tenants. This
has important social consequences, which have explicit spatial
features. Houses offer more than basic shelter; they also condition
access to jobs, shops and services, and influence segregation and
social balance in rural communities.

In Britain, local authorities (councils) have usually built
dwellings for direct renting rather than for sale to the public.
Council dwellings have only been sold to their tenants sporadically
and in a few areas. However, the sale of local authority houses in
the United Kingdom has received unprecedented attention in recent
years as a result of the political controversy which has surrounded
the Housing Act 1980 which gave most tenants the legal right to buy
their homes. For once, the 'rural voice' has been as emphatic and,
possibly, as effective in this debate as its urban counterpart.
Intense lobbying attended the passage of the Act through Parliament
and this resulted in some safeguards being incorporated into the
sales procedures to be adopted in some rural areas. This is

surprising because traditionally in Britain there has been far
greater recognition of the housing needs of urban than rural areas.
The reasons for this will be considered in the final sector of the
paper. The emphasis on sales may have been rather misleading, since
their real significance lies in their relationship to the low level
of provision of public sector housing in rural areas. Sales,
possibly at above average rates, are likely to deplete rural housing
provision which is already below average. Although passionate
emotions have been aroused in this debate, there is a surprising
dearth of data about the level of public housing provision and sales
in rural areas. The main aim of this paper is to outline some of
the distributional features of both the stock and the sales of
council houses. The analysis focuses on two levels, the aggregate
national picture of rural-urban differences, and detailed case study
material from the authors' own work in South Hams District in Devon
(Phillips and Williams, 1982a).

The importance of the debate over council house sales lies in
the social implications of reduced housing provision in rural areas.
These are particularly important since there is a growing awareness
of the pressures on housing in many rural areas. These pressures
have been most acutely observed in some of the more attractive
tourist areas such as North Wales, the South West and the Lake
District. In response to the sales of houses to 'outsiders' or to
holidaymakers in these areas, several authorities have attempted to
restrict the sales of new private sector dwellings to local residents. However, the Secretary of State has usually insisted upon
weakening any specific controls which these authorities have
attempted to impose on the sale of private dwellings (Clark, 1981).
The most notable example has been the decision to veto the use of
'Section 52 agreements' by the Lake District Special Planning Board
(see the paper by Clark in this book). In contrast to this policy
of non-intervention in the public sector, the Secretary of State did
allow certain safeguards to be introduced into the sales of council
houses in rural areas. As one of the main arguments used in favour
of such safeguards was the fear that dwellings would be resold to
'outsiders', the government seems to have made two rather contradictory decisions with regards to rural housing.

LOCAL AUTHORITY HOUSING PROVISION: A NATIONAL PERSPECTIVE

There is general agreement that there is a lower level of provision
of local authority housing provision in rural than urban areas in
Britain and a number of reasons have been advanced to explain this.
Building costs have usually been higher in rural areas and, although
there have been higher subsidies to counterbalance this, they have
not usually been adequate to bridge the difference in costs (Dunn
et al., 1981). Central government has also been more ready to
subsidise building in urban than in rural areas, in response to the
more immediately apparent social crisis in these areas (Lansley,
1979). Housing protest groups have been far less well organised and
less able to press for improvements in rural areas than in urban
areas (Short, 1982). Historically there has also been a degree of
unwillingness by some rural councils to undertake council-house
building. Newby (1979) has shown that this can be linked to the
nature of political control in such areas. Some rural councils have
been dominated by farmers and 'ratepayers' who have deliberately
sought to restrict public-sector houses in order to reduce the level
of rates (local property taxes) and to maintain paternalistic control over a workforce that would be forced to live in agricultural
tied cottages (see also Irving and Hilgendorf, 1975). Ideological
opposition to the public sector has also been an important
consideration as most local councils in rural areas have been

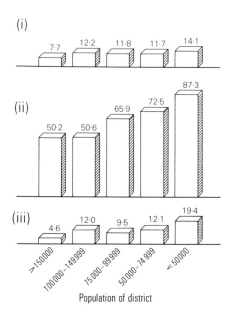

(i)

7·7   12·2   11·8   11·7   14·1

(ii)

87·3

72·5

65·9

50·2   50·6

(iii)

19·4

4·6   12·0   9·5   12·1

>150000   100000-149999   75000-99999   50000-74999   <50000

Population of district

Figure 32.1   Local authority housing in England:
              (i)   number of persons per council house
              (ii)   claims to 'right to buy' in ratio to the stock
                      of dwellings
              (iii)   sales in ratio to the stock of dwellings

dominated by the Conservative Party, or at least by conservative
interests (Bracey, 1959; Butler and Stokes, 1971).

    Whatever the exact reasons for the low rate of public housing
construction in rural areas, it is widely accepted that provision is
at a lower level than in urban areas. However, there is relatively
little documentation of the precise extent of this underprovision.
Dunn *et al.* (1981) have used 1971 census data for England to show
that the proportion of local authority rented dwellings is only 21.1
per cent of the rural housing stock, compared to 28.0 per cent in
the country as a whole. Shucksmith (1981) has also shown that the
rate of council-house building in urban areas was more than double
that in rural areas between 1968 and 1973 and that in 1979/80 the
average English local authority spent £62 per capita on housing
compared to only £53 in rural areas (see the paper by Van Bemmel in
this book). In order to illustrate further these rural-urban
differences, the ratio of the local authority housing stock to total
population has been calculated for different types of district
councils (classified into five population groups) (Figure 32.1(i)).
Population size is taken as an admittedly crude indicator of
rurality. The middle categories in this classification tend to be
rather mixed including some obviously urban areas and some eminently
rural ones. However, at the extremes, this classification is useful.
Category 1 (over 150 000) is composed mostly of Metropolitan
Districts and large cities, while category 5 (population fewer than
50 000) is composed of areas which would be defined as 'rural'
according to almost any indicator. The population statistics are
drawn from the Registrar General's estimates for mid-1980, and the

numbers of council houses are based on returns in the Chartered Institute of Public Finance and Accountancy's *Housing Revenue Account Statistics, 1979-80*.

The ratio of population to the number of council houses has its limitations. It neither measures need (whether of the homeless or those in dwellings in poor condition or overcrowded), nor does it take into account alternative forms of housing tenure (see the paper by Richmond in this book).

Even so, it does help to illustrate different levels of provision. Figure 32.1(i) shows that the ratio of total population to council houses in more rural areas (category 5) is almost double that in more urban areas (category 1), while the other categories are at intermediate levels. At this scale of measurement this represents a gross inequality, especially as there is corroborative evidence that the rural poor may be less well housed than their urban counterparts and that waiting lists for council houses may be at least as long in rural areas as in the cities. Moreover, the aggregate level of analysis does tend to conceal the true extent of under-provision for there are substantial variations within 'rural' areas (Larkin, 1978a).

LOCAL SCALE PROVISION: A CASE STUDY OF SOUTH HAMS

At the local level council houses tend to be concentrated in the main settlements, so that the larger the village or town, the greater the proportion of its housing stock which is likely to be in local authority ownership. There are a number of reasons for this, including the restriction of land supply by planning policies (see the paper by Gilg in this book), the higher building costs associated with using local materials in some villages, the economies of scale of building larger estates, and the effects of government subsidies and construction guidelines which favour the development of large, high-density estates. Recently imposed restrictions on local authority expenditure have also left many authorities with sufficient resources for only a single new building project, which inevitably tends to be located in the main settlements (Phillips and Williams, 1982a). It should be added that, in practice, local authority construction programmes have not always followed the policies of concentration specified in structure plans (Cloke, 1979; Phillips and Williams, 1983) but, in general terms, land-use planning policies have militated against construction in the smaller villages and hamlets. This has been reinforced by the failure of statutory public bodies, such as the water authorities, to provide adequate infrastructure for new housing in many rural zones.

This can be illustrated by reference to South Hams District, one of the category 4 districts (population 50 000-75 000) which were analysed earlier. This is a rural area with a fairly dispersed settlement pattern, and substantial employment in tourism and agriculture. The housing stock in some parts of the area is under considerable pressure, notably in the west from commuters to nearby Plymouth, and in the coastal zones from the demand for second or holiday homes. The main foci of population are four towns with populations of between 4000 and 6000; these are Dartmouth, Ivybridge, Kingsbridge and Totnes. At the other extreme there are 19 parishes with populations of fewer than 400 persons, and these have very dispersed settlements. Using a threefold classification of parishes by population, the ratios of council houses to population have been calculated and are shown in Figure 32.2(i).

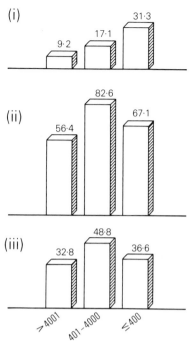

(i)

9·2   17·1   31·3

(ii)

56·4   82·6   67·1

(iii)

32·8   48·8   36·6

>4001   401-4000   ≤400

Population of parish

Figure 32.2   Local authority housing in South Hams:
(i)   number of persons per council house
(ii)   claims to 'right to buy' in ratio to the stock
         of dwellings
(iii)   sales in ratio to the stock of dwellings

There are two points to note with regard to these data. First,
there is enormous variation in provision within South Hams, 31.3
persons per local authority dwelling in the smallest settlements and
9.2 in the four major towns. Secondly, the ratios in all the rural
areas outside the main towns are greater than those observed earlier
for rural districts at the national scale. The greatest contrast is
therefore between people living in the most rural areas and most
of the remainder of the country where the ratio is only a quarter
to a half as great. The inequity of provision is further illustrated
by considering the absolute numbers of council houses. Figure 32.3
shows that there is a highly uneven spatial distribution of council
houses and that there are some large areas with hardly any local
authority dwellings. Eighteen parishes have fewer than 15 council
houses and three parishes have none. The distribution of dwellings
seems even more uneven if account is taken of their characteristics.
In South Hams it has been shown that there is considerable variation
in the proportion of bungalows, houses and flats of various sizes
which are available at parish level (Phillips and Williams, 1982a).
This further exacerbates the mismatch between specific local needs
and provision.

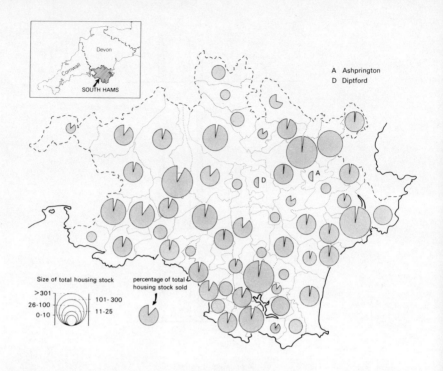

Figure 32.3.   Council housing stock and proportion sold in South
Hams

Rural areas have very low levels of provision of council housing
and the planned reduction by 50 per cent of public expenditure on
housing between 1979-80 and 1983-84 makes it unlikely that this will
be substantially modified in the near future.   Instead, the
acceleration of council house sales following the Housing Act 1980
is likely to reduce the level of provision.

COUNCIL HOUSE SALES

The sale of council houses is not a new feature of housing policy in
Britain.   The Housing Act 1936 gave powers to local authorities to
sell council dwellings although ministerial permission had to be
sought.   This was reinforced later by a number of measures,
especially Circular 64/52 in 1952 which allowed local authorities to
impose resale conditions on all council house sales. Under the
Labour Governments of the 1960s, attempts were made to restrict
sales to take account of local needs, but permission to sell was
never actually withdrawn.   This basic approach was continued
throughout the 1970s except for the brief period of Conservative
control between 1970 and 1974 when an active sales policy was pursued.
These changing attitudes are reflected in the figures for the sales
of council houses which increased from 2000-5000 per annum in the
early and mid-1960s to 17 000-45 000 per annum under the Conservatives in the early 1970s, before falling back to 2000-13 000 per
annum during the remainder of the 1970s.

332

After the election of the Conservative Government in 1979, the emphasis shifted firmly to favouring sales. The main vehicle for this was the Housing Act 1980 which included a wide range of measures, the most pertinent being that most tenants were given the right to purchase council houses irrespective of whether or not the local authority wished to sell the dwellings. In effect, the *power* to sell had been transformed into a *duty* to sell and the *privilege* of being able to purchase has been made into a *right*. The details of the Act are described elsewhere (Crouch, 1980; Schifferes, 1980; Smith, 1981) but, basically, the right to buy was given to 'secure tenants' of three years' duration, at a discount on open-market prices ranging from 33 per cent to a maximum of 50 per cent for those with 20 or more years' residence. The actual passage and review of the Act saw enormous pressure being exerted by rural councils and other pressure groups which resulted in some safe-guards for rural areas in recognition of their small stocks of dwellings and the severity of pressures in some local rural housing markets. The basic right to buy remained and there were no actual restrictions on sales. However, in Areas of Outstanding Natural Beauty (A.O.N.B.s), National Parks and 'designated rural areas', resale conditions were attached to council house sales. In these areas, the council must either be allowed first refusal on resale during the first ten years, or the houses can only be resold to someone who has lived or worked in the locality for at least three years. The extent of the designated rural areas in England is in fact very limited as can be seen from Figure 32.4. In South Hams, for example, the four major settlements and, consequently, one-half of the council housing stock, were excluded. Therefore, only some rural areas have any protection at all, which means that resale conditions are, in effect, quite limited.

The problems with the resale conditions are twofold. First, the high prices of dwellings in some rural areas, especially in the more attractive villages, make it very unlikely that a council could exercise its right to repurchase. This is especially so as houses are sold at a discount and there are also severe restrictions on local authority expenditure. Second, there are a number of ambiguities over the definition of 'locality' in the resale condi-tions. Essentially, this seems to be the county or, if applicable, other parts of the National Park or A.O.N.B. within which a house is located. As an example, the result may be that 'somebody selling a home in the Staffordshire part of the Peak National Park can sell to anybody from any other part of the Park (which straddles five counties), or to any other Staffordshire resident, including areas like the Potteries towns and the fringes of Birmingham. Yet these are the very areas that have created the pressure on the Peak housing market!' (Clark, 1981). The fear amongst rural councils is that sales will deplete their limited stocks, they will lack the financial resources to apply repurchase conditions and, more importantly, they will be unable to undertake new building programmes.

The data which are available on sales confirm some of the apprehensions of the rural councils, and this is the case if either applications to buy or actual sales are examined. It is useful to examine claims alongside sales because, for mainly political reasons, many councils have been reluctant to approve applications and to complete sales. The data on sales and applications are drawn from the Department of the Environment's *Local Housing Statistics,* and they cover the period April 1980 to October 1981. Figures 32.1(ii) and 32.1(iii) show the ratios of 'claims to buy' and of actual sales to the number of local authority dwellings in the different cate-gories of districts. The number of claims to buy per 1000

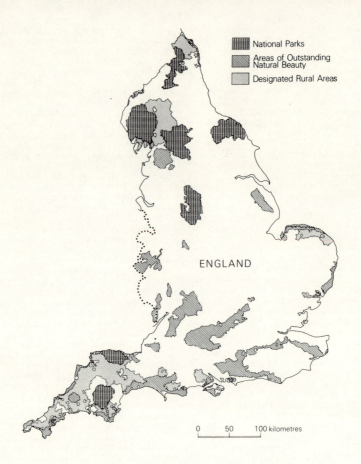

National Parks

Areas of Outstanding Natural Beauty

Designated Rural Areas

ENGLAND

0   50   100 kilometres

Figure 32.4   Areas in which resale conditions apply to
                council house sales

dwellings shows a strong correlation with the degree of 'rurality'.
The highest ratio of applications is in the smallest and more rural
areas and this number falls steadily as the population size
increases. The same general pattern can be observed with respect to
sales.

South Hams mirrors these trends. It has been argued that,
because of ease of resale by their purchasers, dwellings which are
more likely to be sold are those which are located in very small
clusters, whilst houses in the middle of large estates are least
likely to be sold (English, 1977). It is therefore to be expected
that, within South Hams, sales will be greatest outside the four
major settlements where the main estates are located. The data
presented in Figures 32.2(ii) and 32.2(iii) confirm that this is
the case. The claims to buy and the actual sales are greater outside the four main towns. However, there are higher claims and
sales ratios in the parishes with intermediate population levels

334

than there are in the smallest parishes. This is an interesting aspect for which there is no immediately apparent or obvious reason, but it may be related to general levels of accessibility.

The question of the distribution of sales within South Hams is worth pursuing further. Figure 32.3 shows the distribution of sales according to the size of the council housing stock rather than to the size of the population in each parish. This reveals a rather uneven spatial patchwork which could only be explained by reference to specific details of the local housing markets and to social and political structures. However, there are two important features which can be emphasised. First, it is the parishes with the smallest council house stocks which have experienced the highest rates of sales. Some 87 per 1000 dwellings have been sold in parishes with fewer than 15 council-owned properties, contrasting with 37 per 1000 in the main towns and 51 per 1000 in the other parishes. Therefore, it is the parishes with the smallest council stocks which have suffered the most severe depletion. Secondly, the position in particular parishes can be very acute; for example, in Ashprington and Diptford, each of which had fewer than 10 council houses, half of the stock has been sold and it seems only a matter of time before all the houses in such villages are sold.

SOME IMPLICATIONS AND CONCLUSIONS

It can be argued that the sale of council houses offers individuals a share in the financial benefits of house ownership, freedom from bureaucratic restrictions, and the scope to improve their homes according to their own preferences. However, these are consumption advantages for individual tenants and the advantages for the remainder of the community are more dubious. Sales will provide funds for the building of new local authority dwellings, but the costs of new development are likely to outstrip the revenue received from sales (Kilroy, 1980).

By contrast, a number of disadvantages accrue to the community especially in rural areas. In 1980, the House of Commons Environment Select Committee on Housing called for verbal evidence on housing sales from Allerdale and South Lakeland rural councils in the Lake District. It is significant that they were told that most councillors in these areas (which had majorities of Conservative and Independent members) were opposed to the 'right to buy' legislation. Some of their arguments were similar to those advanced by urban councils; both the types of purchasers and the properties sold will be selective, so that the poorest tenants will be left with the least desirable properties.

The local councils also argued that there were particular adverse implications for rural areas. Sales are more likely to occur in the more rural and more attractive areas with small housing stocks, where they will represent substantial reductions in the levels of public sector dwellings. The result will be ever-longer waiting lists and queues for a dwindling number of council houses. In South Hams, for example, there were only 120 to 150 lettings a year to those on the waiting list of applicants which numbered more than 1000 persons. In some villages there were no lettings of council houses over a two-year period (Phillips and Williams, 1982b). The outcome of this may be twofold: either local residents will have to move from their local area in order to improve their housing conditions, or they may be forced to accept inferior accommodation within their villages (Phillips and Williams, 1982c). Those who move out of the parish may, in turn, be faced with long journeys to work or for visits to friends and families. Moreover, the implications for the community may be even more profound than for individuals. The reduced supply of low-cost accommodation may lead to young families having to leave the area, and their replacements (either permanent or temporary second-home residents) may be older and more prosperous people. The results of such social changes are well known, and the run-down of shops, schools and bus services may follow. The point to be emphasised here is that the resale conditions of the Housing Act 1980 will not prevent this occurring. Clearly sales are further exacerbating already very low levels of provision in rural areas.

At another level of analysis, attention must be drawn to the changing role ascribed to rural areas in terms of production and consumption within Britain. Historically, production in rural areas meant agriculture and there was little need for the state to intervene in housing to guarantee the reproduction of labour power or to reduce the costs of labour, as was the case in urban areas (Short, 1982). Farmers were able to use low levels of council housing and the tied-cottage system to guarantee the reproduction of their labour force. In addition, rural areas in the twentieth century were a locus for individual consumption (most spectacularly seen in private house building in inter-war suburbanisation) and state activity was restricted to limiting the environmental impact of the expansion of owner-occupation (Blacksell and Gilg, 1981). There was relatively little state investment in housing in rural areas. It is urban areas rather than rural areas which have acted as the focus for collective consumption of which public housing is a major component (Castells, 1977).

However, in the post-war period there has been a change in the territorial division of labour in Britain so that there has been a net shift of manufacturing and service employment away from metropolitan areas (Massey, 1979). This both resulted from, and contributed to, a parallel shift of population to smaller towns and the more rural regions (Spence, *et al.*, 1982). At the same time, the continuing demands of individual consumption continued to increase pressure on rural housing markets. This operated both directly, as in the purchase of second or retirement homes, and indirectly, through the planning restrictions placed on rural development in order to preserve the rural environment. Considerable pressure was therefore exerted on rural housing markets leading in some areas to a real housing crisis. From the 1960s, however, the ability of the state to intervene in rural housing markets was limited by the increasing attention being directed at the inner-city areas, especially in response to social protests (Larkin, 1978b). More recently, any attempt to remedy the housing crisis in rural areas through public-sector construction has been forestalled by the aims of the Conservative Government which has been committed to privatisation of consumption including housing, the background of which is described by Gough (1979). The Housing Act 1980 was the main vehicle for this policy but, like most housing policies, it seems initially to have been drafted with no account being taken of the special needs of rural areas. This is why there was considerable opposition to the Act even from Conservative councillors in rural areas. It can be argued that the amendment concerning resale conditions in rural areas was always likely to be inadequate because any stronger measures would have undermined the aim of the privatisation of consumption.

REFERENCES

Bielckus, C.L., Rogers, A.W. and Wibberley, G.P. 1972. *Second homes in England and Wales.* Studies in Rural Land Use 11, (Wye College, University of London).

Blacksell, M. and Gilg, A. 1981. *The countryside: planning and change.* (George Allen and Unwin, London).

Bracey, H.E. 1959. *English rural life: village activities, organisations and institutions.* (Routledge and Kegan Paul, London).

Butler, D. and Stokes, D. 1971. *Political change in Britain: forces shaping electoral choice.* (Penguin, Harmondsworth).

Castells, M. 1977. *The urban question: a Marxist approach.* (Edward Arnold, London).

Clark, D. 1981. *Rural housing initiatives.* no.11, (National Council of Voluntary Organisations, London).

Cloke, P.J. 1979. *Key settlements in rural areas.* (Methuen, London).

Crouch, D.J. 1980. A guide to the main provisions of the Housing Act 1980. *Housing,* 16, 4-5.

Dunn, M., Rawson, M. and Rogers, A. 1981. *Rural Housing: competition and choice.* (George Allen and Unwin, London).

English, J. 1977. Council house sales: what will be left? *Housing Review,* 26, 136-9.

Gough, I. 1979. *The political economy of the welfare state.* (Macmillan, London).

Irving, B. and Hilgendorf, L. 1975. *Tied cottages in British agriculture.* Working paper no.1, (Tavistock Institute of Human Relations, London).

Kilroy, B. 1980. From roughly right to precisely wrong. *Roof,* 5, 16-17.

Lansley, S. 1979. *Housing and public policy.* (Croom Helm, London).

Larkin, A. 1978a. Rural housing - too dear, too few and too far. *Roof,* 3, 15-17.

Larkin, A. 1978b. Inner-city infatuation - rural areas must fight it. *Municipal and Public Services Journal,* 86, 1277-8.

Law, C.M. and Warnes, A.M. 1976. The changing geography of the elderly in England and Wales. *Transactions of the Institute of British Geographers,* N.S.1, 453-71.

Massey, D. 1979. In what sense a regional problem? *Regional Studies,* 13, 233-44.

Newby, H. 1979. *Green and pleasant land?* (Hutchinson, London).

Phillips, D.R. and Williams, A.M. 1982a. *Rural housing and the public sector.* (Gower, Aldershot).

Phillips, D.R. and Williams, A.M. 1982b. Local authority housing and accessibility: evidence from the South Hams, Devon. *Transactions of the Institute of British Geographers,* N.S.7, 304-20.

Phillips, D.R. and Williams, A.M. 1982c. The need for rural council houses. *Housing,* 18, 16-19.

Phillips, D.R. and Williams, A.M. 1983. Rural settlement policies and local authority housing: some observations from a case-study of South Hams, Devon. *Environment and Planning A,* 15, 501-13.

Phillips D.R. and Williams, A.M. 1984. Rural Britain: a social geography. (Blackwells, Oxford).

Randolph, W. and Robert, S. 1981. Population redistribution in Great Britain, 1971-1981. *Town and Country Planning,* 50, 227-30.

Schifferes, S. 1980. Housing Bill 1980: the beginning of the end for council housing. *Roof,* 5, 10-15.

Short, J. 1982. *Housing in Britain: the post-war experience.* (Methuen, London).

Shucksmith, M. 1981. *No homes for locals?* (Gower, Aldershot).

Smith, P.F. 1981. *Housing Act 1980.* (Butterworths, London).

Spence, N., Gillespie, A., Goddard, J., Kennet, S., Pinch, S. and Williams, A. 1982. *British cities: an analysis of urban change.* (Pergamon Press, Oxford).

Chapter 33

Alternative tenures in rural areas:
the role of housing associations

Patricia Richmond

INTRODUCTION

The rural housing market has an unusually low proportion of state-provided housing (21 per cent compared with 28 per cent, at least, in urban areas) and in some cases the effects of the current sale of council houses will be severe. Added to this is the anomaly of a large but rapidly declining 'tied' agricultural sector, and the pressure for second homes and tourist accommodation which forces unrealistically high prices for houses in favoured areas. A recent study by Phillips and Williams (in this book) showed that the distribution of housing in one rural local authority area in 1971 was as follows - 55 per cent owner-occupied; 27 per cent privately rented and only 19 per cent local authority, and these figures are by no means exceptional.

This paper examines one area of housing provision in the more rural areas of the country as a balance to the wealth of urban-oriented research on housing. It looks at the housing association sector which includes a variety of voluntary, non-profit-making bodies, differing in size, objectives and operations according to the nature of their constitution and the period of their formation. The reasons why attention is focussed on this sector are threefold.

Firstly, the housing association sector, or voluntary housing movement as it may be called, plays a significant but often over-looked role in the housing system. In 1980, 11.91 million British houses were owner-occupied, 6.82 million were provided by a local authority or New Town Development Corporation, and 2.81 million were privately rented (Department of the Environment, 1980). Of this total of 21.54 million dwellings, some 500 000 were owned and managed by a housing association (Housing Corporation, 1982).

Secondly, as part of the public sector the voluntary housing movement has been ear-marked for a continuing share of the public purse. The total budget of the Housing Corporation, the main source of finance to the movement, is to be £556 million for the U.K. in 1982-83. In real terms this is at, or indeed above, allocations for the last financial year (Housing Corporation, 1982). With an already small and rapidly diminishing public sector in most rural localities, a more detailed look at the provision of the quasi-public sector or 'third arm' would appear of prime interest.

Thirdly, disadvantages in housing provision are exacerbated for rural populations by inaccessibility, lack of employment prospects and the higher costs of social service provision (Neate, 1981). The successes of many community self-help initiatives, amongst which have been the setting up of village housing associations (Clark, 1981) suggest that the voluntary housing movement is suited to alleviating at least some of the rural housing problems. As Bourne (1981) stressed, 'the prevailing view is no longer of a single national housing problem to which a uniform national policy solution is either necessary or appropriate, but of a series of localised problems which differ in nature and extent by community and which are best dealt with by local governments and by more spatially sensitive and precise policy instruments'. Housing associations, by their nature and organisation, attempt to fulfil such a specification.

Nonetheless, within geography housing associations remain relatively unresearched. Recent texts dealing with the peculiarities of rural housing markets (Dunn et al., 1981; Shucksmith, 1981; Phillips and Williams, 1982) recognise the potential within the voluntary housing movement, but as yet there is little evidence of its actual contribution. There have been a number of studies of housing association activity in Housing Action Areas and inner-city rehabilitation schemes (Page, 1971; Thomas, 1979). In comparison there is a need to evaluate their work in non-metropolitan areas and especially those rural areas where incomes are low and property is in a poor state, in essence, a similar crisis situation as that identified in the inner-city housing market (Basset and Short, 1980). An analysis of schemes in a mainly rural county is intended to fill this gap.

THE VOLUNTARY HOUSING MOVEMENT

In studying the voluntary housing movement two approaches may be adopted. Firstly, the emergence of the sector may be seen in the light of legislative and political developments from its foundation in the early nineteenth-century to its role in the post-1980 U.K. housing system. This reveals fundamental shifts in the ideology behind the movement and a growing dependence on centralised control with the establishment of the Housing Corporation in 1964 and its extension in 1974.

There have been essentially three stages in the growth of the voluntary housing sector in this country. The first, covering a period from the early nineteenth century to the introduction of direct legislation in 1890, saw the foundation of both small, charitable and co-operative-minded housing groups, and of large societies such as the Sutton Trust and Peabody Trust associated with philanthropists (Figure 33.1). Many of these continue to play a considerable role in the provision of housing for those in need. In the second stage of their history, their powers and authority were extended by the housing legislation of the early 1920s. However, housing associations or the public utility societies, as they were then known, remained insignificant in contrast to the life and vigour enjoyed by similar voluntarily-based housing abroad. This is due in part to the large-scale stimulation of local authority provision, especially by Labour governments, and later by the post-1951 Conservative emphasis on home-ownership.

The present-day image of the housing association movement has evolved since 1960. Under earlier legislation, the emphasis was on the formation of co-ownership societies offering housing on a share-ownership basis, and on cost renting. The latter was replaced

Figure 33.1 A guide to the voluntary housing movement in the U.K.

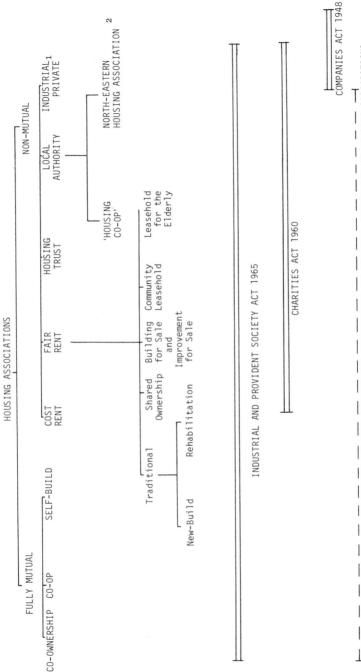

by Fair Rent Associations in the 1970s, providing property to rent at costs fixed by the local Rent Officer and supported by Department of the Environment loans. This has lately been superseded by a shift to wider home-ownership, and since the Housing Act of 1980 the trend has been towards encouraging schemes for shared ownership, improvement and building for sale, and leaseholds for the elderly. Fair-rent provision is considered less of a priority except for special needs such as the disabled and mentally handicapped, and the original co-ownership schemes which proved, for the most part, costly and impractical, have been largely disbanded.

In summary, there has been a noticeable shift away from the early role of the quasi-public sector - the provision of homes to rent for low-income groups - to that of providing schemes aimed at the lower end of the property-buying market.

At present the Housing Corporation has a total of 3020 Associations on its register. Last year (1981) it advanced some £514 million resulting in 26 601 completions and the approval of over 28 000 schemes nationwide. This is a noticeable decrease on the 1979-80 house-building programme (24 601 completions and 68 000 approvals) but reflects only the net decline of all public-sector provision. The 1980 house-building figures show that out of a total of 59 000 houses started in the public sector, almost 15 000 were by housing associations compared with 37 500 by local authorities and some 6000 by New Town Development Corporations. This reflects the increasing importance of housing associations which have risen from 7.6 per cent to 25 per cent of all public-sector housing since the early 1970s.

The second approach to studying the voluntary housing movement takes a more geographical perspective, focussing attention on the externality effects of locations, such as the countryside, and the subsequent differentiation of housing provision between the appropriate agents. Thus housing associations can be seen as a necessary means of satisfying demands for a particular category of housing such as that for the elderly, local people or a rural low-income group.

In a government circular in 1974, the role of the housing association movement was defined as 'provision in areas of particular need' such as designated Housing Action Areas (H.A.A.) and General Improvement Areas (G.I.A.), and also for the provision of special-needs housing. Associations were acclaimed as the most fitting replacement for private landlords and as providers of housing for key-workers. Since then there has been an emphasis on the revitalisation of inner cities. Priority in housing association funding has been given to schemes in H.A.A.s and areas of housing stress. In Birmingham, for example, 33 per cent of allocations in one year were in G.I.A.s, and 25 per cent in H.A.A.s, representing 2748 out of a total of only 4000 developments in the entire area (Page, 1971).

However, the Secretary of State specified that Housing Corporation allocations for 1981-82 should give special attention to the needs of rural areas. The 1979-80 Annual Report stated that the Corporation would be glad to fund well-judged schemes in rural areas 'where the numbers of people without a decent home may be small but the need acute'. This comment was reiterated by the Chairman of the Housing Corporation as recently as June of this year. To what extent has this approach been implemented?

A CASE STUDY OF THE VOLUNTARY HOUSING MOVEMENT IN DEVON

It is an immense task to identify those associations operating in rural locations, especially on a national scale. For one thing, many of the larger associations are based in London but have projects in a variety of places, including both cities and small settlements. Few have a specific rural policy unless they have been formed locally with a village bias. Of 19 registered housing associations providing rented accommodation within Devon, half have headquarters located outside the region, predominantly in London and the South East.

Furthermore, the delineation of 'rural' presents immediate problems, and cannot be limited to settlements of below a certain population level. 'Local' need may cover the area of a district council or just a single parish, and planning regulations may be such as to locate schemes, involving new development, only within designated settlements. Evidence of this has been found particularly among those associations providing for the elderly or handicapped where accessibility is an essential factor.

Finally, the constitutional nature of the voluntary housing movement makes classification extremely complex. Baker (1976) gives some idea of the enormous variation in the type and make-up of organisations found under the broad umbrella of the movement, which he describes as 'multiform' or 'heterogeneous' (Figure 33.1).

This study is concerned with those housing associations registered with the Housing Corporation in January 1982. It recognises, however, the existence of associations outside this category - those that prefer not to register with the Corporation - but the contribution of these groups is of minimal importance as regards both present and future developments since they are not eligible for Housing Corporation funding.

The data used have been compiled from information obtained from the Housing Corporation Regional Office for the West of England. Re-organisation between the period December 1980 and March 1981 resulted in changes in the regional boundaries governing Housing Corporation administration. This has left some inconsistency in available statistics and is one reason for concentrating on data for Devon only.

A number of features of the geographical distribution of housing association stock within the study area are worth commenting upon (Table 33.1). Firstly, there is noticeable unevenness in the distribution of housing association schemes. The location of schemes funded by the Housing Corporation and by district councils since the early 1970s is shown in Figure 33.2. It is clear that Housing Association provision is located in the main urban areas with a pronounced concentration in the larger towns of Barnstaple, Newton Abbot, Honiton and Exmouth. The three major cities of Devon - Exeter, Torbay and Plymouth - were excluded from the study. Nevertheless, some small towns and villages in remote localities have benefited from Corporation allocations quite out of proportion to the populations involved. For example, Uffculme in mid-Devon has a population of under 500 and has some 62 houses planned for completion in 1983. This point is highlighted for a single district council in Figure 33.3 in which the number of houses per parish provided by housing associations is compared with 1981 population statistics. Once again predominantly large centres of population have been catered for, namely Honiton (6567) Sidmouth (12 446) and Exmouth (28 787), although Ottery St Mary (7069) and Teignmouth (11533) are not covered.

343

Table 33.1   Housing stock in Devon by District Council areas

DISTRICT COUNCIL HOUSING STOCK IN NO. OF UNITS

District	Total Population (1981)	Council	Private	Housing Associations			Area (ha)	Population (per ha)
				Fair Rent	Other	Total		
Teignbridge	95 665	5515	33 443	168	104	272	67 498	1.4
South Hams	67 861	4541	24 118	136	38	174	88 695	0.8
Mid-Devon	58 057	5515	16 258	218	76	294	91 561	0.6
North Devon	78 728	4790	25 186	233	20	253	108 614	0.7
Torridge	47 275	2956	16 342	61	12	73	98 492	0.5
East Devon	81 656	6496	38 299	197	214	411	81 656	1.3
West Devon	42 996	2219	*	49	0	49	115 973	0.4
Torbay	115 582	5009	*	775	316	1091	6 282	18.4
Plymouth	243 895	24 378	*	2813	49	2862	7 929	30.8
Exeter	95 621	8747	24 791	853	201	1054	4 388	21.8
Totals	927 336	70 166	178 437	5503	1030	6533		

* Insufficient data

Sources: Census 1981

Housing Rent Statistics April 1981 CIPFA

Local Authority Housing Investment Programme Reports 1981; 1982

Housing Corporation Records

Figure 33.2   Location of housing association projects in Devon

Secondly, Phillips and Williams (1982) revealed that the uneven distribution evident from council stock figures is further exacerbated when the character of dwellings is considered. By breaking down total housing association stock into types of house, it can be seen that their contribution is made even more specific by limited eligibility. Of 1062 houses for rent in Devon, 675 (63.5 per cent) are for the elderly alone, and 155 (14.6 per cent) are for special needs of other kinds, such as the handicapped or single-parent families. Again distribution bears little relation to expected demands. There may, however, be some validity in the claim by housing associations that they supplement local authority provision, for Figure 33.4 shows that there is a close correlation between the two types of stock. In view of recent trends, however, it must be pointed out that of the 41 associations receiving Housing Corporation funding to January this year, 19 of these were, in fact, self-build and co-ownership societies. By serving a more middle-of-the-road purpose, that is, of first-time buyers or half-and-half home-ownership, these societies are not directly comparable with the traditional, rented, local authority houses which cater for the needs of low-income families.

Thirdly, in the majority of cases allocation of housing by and among housing associations appears to follow no standard pointing

345

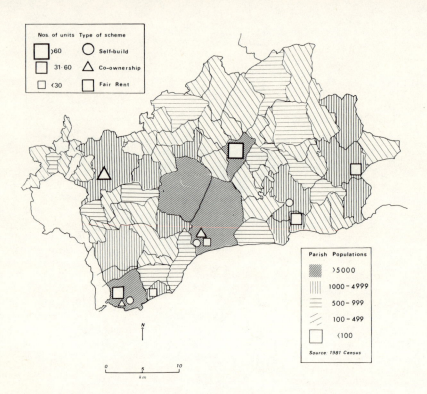

Figure 33.3   Housing association projects in East Devon

system or official waiting list; such organisation is often
considered unnecessary given the scope and scale of activity undertaken.   Depending on their constitutional aims, many associations
provide only for residents of the parish.   This may result in the
availability of a substantial number of units for a limited number
of inhabitants in villages fortunate to have their own association
scheme.

     In other cases, housing may be provided for a larger catchment
area but with the proviso of a recognised 'need' and perhaps with
archaic stipulations that allocation include a specified category
of person (e.g. widows of seamen).   In comparison to the strict
eligibility measures controlling the allocation of council housing
(see the paper by Phillips and Williams in this book), access to
a housing association property may rest largely on chance
circumstances or in more dubious instances, a local network of
social relations.

SOCIAL IMPLICATIONS AND CONCLUSIONS

The implications of this initial study of rural housing provision by
the voluntary housing movement are far-reaching and open up a
number of avenues for further research.

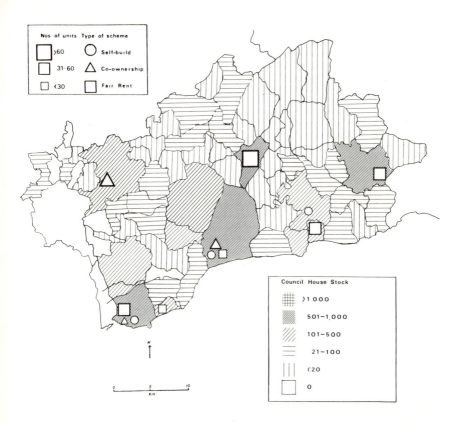

Figure 33.4   Housing association projects and council housing
stock in East Devon

Firstly, there is the difficulty of defining need and thus the
appropriate levels of provision for any area. The Abbeyfield
Societies, which provide accommodation to let solely to the elderly,
have defined a target of 1 scheme for 10 000 head of population.
Such targets must take into consideration local factors such as
council house stock and population structure, which can only be
studied at a local scale, and are best supported by the opinions of
those with an intimate knowledge of the communities under consideration, for example, parish councils.

Secondly, Shucksmith (1981) suggested that without government
intervention the rural housing market cannot hope to be efficient
or equitable in its provision. This applies to the management of
housing associations no less than to local authority stock. At the
managerial level, therefore, more knowledge is required of allocation methods and organisational structures within individual
associations. In particular, where restrictions are imposed, for
example of 'local' need, how is this assessed and does it place
unnecessary restraints on the utilisation of the housing stock?
Whilst such criticisms have regularly been made of local authority
procedures and public-sector housing allocations, there has been
little or no observation of housing association activities at the

347

regional level. In Housing Action Areas and the inner cities, the Housing Corporation controls the spatial operations of individual associations by means of a zoning policy. Over a wider rural arena such as Devon no such monitoring system exists. There are instances without doubt, where there is scope for voluntary housing activity in south-west England but where no suitable nationally-based association has penetrated or is willing to do so perhaps for economic reasons. Alternatively there is the situation where the siting of a housing association scheme and the manipulation of public funding has depended solely on the dynamism and initiative of a minority of the local community. Examples found in Devon include the Dartington Housing Trust, attached to Dartington Hall near Totnes, which provides homes for workers on the estate and for local people. Is it possible that districts with a well-organised middle class receive more help than those with a larger, more needy working-class population?

Thirdly, greater equality of provision need not involve over-centralised control, which may be in danger of destroying the very ethics upon which the voluntary housing movement was founded, and to which many of the smaller associations lay claim. The movement thrives, in its traditional mode, on a philosophy of self-help and non-profit-making, maintaining flexibility to tackle new challenges and experiment with housing ideas. Above all, associations like to feel they offer a more humane and sensitive approach to local housing problems. Many are likely to be deterred from development by the plethora of paperwork required by the Housing Corporation, leaving the field open for the domination of large, national bodies with full-time staff. It is possible that with unrelenting control over every aspect of their activities, housing associations, like the building societies from which they grew, are in danger of becoming expanding financial institutions characteristically inflexible and insensitive to local needs. Since local authorities and housing associations are apparently working together within the public sector, greater control should perhaps be exercised at the district council and regional levels rather than by central govern-ment legislation and Housing Corporation policy. A frequently cited failing within the field of housing provision is in the communica-tion and liaison between local housing departments and associations operating within the area. A major problem is to determine how the most effective use of public finance can be ensured.

It has been suggested that housing associations, by providing housing to rent where there is a recognised demand, present an alternative to council or private-rented accommodation and thereby a means of maintaining diversity of tenure in an increasingly limited housing stock. This is considered particularly relevant to rural communities. The above analysis shows two things. The number of units provided by the voluntary housing movement in one form or another in remote rural areas is very limited and follows no systematic pattern in comparison with population levels or alternative rented stock. Secondly, access to this stock is by no means open but confined by the objectives of each individual association.

Finally, attention is shifted to consider the outcomes of the 1980 Housing Act, namely, the unprecedented push towards higher levels of home-ownership and the inclusion of some of the Housing Association sector under the 'right to buy' provision. In addi-tion, housing associations along with local authorities are involved in the still experimental provision of self-build and shared owner-ship type of schemes. The 1981-82 figures (Housing Corporation, 1982) reveal already a total of 322 completions in housing of this kind. Rented property, for general purpose needs, is likely to be

in shorter supply in the future. Housing associations have an
important role to play in satisfying this demand and perhaps particularly in the re-use of redundant dwellings in conversion and
rehabilitation schemes. For rural areas, such as the greater part
of Devon, with the external pressures mentioned becoming more acute
in post-industrial rural society, the spatial allocation of resources
demands attention for the real and future needs of the communities.

REFERENCES

Baker, C.V. 1976. *Housing associations.* (Estates Gazette, London).

Basset, K.A. and Short, J.R. 1980. *Housing and residential
structure: alternative approaches.* (Routledge and Kegan Paul,
London).

Bourne, L.S. 1981. *The geography of housing.* (Edward Arnold,
London).

Clark, D. 1981. *Rural housing initiatives.* (National Council of
Voluntary Organisations, London).

Department of the Environment, 1980. *Housing and construction
statistics.* (H.M.S.O., London).

Dunn, K., Rawson, M. and Rogers, A. 1981. *Rural housing: competition and choice.* (Allen and Unwin, London).

English, J. (ed) 1982. *The future of council housing.* (Croom Helm,
London).

Gray, F. 1976. Selection and allocation in council housing.
Transactions of the Institute of British Geographers,
NS 1, 34-6.

Housing Corporation, 1980. *Annual Report 1979/1980.* (London).

Housing Corporation, 1982. *Annual Report 1981/1982.* (London).

Kirby, A. 1977. *Housing action areas in Great Britain.*
Geographical Papers no. 60, (Department of Geography, Reading
University).

Neate, S. 1981. *Rural deprivation: an annotated bibliography.*
(Geo Books, Norwich).

Page, D. 1971. *Housing associations - three surveys.* Research
Memorandum 7, (Centre for Urban and Regional Studies,
University of Birmingham).

Phillips, D. and Williams, A. 1982. *Rural housing and the public
sector.* (Gower, Farnborough).

Shucksmith, M. 1981. *No homes for locals.* (Gower, Farnborough).

Thomas, A.D. 1979. *Area-based renewal: three years in the life of
a housing association.* Research Memorandum 72, (Centre for
Urban and Regional Studies, University of Birmingham).

This book is to be returned on or before the last date stamped below.